Principles of Horticulture

Gardening and horticulture generally are essentially practical activities much enhanced by an understanding of how plants grow. This colourful guide will introduce you to the fundamentals of horticulture. It is written in a clear and accessible style and covers the principles that underpin growing plants for the garden and allotment, with reference to how these are tackled by professionals.

With highlighted definitions, key points and illustrations in full colour, this book will be a useful companion as you progress in the study and practice of horticulture. The book covers topics such as classifying and naming plants, the plant life cycle, ecology and garden wildlife, soils, composts, hydroponics, weeds, plant nutrition, plant pests, and plant diseases and disorders. The new edition has been updated to reflect changes in legislation and the modernization of horticultural practices. It is also fully reflective of the changes in the new syllabuses for horticulture at Level 2.

Principles of Horticulture is a valuable resource whether you are taking a Level 2 RHS, City and Guilds, Teagasc or SNQ course, or are a keen amateur or seasoned gardener.

The book is accompanied by ancillary materials including essential and extended information on horticultural principles and downloadable instructor resources.

Charles Adams is Fellow of the Chartered Institute of Horticulture, formerly Vice-Principal of Capel Manor College and Head of the Science Department at Oaklands College. He is an examination moderator of horticulture examinations and a member of the Royal Horticultural Society Qualifications Advisory Group.

Jane Brook is a freelance lecturer in plant science and horticulture, formerly at Capel Manor College, Middlesex University and the University of Hertfordshire, UK. She is also an examiner in horticulture at the Royal Horticultural Society, UK.

David Francis is a professional associate of the Royal Horticultural Society and lives in Japan. He was formerly a lecturer in horticulture at Capel Manor College in London.

Mike Early was a lecturer in horticulture science at Oaklands College, UK, and a landscape gardener.

Principles of Horticulture

Eighth Edition

Charles Adams, Jane Brook,
David Francis and Mike Early

LONDON AND NEW YORK

Designed cover image: *Rosa* 'Francis E. Lester'

Eighth edition published 2025
by Routledge
4 Park Square, Milton Park, Abingdon, Oxon, OX14 4RN

and by Routledge
605 Third Avenue, New York, NY 10158

Routledge is an imprint of the Taylor & Francis Group, an informa business

First edition published by Elsevier Ltd. 1984
Seventh edition published by Routledge 2014

British Library Cataloguing-in-Publication Data
A catalogue record for this book is available from the British Library

ISBN: 978-1-032-94693-1 (hbk)
ISBN: 978-1-032-94690-0 (pbk)
ISBN: 978-1-003-58126-0 (ebk)

DOI: 10.4324/9781003581260

Typeset in Univers LT Std
by Apex CoVantage, LLC

Access the Support Material: www.routledge.com/9781032946900

For Product Safety Concerns and Information please contact our EU representative:
GPSR@taylorandfrancis.com
Taylor & Francis Verlag GmbH, Kaufingerstraße 24, 80331 München, Germany

Contents

Preface

This book is for those who are keen to improve their knowledge of cultivating plants whether it is to improve their gardening and growing or because they are intending to work professionally in horticulture. It provides the appropriate principles of horticulture for those studying the subject to Level 2 with some insight into a higher level to encourage the continued study of what can be a life-long interest in the world of plants and associated subjects.

Horticulture involves the growing of plants: from the production of flowers, fruit and vegetables outdoors to houseplants and the more tender plants under protection. It includes the establishment and maintenance of plants for our enjoyment through the landscape industry and the provision of sports turf. Gardening for leisure and as a professional activity is the practical and artistic expression of horticultural science; 'gardening is to horticulture as theatre is to literature'. Those beginning their study of horticulture and plants are usually familiar with growing in gardens in order to create an attractive area around them and to provide leisure facilities such as lawns and sitting out areas. Many embark on the outdoor production of vegetables and fruit in their garden or on an allotment. The use of greenhouses or other protected structures enables the enthusiasts to extend the growing season and grow tender plants not normally possible in their garden. This background makes an appropriate starting point for an introduction to the principles of horticulture.

There are many techniques involved in horticulture and some familiarity with them is gained through our own experience of gardening. By studying the **principles of horticulture**, we are able to learn how plants grow and develop. In this way, a better understanding of the plant's requirements and its responses to various conditions enables us to grow plants more effectively. In the end, the horticulturist wants to be able to manipulate the plant for many reasons including optimizing its growth, fitting it into a pleasing planting scheme or decorative arrangement, or benefitting wildlife.

In the introductory chapter, the many and varied sectors of the horticultural industry serving the gardener are introduced. Gardens do not exist in isolation, so some of the current issues surrounding **sustainable practice** are explored and the importance of **conservation** of our garden plant heritage, including some of the organizations which help to bring this about, are described. Gardens are increasingly seen to have an important role in conserving and increasing biodiversity, supporting and enhancing communities, providing environmental benefits especially in urban areas and improving health and well-being. In effect, horticulture can make a significant contribution to the three goals of sustainability: environmental, economic and social. In addition, the impact of a changing climate and its relationship to horticulture are considered.

The early chapters introduce the biology of plants to the reader. The gardener currently selects from an enormous range of plant species that have been collected from all over the world and are adapted to a diversity of climates and habitats. The challenge of growing these introduced plants in situations local to us is explored. Through the study of **ecology**, the interrelationships between garden organisms is outlined to better understand how garden design, planting and management can benefit both the garden and the wider environment. The processes of **classifying** and **naming** plants, which are important in understanding the complexity of the plant kingdom, communicating accurately with others and correctly identifying plants, are described. The **internal** and **external structural features** of the plant and the many and varied ways in which they are adapted to different environments are detailed in order to inform accurate and successful plant selection and the concept of 'Right Plant, Right Place'. The significance of these plant features is explained in relation to plant processes such as **photosynthesis**, **respiration** and **transport**, an understanding of which helps create opportunities for optimizing and manipulating plant growth, development and behaviour to the grower's advantage. With the details of plant reproductive methods including **pollination** in its many forms and **fertilization**, the reader is given the background to **plant propagation** whereby plant stock can be increased by seed or vegetative means. Pollination and its importance in fruit and seed production is also explored.

Plants have been traditionally grown in soil, but now the range of **growing media** has expanded enormously with soil substitutes such as composts, aggregate culture and hydroponics. Usually, the plant's water and mineral requirements are taken up from the growing medium by roots. Active roots need a supply of oxygen, and so **the root environment** must be managed to include aeration as well as to supply water and minerals. The growing medium must also provide anchorage and stability. The physical characteristics of soil are described to help explain how satisfactory root environments can be produced and maintained. **Organic matter**, **water** and **nutrients** are considered in detail because they play such an important part in the management of productive soils along with **soil pH**, which has a major effect on the availability of nutrients. Soil conditions are modified by cultivations, irrigation, drainage and liming, while fertilizers are used to adjust the nutrient status to achieve the type of growth required. **Alternative growing media** and the management of plants grown in pots, troughs, peat bags and other containers are discussed in the context of the restricted rooting volume that makes it significantly different to growing in soil.

In growing plants for our own needs, we create a new type of community that introduces competition for environmental factors between one plant and another of the same species, between the crop plant and a **weed**, or between the plant and a **pest** or **disease-causing organism**. These bring about the need to address the challenge of maintaining plant health. It is only by the identification of the competitive organisms (weeds, pests and diseases) and an understanding of their life cycle and biology that we may select the correct approach to keeping them under control and understanding their wider roles in the gardens or crop ecology. With the larger pests there is little problem of recognition, but the smaller insects, mites, nematodes, fungi and bacteria are difficult to see with the naked eye. In this situation **signs** of a problem (physical evidence of the pathogen) are not always evident, so the grower must rely on the **symptoms** produced (usually the type of damage). In this section the symptoms of physiological disorders such as frost damage, waterlogging and mineral deficiencies that may be confused with pest or disease damage are addressed.

This edition provides the ideal support for those studying horticulture up to Level 2, including the current Royal Horticultural Society's 'Certificate in the Principles of Plant Growth and Development', and provides the principles underpinning RHS Level 1 Award in Practical Horticulture and RHS Level 2 Certificate in Practical Horticulture. In addition, it covers the plant science, plant/crop protection and soils components of the City & Guilds Level 2 Certificate, Extended Diploma, Diploma and Work-Based Diplomas in Horticulture, Teagasc Certificate in Horticulture Award and Advanced Certificate in Horticulture and in Scotland the NQ Introduction to Horticulture with Gardening Skills and the SVQ Level 3 Parks, Gardens and GreenSpace. For those studying to gain a specific qualification, the book should be read in conjunction with the relevant syllabus/qualification specification. It is also intended to be a comprehensive source of information for those taking City & Guilds Land-Based Services Certificate in Gardening modules and for any keen gardener who wishes to expand their knowledge and understanding.

The indexing and key word cross-referencing is to help the reader integrate the subject areas and to pursue related topics without laborious searching. It is hoped that this will enable readers to start their studies at almost any point, although it is recommended that an overview of the subject is gained by reading the early chapters first. Essential definitions are picked out in orange boxes alongside appropriate parts of the text. Further details of some of the science associated with the principles of growing have been included in the blue boxes and specialist areas of the horticulture industry are picked out in green boxes. Each chapter concludes with further reading on the subjects covered. The **Support Material** accompanying the book provides further information and illustrations and takes the reader towards the next level of study. This can be accessed by the QR code on the cover or at the end of each chapter.

Charles Adams
Jane Brook
David Francis

The online material is accessible via the QR code and includes further information on many of the topics in the book, such as

- The plant collectors
- Climate and microclimate
- Plant propagation
- An introduction to genetics

Horticulture and gardening

Figure 1.1 A community orchard

This chapter includes the following topics:

- The nature of horticulture and gardening
- The horticultural industry
- The plant
- Climate and microclimates
- Sustainable gardening
- Invasive non-native species
- Waste management
- 'Greenhouse gas' emission
- Organic gardening
- Conservation

DOI: 10.4324/9781003581260-1

Figure 1.2 A garden idyll for somebody

The nature of horticulture and gardening

We become familiar with gardens, whether through visiting the many public gardens, involvement in community projects, growing fruit and vegetables on an allotment, tending houseplants or gardening at home.

Around us are examples of gardens that are little more than surrounds to houses with the priority being to provide standing room for the car, utility areas and a place to sit out when the weather is good enough; the emphasis is probably on minimizing the workload with much hard surface. For others, the area is an opportunity to provide an attractive view, to enhance the look of the property or to have a safe playing area, but without wanting to be more involved with the plants than necessary. Again, the emphasis is likely to be on minimizing their input, with someone brought in considered to be a good solution to dealing with the time required and complications to keep good order. In contrast, gardeners consider a garden to be where they find deep satisfaction in working with nature, fulfil their wish to work with plants and seek to create their 'paradise on earth' (see Figure 1.2).

We learn quickly that gardening, wherever it takes place, is not a simple process because it is a dynamic situation that we face. Plants and plantings change over time. Over the year there are seasonal changes; and as time passes the plants grow in size. Both of these have significant consequences in achieving and maintaining what we wish to achieve. We can choose between rigorously maintaining our planned garden and allowing it to evolve. There are many skills and techniques associated with both ways forward: planting and replanting; controlling the size of the plant; allowing plants to spread by seed or vegetative means such as runners or placing plants according to carefully devised plans; but, probably above all, selecting the right plant for the situation in the first place. It is an ongoing job for gardeners to identify and hold at bay the undesirable plants whilst protecting their chosen ones from the attack of pests and diseases.

Gardeners benefit from knowing about what can improve or harm their plant's growth and development. The main aim of this book is to provide an understanding of how these factors contribute to the ideal performance of the plant in particular circumstances. For many their intention is to apply the knowledge to growing plants better, whether it is in their own garden, the window box, allotment, community garden or the private or public gardens where they volunteer their help. Others are intending to become professional horticulturists.

What goes on in our garden can give an insight into the wider world of horticulture.

Horticulturists as defined by the Chartered Institute of Horticulture

Horticulturists apply the knowledge, skills and technologies used to grow intensively produced plants for human food and non-food uses and for personal or social needs. Their work involves plant propagation and cultivation with the aim of improving plant growth, yields, quality, nutritional value and resistance to insects, diseases and environmental stresses. They are also involved in the application of post-harvest technologies, supply chain management and the economics, management and marketing of quality horticultural products and services to customers and consumers. They work as gardeners, growers, therapists, designers, operatives, technical advisors, educators, managers and business owners in the food and non-food sectors of horticulture.

Figure 1.3 Three-dimensional bedding

The horticultural industry

Horticulture may be described as the practice of growing plants in a relatively intensive manner. This contrasts with agriculture, which, in most Western European countries, relies on a high level of machinery use over an extensive area of land, consequently involving few people in the production process. The boundary between the two is far from clear, especially when considering large-scale **outdoor production**. When vegetables, fruit and flowers are grown on a smaller scale the difference is clearer cut and horticulture is characterized by a large labour input and the grower's use of technical manipulation of plant material. **Protected culture** is the more extreme form of this where the plants are grown under protective materials such as polythene or in greenhouses (clad in glass or modern plastics).

Service horticulture is the development and upkeep of gardens, parkland and landscape for amenity, cultural and recreational purposes. Increasingly horticulture can be seen to be involved with social well-being and welfare through the impact of plants on human physical and mental health; **horticultural therapy** is a valued means of helping many people through working with plants.

Horticulture encompasses large- and small-scale landscape design and management. Those involved will be engaged in **plant selection, establishment and maintenance**; many will be involved in aspects of garden planning such as surveying and design. Inevitably this may take horticulturists into environmental protection and conservation. Parts of the professional world come very close to what we do in our gardens such as the work of **self-employed gardeners**.

There may be some dispute about whether **countryside management** belongs within horticulture, dealing as it does with the upkeep and ecology of large semi-wild habitats. In a different way, the use of alternative materials to turf seen on **all-weather sports surfaces** tests what is meant by the term horticulture; many working 'in horticulture' have responsibilities for much beyond the growing of plants.

This book concerns itself with the principles underlying the growing of plants in the following sectors of horticulture:

- **Professional and heritage gardening** covers the use of plants in both public and private gardens and embraces many aspects of horticulture. It often includes both the decorative and productive sides of horticulture as seen in many of our stately homes open to the public. Bedding in the Victorian gardens of Waddesdon Manor in Buckinghamshire, for example, is provided on a grand scale and has utilized innovative techniques to provide carpet and three-dimensional bedding (Figure 1.3).
- There are numerous **specialist gardens** and **arboreta**, many of which are open to the public such as the National Trust; Royal Horticultural Society's Gardens; English Heritage Gardens; Westonbirt Arboretum; National Fruit Collection, Brogdale; Royal Botanic Gardens, Kew, in London; National Botanic Garden, Glasnevin, Dublin; Botanic Gardens, Belfast; National Botanic Garden of Wales; and the Royal Botanic Garden, Edinburgh.
- **Landscaping and garden construction** require the skills of construction (**hard landscaping**) together with the ability to develop planted areas (**soft landscaping**) and the use of water (**aquatic gardening**).

- Closely associated with this sector is **grounds maintenance** of local authority green areas, the maintenance of trees and woodlands (**arboriculture** and **tree surgery**).
- **Interior landscaping** is the provision of semi-permanent plant arrangements inside conservatories, offices and many public buildings and involves the skills of careful plant selection and maintenance. It is considered desirable for the health of workers to maintain indoor landscapes.
- **Turf culture** includes decorative lawns and sports surfaces for football, cricket, golf and the like. Turf management is very much a specialist horticultural profession which is an essential part of the sports industry. On golf courses, for instance, expertise is needed to provide and maintain a range of turf types and, more recently, other areas of golf courses are being managed for biodiversity too (Figure 1.4).
- **Nurseries and the hardy stock industry** are concerned with growing and supplying the other sectors of horticulture including the production of soil-grown or container-grown shrubs and trees (see Figure 1.5), bedding plants and the young stock of **soft fruit** (strawberries etc.), **bush** and **cane fruit** (blackcurrant, raspberries etc.) and **top fruit** (apples, pears etc.).
- **Garden centres** provide plants for sale to the public, which involves handling plants, maintaining them and providing horticultural equipment, chemicals, materials and advice. A few have some production on site, but stock is usually bought in.
- **Production horticulture**. The orchards, fields of vegetables, bulbs and flowers that we see on our travels give us some idea of the area over which **outdoor production** is undertaken (see Figure 1.6). Even from the road we can see, in general terms, the work being done over the year and the wide range of equipment being used in the fields. Less obvious is what exactly is being done and the technology involved, especially with crops grown in protected culture.

The huge blocks of polythene tunnels or greenhouses indicate that **protected cropping** is being undertaken on a scale very different to the greenhouse in the garden (one of the largest tomato producers in Britain and Ireland has a block of 18 hectares, i.e. it could contain 25 football pitches). Protected cropping enables plant material to be supplied outside its normal season and ensures high quality, for example chrysanthemums all the year round (see Figure 1.7), tomatoes to a high specification over an extended season, and

Figure 1.5 A hardy nursery stock area

Figure 1.4 Different areas of turf on a golf course and a range of habitats designed to increase biodiversity

Figure 1.6 A bulb field that shows the scale on which familiar plants are grown in commercial production

Figure 1.7 An area of year round chrysanthemums (YR) in a heated wide span greenhouse with lighting and shading to create daylengths needed to ensure flowering and strong stems throughout the year

cucumbers from an area where the climate is very different. Also out of sight is all the work being done with specialist equipment in the packing sheds where so much is undertaken, including the processing, grading and storage as well as the packaging of what is seen in the shops.

- **Plant propagation** uses a variety of structures to provide seedlings and cuttings for outdoor growing as well as for the other parts of protected culture and retail outlets like garden centres.
- The horticultural industry is supported throughout Britain and Ireland by horticulturists working in research and development, trials gardens, teaching and training organizations.

The plant

At the heart of our garden is plants. In order to discuss them we need to have an unambiguous means of naming and ideally a way of seeing how they relate to each other (see Chapter 2). Whilst they come in a great variety of shapes and sizes, they do have some fundamental similarities in their life cycles (see Chapter 3) and how they grow (Chapter 4).

It is important to have a clear idea of what a **healthy plant** is like at all stages of its life. The appearance of abnormalities can then be identified at the earliest opportunity and appropriate action taken. This is straightforward for most plants, but it is rather important to be aware of those plants whose healthy leaves are not normally green i.e. variegated, yellow, silver, purple, etc. Many 'stunted' plants are often expensive, dwarf forms such as *Betula nana, Berberis thunbergeii* 'Atropurpurea Nana' (the clue is in the name. Some appear 'monstrous', such as those with contorted stems such as *Salix babylonica* var. *pekinensis*

Figure 1.8 *Corylus avellana* 'Red Majestic' (corkscrew hazel) with insert showing detail of the contorted twigs

'Tortuosa' and the corkscrew hazel *Corylus avellana* 'Red Majestic' (see Figure 1.8).

It should be noted that **physiological disorders** account for many of the symptoms of unhealthy growth which includes conditions caused by nutrient deficiencies or imbalance (see Chapter 13). Damage may also be attributable to environmental conditions. These disorders are examined in Chapter 18 along with the problem of **plant diseases**, the symptoms of which may look similar. The problems that **plant pests** bring and how to deal with them is the theme of Chapter 17.

The incorrect functioning of any one factor may result in undesired plant performance. Factors such as the soil conditions, which affect the underground parts of the plant, are just as important as those such as light, which affect the aerial parts. The nature of soil is dealt with in Chapter 10. Increasingly plants are grown in alternatives to soil such as green waste, bark, inert materials and water culture, which are reviewed in Chapter 14.

Climate and microclimates

Climate can be thought of as a summary of the weather experienced by an area over a long period of time. More accurately, it is the long-term state of the atmosphere. Usually, the descriptions apply to large areas dominated by atmosphere systems (global, countrywide or regional). Britain and Ireland, surrounded as they are by large bodies of water, are considered to have a maritime climate, experiencing much less extreme temperatures than the landmasses of the continent. This low annual range of temperature plus high rainfall and humidity makes the natural vegetation of the western side of Britain and Ireland temperate rainforest. However, there are only small areas of the natural vegetation left after centuries of destruction.

Local climate reflects the influence of the topography (hills and valleys), altitude and large bodies of water (lakes and seas). Growers are aware that even their regional weather forecast does not do justice to the whole of the area because of these factors. The features of the immediate surroundings of the plant can further modify the local climate to create the precise conditions experienced by the plant. This is known as its **microclimate**. More details of these and the following topics are given in the Support Material.

- climate of Britain and Ireland
- world climate
- local climate
- microclimate
- growing seasons
- plant hardiness
- measurement of rainfall, temperature, humidity, wind, sunlight.

Climate change is having a significant effect on choice of plants and especially long-term planning (see Webster, E. et al).

Predictions include:

- drier summers with increasing periods of drought
- wetter winters, heavier and more intense rainfall, with greater risk of flooding
- higher and more variable temperatures and a reduction in ground frost
- possible increases in the number and intensity of storms
- increase in solar radiation due to reduced cloud cover.

Leading to:

- drier soils (increasing soil moisture deficits) in spring, summer and autumn and the development of arid areas (see p. 178)
- many impacts on the introduction, behaviour and spread of pests and diseases (see p. 236)
- changes in phenology (the timing of stages in plant and animal life cycles) leading to potential mismatch between flowers and their pollinators (see p. 104) and impacts on pollination (see p. 105)
- reduced cold treatments necessary for seed and germination and chill requirements for bud break and flowering (see p. 35)
- physiological changes affecting photosynthesis, respiration and transpiration and the balance between them (see p. 121)
- damage to soils (see p. 168).

Responses to climate change in the garden could be:

- a need to change plant selection, choosing a greater diversity of plants with improved resilience to potential climate change impacts such as higher temperatures, drought and wetter winters (see plant adaptations p. 68)
- better water management and irrigation practices using recycled water, mulching and attention to good soil structure (see p. 167)
- adoption of garden practices and purchase of products with lower carbon footprints (see sustainable gardening next).

Sustainable gardening

There are many ways to define the word '**sustainable**' but all agree in essence that, if an action or process is sustainable, it provides the best for the environment and people, socially and economically both now and indefinitely. Whilst most of us are familiar with the idea of 'environmental sustainability', we should not forget 'social sustainability' which focusses on the physical and mental health of people and communities and 'economic sustainability' (providing long-term economic benefit whilst protecting people's well-being and the environment). Horticulture can contribute to all of these by providing suitably rewarded employment by improving wellbeing and by adopting horticultural practices which do not damage the environment.

Environmental sustainability means making decisions and taking actions that do not degrade the natural world irreversibly. The end result of most environmentally unsustainable practices is **loss of biodiversity**, that is, a reduction in the number of habitats and species present in the wild and also less genetic variation in wild species and fewer cultivated species. Threats to biodiversity include habitat destruction, pollution, introduced species, global warming and overexploitation of natural resources.

When focussing on environmental sustainability, there are many issues that have relevance for gardeners and horticulturists. Some of these are:

- invasive non-native species
- 'greenhouse gas' emissions
- waste
- removal of rare species from the wild
- peat usage. Peat has been much valued as

an excellent component of composts, but for some time there have been plans to ban its use. However, this has been delayed for amateur as well as professional use for the time being. Much progress has been made to provide alternatives to incorporate in composts and gardeners are encouraged to adopt them; some of these are discussed in Chapter 14.

Peatlands are valuable as:

- a unique habitat supporting many rare species
- an important carbon sink, locking up about a third of the world's soil carbon
- an archive containing archaeological and geochemical historical information going back hundreds of years
- a habitat which plays an important role in the water cycle, contributing to flood prevention, water quality and quantity of freshwater.

They develop at a rate of just 1 mm a year so a10 m deep layer of peat has taken around 10,000 years to form. Their development is described in more detail in the Support Material.

Sustainable gardening practices

Wildlife gardens help to conserve biodiversity both in the garden itself and in the wider area (see Chapter 9).

Timber for garden structures should be produced from sustainable sources such as those certified by the Forestry Stewardship Council (FSC).

Discourage habitat destruction. It is all too easy to buy products which only become available following the destruction of habitats. On a world scale this involves deforestation (see previous item 'timber'), peatland (see earlier), removal of rainforests for plantations (e.g. for palm oil) and land clearing and drainage for agriculture such as soya and large-scale greenhouse production such as tomatoes. Nearer to home, limestone pavement (see Figure 1.9) is a natural rock formation which provides a habitat for many rare species and has been popular for making rock gardens, and although it is now protected in Britain and Ireland its removal still occurs.

Invasive non-native species

These include many weeds, pests and diseases which have arrived in this country from elsewhere. Their impact is second only to habitat destruction

Figure 1.9 Limestone pavement is now protected from removal from the wild

in reducing biodiversity. In the heyday of **plant collecting** little thought was given to the effects on native wildlife of foreign plants (more details of plants from across the world and the plant collectors can be found in the Support Material). Whilst most were well behaved, some garden escapees are now causing increasing problems for **native plants** and habitats. They also pose a challenge to forestry, tourism, agriculture and construction. They were unaccompanied by their natural diseases and predators which might have kept them in check. Three examples of invasive plant species are:

- ***Reynoutria japonica*** (**Japanese knotweed**) which was introduced to Britain via Kew in 1850 by Siebold as an attractive ornamental (Figure 1.10a). It can grow at an alarming metre or more a month, penetrating tarmac and concrete, and is highly resistant to weedkillers. It can add enormously to the preparation of sites for construction.
- ***Rhododendron ponticum*** is a familiar weed in woodlands with attractive purple flowers in spring (Figure 1.10b). It is one of nature's most successful plants, spreading readily by seed and rooted branches, and its dense evergreen canopy cuts out the light and produces poisons and a deep leaf litter which suppress the growth of everything beneath it including woodland tree seedlings. It was introduced as an ornamental garden shrub in 1763 and also used as a rootstock for grafting. It was first recorded in the wild in 1893 and has spread to woodland and moorland particularly on acid soils. The toxins in its leaves mean that it is undamaged by insects, and it can also harbour fungal diseases such as *Phytophthora ramorum* and *P. kernoviae* which can spread to oaks, beech and nursery stock.
- ***Impatiens glandulifera*** (**Himalayan balsam**) can grow to 3 metres tall and is a serious problem along river banks (Figure 1.10c). It was introduced in 1839 from Northern India and

Figure 1.10 Three invasive non-native plant species; (a) *Reynoutria japonica* (Japanese knotweed); (b) clearing *Rhododendron ponticum* from a woodland, note the lack of other plant species in the cleared area; (c) *Impatiens glandulifera* (Himalayan balsam) along a river bank

grown for its attractive pink and white slipper-shaped flowers. Its seed pods can project the seed explosively up to 7 metres away, and seeds are carried downstream to colonize new areas. The dense stands shade out other species, and it produces more sugary nectar than any other European plant species, attracting pollinating insects such as bees away from native plants. When it dies back in late autumn it leaves river banks exposed and liable to erosion.

Invasive species are particularly troublesome along and in waterways where it is difficult to use chemical control. Aquatic species such as water fern, Australian swamp stonecrop, floating pennywort, Canadian pondweed and parrots' feather are particularly invasive. Others include *Buddleja*, *Cotoneaster*, some broad-leaved bamboos and some exotic honeysuckles. Gardeners can best prevent the spread of these damaging weeds by knowing precisely which plants they are buying, choosing non-invasive species and disposing of plant material carefully.

The global trade in plants has also led to the introduction of many new **'alien' pests and diseases** (see Chapters 17 and 18). Ash dieback (*Hymenoscyphus fraxineus*) is just the latest in a series of fungal diseases affecting our trees (p. 236).

Never collect plants from the wild, and ensure that purchased plants are sustainably sourced. The removal of rare species from the wild is a result of the demand for new and ever more interesting plants. Overcollection and illegal importation of wild bulbs has become an issue. Many are now protected to some extent by CITES (Convention on Trade in Endangered Species) which monitors and issues quotas for imports and exports of threatened species. CITES produces a checklist which can be downloaded from its website; most of the plants listed are orchids, cacti, cycads (Figure 1.11), bulbs, some succulents and some carnivorous plants.

Reduce, Reuse, Recycle

Waste is an issue because of the damage to the environment from landfill and the energy used to collect, dispose of and recycle it. Horticulture uses many types of plastic from trays and pots to polythene used to cover polytunnels, fleece to insulate crops and packaging for sale of fruits and vegetables, cut flowers and plants. Growing also generates green waste such as plant material, much of which can be composted and can be used as a source of energy.

Good practice can be summarized as **Reduce, Reuse, Recycle**.

Figure 1.11 *Encephalartos ferox*, a **cycad** with female cones (leaves removed on one side). All cycad species are on the CITES Red Data List of endangered species

Reduce. Gardeners can help reduce their environmental impact in many ways by reducing the use of:

- **Water** by planting drought-tolerant plants and mulching (see p. 188). Use rainwater to water plants (purification of mains water is energy intensive) or reuse household water where appropriate. Water only when necessary and do this efficiently, that is, thoroughly and less frequently (see Chapter 11).
- **Energy**, for example heating and lighting in greenhouses or power tools such as air blowers outdoors. Insulate greenhouses as an alternative to heating. Source garden materials and plants locally if possible, to reduce fossil fuels used in transport.
- **Fertilizers** and **pesticides** along with the many other garden chemicals. Only treat pests and diseases when necessary and follow the instructions carefully, using only approved products. Incorrect and excessive use is likely to kill beneficial organisms and pollute soil, ponds and groundwater.
- **Waste**. Within the overall need to reduce waste, a major consideration is avoiding excessive packaging.

Reuse and repurpose items such as plastic pots and trays which should be thoroughly cleaned and sterilized before reuse some garden centres have return schemes for pots and trays. Hard landscaping materials such as slabs, rocks, building materials and timber can be reused, which also avoids further damage to habitat. Woodchip for paths can be sought from arborists or straw for insulation (rather than composted, dug-in or landfilled). Cardboard packaging can be repurposed as a mulch on vegetable beds or weed suppressant. Bubble wrap can be used to insulate tender plants outdoors or in the greenhouse.

Recycle items that cannot be reused, although many horticultural plastics such as seed trays and polystyrene can only be landfilled or incinerated. Garden waste can be composted (see Chapter 12). Homeowners can do this, which avoids energy being used for collection and disposal as well as providing a valuable soil improver. Organic waste in landfill produces methane which is 25 times stronger than carbon dioxide as a greenhouse gas.

'Greenhouse gas' emission

'Greenhouse gas' emission, largely due to due to the burning of fossil fuels and conversion of land for agriculture (landuse change), has been identified as the major cause of global warming. Fruit, vegetables and flowers all have a **carbon footprint** made up of the 'greenhouse gas' emissions incurred in their production and supply. Fossil fuels are used directly for cultivation, heating and transport of produce and plants. They are also used indirectly to supply energy for lighting, refrigeration, water treatment and manufacture of materials like plastics, fertilizers and pesticides. Issues such as the sale of all year round produce, which involves heating and lighting in winter and long distance transport ('air' and 'road miles'), are all hotly debated. Growers can reduce their carbon footprint by reducing their energy use or using alternative 'green' energy. For example, combined heat and power generated from green waste can be used for tomato production in the winter and wood chip burners instead of gas can heat greenhouses in areas. Some major greenhouses are located beside power stations to make use of 'waste' heat. A massive greenhouse production area generates all its energy from crop waste and exports the remainder to the National Grid. Another successful scheme uses 'waste' carbon dioxide produced in an adjacent factory to boost photosynthesis (see p. 113) in its greenhouse-grown tomatoes, thus reducing the amount lost to the atmosphere.

Organic gardening

There are many different ways of gardening that make use of the principles of horticulture outlined in the following chapters. Most of the book deals with the nature of plants and soils common to all growers, but those growing 'organically' will focus on some specific parts of the text, such as soil health, and not make use of others such as the use of artificial fertilizers and most pesticides. The degree to which the gardener adopts an 'organic' approach can range from the strict adherence to the organic philosophy (as required in commercial horticulture where growers have to adhere to strict criteria) to those who are inclined to 'no chemicals' but are less concerned about obtaining organically acceptable plants, propagation material, manures and so on. **Organic growing** is described in more detail in the Support Material.

Conservation

Conservation is the management of animals, plants and other organisms to ensure their survival as a resource for future generations. It focusses on reducing threats to biodiversity in the wild but is also concerned with conserving cultivated plants and their wild ancestors. These are the gene pool on which future plant breeders can draw for further improvement of plant species.

***In situ* conservation** involves the preservation (and sometimes the creation) of natural reserves where they occur to conserve habitats and the wild species they contain, for example the work done by The Wildlife Trusts or National Trust. Sometimes cultivated plants are also conserved *in situ*; for example the UK Orchard Network is a partnership of organizations conserving orchard fruit and nut trees and their cultivars, from individual trees to whole orchard habitats.

***Ex situ* conservation** includes whole plant collections in botanic gardens, arboreta, pineta and genebanks where seeds, vegetative material and tissue cultures are maintained, for example the Millennium Seed Bank of the Royal Botanic Gardens, Kew, at Wakehurst, Surrey. The botanic gardens are coordinated by Botanic Gardens Conservation International (BGCI), a global organization which is based at the Royal Botanic Gardens, Kew, in London and are primarily concerned with the conservation of wild species.

For cultivated species, large national collections include the National Fruit Collection at Brogdale, Kent, which holds over 3,500 nut and fruit cultivars in its orchards and the Warwick Genetic Resources Group which collects, conserves, documents and researches into a wide range of vegetable crops and their wild relatives. The Garden Organic Heritage Seed Library conserves old varieties of vegetables which were once commercially available, but which have been dropped from the National List (and so become illegal to sell), and operates a seed exchange programme. Plant Heritage (formerly the National Council for the Conservation of Plants and Gardens (NCCPG)) was set up by the Royal Horticultural Society at Wisley in 1978 and is an excellent example of professionals and amateurs working together to conserve stocks of garden plants threatened with extinction. The aim is to ensure the availability of a wider range of plants and to stimulate scientific, taxonomic, horticultural, historical and artistic studies of garden plants. There are over 600 collections of ornamental plants encompassing 400 genera and some 5,000 plants. A third of these are maintained in private gardens, many of which are open to the public. Many are held in publicly funded institutions such as colleges, for example *Sarcococca* at Capel Manor College in North London, *Escallonia* at the Duchy College in Cornwall, *Penstemon* and *Philadelphus* at Pershore College and *Papaver orientalis* at the Scottish Agricultural College, Auchincruive. A *Hamamelis* (witch hazel) collection is held at the Witch Hazel Nursery in Kent.

Rare plants are identified and classified as 'pink sheet' plants. Plant Heritage's 'Threatened Plants Project' aims to identify and assess for heritage the value of all named plant cultivars in Britain and Ireland and to identify rare and threatened plants in its collections. It develops conservations plans with partner organizations fulfilling global targets for conserving the diversity of plants with cultural and socio-economic value.

Further reading

Adams, C., Brooks, J. and Early, M. (2015) *Principles of Horticulture Level 3*. Routledge.

Baines, C. (2000) *How to Make a Wildlife Garden*. Frances Lincoln.

Brickell, C. (ed.) (2006) *RHS Encyclopedia of Plants and Flowers*. Dorling Kindersley.

Chatto, B. (2021) *Green Tapestry Revisited: A Guide to a Sustainably Planted Garden*. Berry & Co.

Clevely, A. (2006) *The Allotment Book*. Collins.

Davies, G. and Lennartsson, M. (2006) *Organic Vegetable Production: A Complete Guide*. The Crowood Press.

Dixon, D. (2019) *Garden Practices and Their Science*. Routledge.

Dowding, C. (2013) *Organic Gardening*. Green Books.

Farrimond, S. (2023) *The Science of Gardening: Discover How Your Garden Really Grows*. Dorling Kindersley.

Greenwood, P. and Halstead, A. (2003) *RHS Pests and Diseases*. 2nd ed. Dorling Kindersley.

HDRA. (2005) *Encyclopaedia of Organic Gardening. Henry Doubleday Research Association (Garden Organic)*. Dorling Kindersley.

Hessayon, D.G. (1958) *Garden Expert Series*. Expert Publications.

Littlewood, M. (2007) *Organic Gardeners Handbook*. The Crowood Press.

Mason, K. (2023) *Gardener's Guide to Adaptive Gardening*. The Crowood Press.

Pears, P. and Sticklands, S. (2007) *The RHS Organic Gardening*. Bounty Books.

Stewart, D. (2023) *Sustainable Gardening*. The Crowood Press.

Thomas, H. and Wooster, S. (2008) *The Complete Planting Design Course*. Mitchell Beazley.

Webster, E., Cameron, R. and Culham, A. (2017) *Gardening in a Changing Climate*. RHS.

www.bbc.co.uk/bitesize/guides/z2m39j6/revsion/1 for Food Chains

www.greensandtrust.org/working-woodlands-project

www./Handlers/Download.ashx?IDM=

www.moormeadows.org.uk

www.plantlife.org.uk
www.rhs.org.uk/advice/health-and-wellbeing/articles/why-gardening-makes-us-feel-better
www.theorchard.org.uk/guides
www.wearebs15.co.uk and www.foodcycle.org.uk for Role of Community Organizations
www.wildlifetrusts.org
www.woodlandtrust.org.uk

Climate change

www.bgci.org/resources/bgci-tools-and-resources/plants-and-climate-change-which-future/
www.rhs.org.uk/science/pdf/climate-and-sustainability/urban-greening/gardening-matters-urban-greening.pdf
www.stockholm.org/research/research-news/2022-04-26-freshwater-boundary-exceeds-safe-limits.html

Plant trials/research

https://archive.ahdb.org.uk/horticulture
www.kew.org/science
www.rhs.org.uk/plants/trials-awards/plant-trials-awards
www.stockbridgetechnology.co.uk

Horticultural statistics

Fruit and Vegetables in the UK – Statistics & Facts | Statista Fruit and Vegetables in the UK – Statistics & Facts
The Global Impacts of Habitat Destruction – National Geographic Society Newsroom
Horticulture Statistics – GOV.UK (www.gov.uk)
www.mordorintelligence.com/industry-reports/united-kingdom-fruits-and-vegetables-market. UK Fruits and Vegetables Market Size
Production of Nursery Stock and Ornamental Plants – Teagasc | Agriculture and Food Development Authority
www.rhs.org.uk/science/ornamental horticulture roundtable
www.rhs.org.uk/science/pdf/industry-growth-report-ohrg.pdf Growing a Green Economy The Importance of Ornamental Horticulture and Landscaping to the UK
Savills UK | UK Garden Centre Spotlight – 2023 UK Garden Centre Spotlight
https://www.statista.com/statistcs/316054/plants-and-flowers-production-value-in-the-united-kingdom-uk/ Plants and Flowers Production Value in the UK
www.teagasc.ie/crops/horticulture/cut-foliage/ Cut Foliage in Republic of Ireland
www.teagasc.ie/crops/horticulture_Horticulture Statistics of Republic of Ireland

The online material is accessible via the QR code and includes further information on many of the topics in the book, such as

- Health and safety
- Climate of Britain and Ireland
- Responding to climate change
- World climate and biomes
- Local climate
- Microclimates
- Weather station instruments
- Peat development
- Organic growing
- Rocks
- Benefits of green spaces
- Historic garden styles
- Plants for a purpose
- Colour wheel
- Basic chemistry

CHAPTER 2

Classifying and naming plants

Figure 2.1 *Berberis × hortensis* – an evergreen, hybrid shrub which has changed its name from *Mahonia × media*

This chapter includes the following topics:

- What is a plant?
- The plant kingdom – mosses and liverworts, ferns and horsetails, gymnosperms and angiosperms
- Monocotyledons and eudicotyledons
- Plant classification – families, genera and species
- Plant nomenclature – the binomial system, hybrids, cultivars and trade designations, plant breeders' rights, plant name changes
- Further classifications of plants used in horticulture
- Fungi

DOI: 10.4324/9781003581260-2

What is a plant?

The world of living organisms can be grouped in many ways. One such classification is the five kingdom system:

- Plantae (plants)
- Animalia (animals)
- Fungi
- Prokaryota (bacteria)
- Protoctista (all other organisms that are not in the other kingdoms, including algae and protozoa).

What distinguishes the plant kingdom from the other four kingdoms? Plants are largely sedentary and live on land (albeit some have adapted to aquatic environments such as ponds and rivers). They are multicellular and their cells have cellulose cell walls and a nucleus. Most importantly, they are **autotrophs**, organisms that are able to convert energy from one form, that is light, into a chemical form stored in organic molecules like sugar and starch through the process of photosynthesis (see p. 112). Animals, among other things, have no cell walls and are **heterotrophs** (the 'consumers' described on p. 138). They rely on eating ready-made organic molecules for their nutrition and feed on plants and other organisms. Fungi do not photosynthesize and have cell walls made of chitin rather than cellulose, and although some bacteria do photosynthesize, they do not have a nucleus. Green algae, from which land plants are thought to have evolved, are included in the plant kingdom by some taxonomists but others exclude them because they are primarily aquatic organisms.

The plant kingdom

The plant kingdom contains a range of plant groups which reflect their evolutionary pathway from simple organisms to more advanced plants. Within the kingdom, primitive '**non-vascular**' plants (mosses and liverworts) have no specialized tissue for uptake and transport of water and minerals. In contrast '**vascular**' plants contain conducting tissue (xylem and phloem) (see Chapter 8). Within this group, '**lower vascular**' plants (including ferns and horsetails) do not produce seeds whereas '**higher vascular**' plants (including gymnosperms and angiosperms) do.

The non-vascular non-seed-bearing plants are often referred to as **bryophytes**, the vascular non-seed-bearing plants as **pteridophytes** and the seed-bearing plants as **spermatophytes.**

Mosses and liverworts

Mosses and liverworts are '**non-vascular**' plants as they have no specialized tissue or organs for conduction of water and minerals. They absorb water and minerals over their entire surface (Figure 2.2) and are without a waterproofing surface layer (cuticle). This, together with the need for water for reproduction (they employ free-swimming sperm), restricts them to damp shady habitats, and their lack of true roots, leaves and stems means they remain low growing. Liverworts are frequently seen on the surface of plant pots containing woody stock in nurseries and may be found in the wild on wet rocky surfaces often next to streams. Mosses are most noticeable in woods in high rainfall areas. Mosses can be a problem weed particularly in shaded and poorly drained lawns, on damp capillary benches and around glazing bars in greenhouse roofs. However, they may also be used as an ornamental feature in Japanese gardens (Figure 2.3) and can be an attractive ground cover in damp shaded areas in the garden. Sphagnum mosses are used in hanging baskets because of their excellent water holding properties although their collection from the wild for this purpose can be damaging and unsustainable. Many mosses and liverworts are pioneer plants (see p. 140) that play an important part in the early stages of soil formation.

Ferns and horsetails

The '**lower vascular**' plants, of which ferns and horsetails are examples (Figure 2.2), do not produce flowers or seeds. They have true leaves, stems and roots containing a vascular system with xylem and phloem and a waterproof cuticle over their surface to reduce water loss. Like mosses and liverworts, they spread by means of spores (see p. 109), and they also require water for reproduction so tend to be found in damp places. Fern leaves (fronds) are generally divided into many pinnae. On the underside of the pinnae are spore-bearing structures called sporangia clustered in sori which often appear as rusty brown spots or stripes (Figures 2.4b and 6.27a).

Ferns provide a wide range of decorative plants in the garden and home with many attractive leaf shapes and forms and a variety of sizes, and they are especially useful for planting in shade, for example, hart's tongue fern (*Asplenium scolopendrium*) (Figure 2.4). They are also grown as houseplants, such as maidenhair fern (*Adiantum cuniatum*). *Dicksonia antarctica* (a tree fern) is unusual in developing a tall stem from overlapping leaf bases which make a striking specimen in a sheltered garden.

Horsetails are primitive plants with an unbranched central stem and much reduced scale-like leaves attached in whorls. They feel rough to the touch because of silica contained in their tissues, which deters herbivores. Some horsetails can be decorative

Figure 2.2 Four plant groups with horticultural significance: (a) moss; (b) liverwort; (c) fern; (d) horsetail (source: Shutterstock, Orest lyzhechka)

Figure 2.3 A moss garden at Hotel Chinzanso in Tokyo, Japan

Figure 2.4 (a) Hart's tongue fern (*Asplenium scolopendrium*) a native British fern; (b) brown sori on the back of frond

Figure 2.5 Bracken (*Pteridium aquilinum*) showing underground rhizomes

such as *Equisetum hyemale*, which can be grown around pond edges, but field horsetail (*Equisetum arvense*) (see p. 253) and the fern, bracken (*Pteridium aquilinum*), which both spread by underground rhizomes, are invasive and are difficult weeds to control (Figure 2.5).

Gymnosperms and angiosperms

The '**higher vascular**' plants, the gymnosperms and angiosperms, are more evolutionarily advanced and are distinguished from lower plants by their reproductive behaviour. These are the **seed-bearing plants** which spread by dispersal of seeds rather than by spores.

Gymnosperms characteristically produce male and female cones which bear only partially enclosed, 'naked' seeds, hence the name 'gymnosperm' (Figure 2.6).

By far the largest gymnosperm group is the conifers, which include many hundreds of species such as the pines, junipers, spruces and yews and form the vast boreal forests of the northern hemisphere. Conifers are also found in Australia, Papua New Guinea and South America. Chilean pine or monkey puzzle tree (*Araucaria araucana*) was once widely planted in Victorian gardens.

Conifers have the following characteristics:

- primarily perennial, woody trees and shrubs
- have male and female cones (dioecious or monoecious) rather than fruits
- may have 'berry-like' arils instead of cones in some species (Figure 2.7a)
- male cones (Figure 6.8) produce prodigious amounts of pollen which is spread by wind
- seeds are usually borne in female cones (Figure 2.7b)
- often found in a limited range of habitats where water is in scarce supply either due to low rainfall or because the ground is frozen much of the year
- frequently display structural adaptations to reduce water loss, for example, needles or scale leaves (Figure 2.8) and branches designed to shed snow
- may contain resin in their wood which protects against pests and diseases and is wound healing
- mostly evergreen to take full advantage of the short growing season in higher latitudes and to avoid expending energy on producing new leaves each year, although some genera such as *Taxodium*, *Metasequioa* and *Larix* are deciduous
- single fertilization (see p. 103) and do not produce an endosperm in their seeds (see p. 106).

There are very many important conifers. Some are major sources of wood or wood pulp, but within horticulture many are valued because of their interesting plant habits, foliage shapes and colours. The cypress family Cupressaceae, for example, include fast-growing species, e.g. ×*Hesperotropsis leylandii* (formerly ×*Cupressocyparis leylandii*), which can be used as windbreaks or hedges, and small slow-growing types very useful for rock gardens, for example *Juniperus procumbens*. The yews are a highly poisonous group of plants that includes common yew (*Taxus baccata*) used in ornamental hedges, topiary and mazes. It is unusual in not bearing cones but encasing its seeds in a fleshy structure called an aril (Figure 2.7a).

Figure 2.6 A cone of stone pine (*Pinus pinea*) showing pairs of seeds exposed just before being shed

Figure 2.7 (a) Yew (*Taxus baccata*) with 'berry'-like arils – a conifer; (b) immature female cones

Figure 2.8 Conifer leaves: (a) needles in *Pinus* spp.; (b) scale leaves in *Thuja*

The **angiosperms**, or flowering plants, encompass the greatest diversity of plant life with adaptations for the vast majority of global habitats. There are estimated to be some 300,000 known species of flowering plant on earth and they represent the most advanced plant life forms. Angiosperms have the following characteristics:

- unique in having flowers, usually hermaphrodite (see p. 89), which are pollinated both by wind and other agents and in many cases these flowers are highly adapted to their specific pollinators
- produce seeds enclosed within a protective fruit which develops from an ovary
- life cycles encompass the full range of ephemerals, annuals, biennials and perennials
- can be both woody and herbaceous in structure
- can be evergreen or deciduous
- occupy the greatest range of habitats
- have double fertilization (see p. 103) and their seeds contain an endosperm (see p. 106).

Many flowering plants are important in gardens as crop plants, ornamentals and weeds. Chapters 2 to 8 will focus mainly on the angiosperms.

Table 2.1 Differences between monocotyledonous and eudicotyledonous plants

Monocotyledons (Monocots)	Eudicotyledons (Eudicots)
One seed leaf (cotyledon)	Two seed leaves (cotyledons)
Parallel veined leaves, usually alternate and sword-shaped with smooth margins	A variety of leaf vein patterns, e.g. reticulate (net) veined and many different shapes and margins
Vascular bundles in stem are scattered Vascular tissue (stele) in the root has many arms	Vascular bundles in stem are arranged in rings Vascular tissue (stele) in the root has up to seven arms
Fibrous root systems	Both fibrous root and taproot (primary) systems
Flower parts (**tepals**, (see p. 88) carpels) usually in threes or multiples thereof, also often three seed chambers in fruit	Flower parts (**petals, sepals**, carpels) usually in fours or fives or multiples thereof, also often four or five seed chambers in fruit
Pollen with one pore	Pollen with three pores
Small non-woody herbaceous plants (except palms and bamboos)	Both small and large, woody and herbaceous species with woody stems undergoing secondary thickening showing annual rings and bark

Monocots and eudicots

For many years, the angiosperm families were divided into two groups, the monocotyledons and dicotyledons, but these have now been reclassified based on the Angiosperm Phylogeny Group (APG) classification of flowering plants. The two largest groups are now the **monocotyledons** (**monocots**) and the **eudicotyledons** (**eudicots** meaning 'true' dicots). Two smaller groups, the **basal angiosperms** and **magnoliids**, are also recognized. The main differences between monocots and eudicots are given in Table 2.1 and are further described in the following chapters.

Monocotyledons include some important horticultural families, for example Poaceae (formerly Graminae), the grasses and bamboos; Alliaceae, the 'onion' family; Liliaceae, which includes bulbous plants such as lilies and tulips and Amaryllidaceae which includes daffodils; and food plants such as Musaceae, the 'banana' family (Figure 2.10).

Eudicots are globally important, providing food, timber, ornamental and other economic crops. They are the most significant group globally with some 416 families and 13,000 genera recognized by the APG IV (see box). Some examples are Fabaceae, the legumes or the 'pea and bean' family; Caprifoliaceae, which includes honeysuckles; Cactaceae, with plants commonly called 'cacti'; Malvaceae, the 'mallow' family; Ranunculaceae, often called the 'buttercup' family; Theaceae, which includes tea; Betulaceae, which includes birches, hazels,

'*Ginkgo biloba* – a living fossil'

Figure 2.9 Maidenhair tree (*Ginkgo biloba*) (a) distinctive leaves; (b) ripe 'fruits'; (c) planted as a street tree showing autumn colour

The maidenhair tree (*Ginkgo biloba*) (Figure 2.9) is a truly unique gymnosperm. It is the only species in its genus, in fact in its whole family Ginkgoaceae. Its characteristics make it hard to place in classification systems. The seeds are exposed, making them a gymnosperm. They are produced in clusters on the surface of the trunk and branches and include a fleshy, foul smelling seed coat (thus resembling a fruit). Their reproductive cycle has sperm-like sex cells which swim down to the egg cells and their leaves have a primitive leaf shape and venation, branching in twos. Ginkgo trees are dioecious in common with many other gymnosperms. Living specimens are little changed from fossils dated back to the Permian period some 240 million years ago and disappeared from Europe only 2.5 million years ago. Ginkgos were 'discovered' in China in the 18th century where they were revered and protected by Buddhist monks because of their longevity; they can live for 1,000 years or more, but they probably no longer exist in the wild. Because they are tolerant of pollution, ginkgos are very useful street trees in urban areas and can survive difficult conditions of compacted soil, limited root runs and irregular water supply. They provide yellow autumn colour and although they can reach 30 m or more, smaller clones and male trees which do not bear smelly seeds are usually planted.

hornbeams and alders; Brassicaceae (formerly Cruciferae), the 'cabbage' family; Fagaceae, which includes oaks, beeches and chestnuts; Solanaceae, often called the 'nightshade' family which includes food crops such as potatoes, aubergines and tomatoes; Crassulaceae, which includes stonecrops and other succulents; and Lamiaceae, the 'dead nettle' family.

Plant classification – families, genera and species

Classification (taxonomy) involves putting objects or organisms into groups based on characteristics which the members of the group share; it is something we do all the time and is very useful. For example, the fresh produce section of the

Figure 2.10 Monocotyledonous angiosperms: (a) fragrant orchid (*Gymnadenia conopsea*); (b) a bamboo grove of *Phyllostachys edulis*

supermarket contains a salad section, a vegetable section and a fruit section. The fruit section in turn is subdivided into smaller groupings such as citrus (oranges and lemons), top fruit (apples and pears) and soft fruit (strawberries, blackberries and raspberries). This enables us to find what we want quickly and to know that what we are buying has particular characteristics.

In the same way, the plant kingdom is classified into major groups or divisions, which are in turn are subdivided into smaller and smaller groups forming a **taxonomic hierarchy**. For example, lettuce 'Little Gem' is a cultivar of the species *Latuca sativa*, which is one of many species within the genus *Latuca*, which in turn is one of many genera in the family Asteraceae. Asteraceae is one of the families in the order Asterales which is in the eudicot group within the angiosperms.

For gardeners, the groupings of plants which are most commonly encountered are **family**, **genus** and **species**.

Plant **families** were first described comprehensively by the Swedish botanist Linnaeus in the 18th century (Figure 2.11). His classification for flowering plants was originally based on flower structure, although nowadays many other factors, particularly genetics, are used in addition to, or instead of, their external features. DNA 'fingerprinting' has become especially useful in establishing evolutionary relationships between plants (see box 'What is the APG?') and this has revealed previously unknown connections leading to many plants being reclassified. For example genetic studies have shown that the sacred lotus (*Nelumbo*) (Figure 3.3) is closely related to plane trees (*Platanus*). Classification is not an exact science and systems and the position of plants in them are constantly being reviewed and updated as more evidence comes to light. Nevertheless, many of Linnaeus's original family names and groupings still stand with flower structure being the most important identifier for the gardener. Members of Rosaceae (the rose family), for example, commonly have five petals and five sepals, whilst the Brassicaceae (the cabbage family, formerly called Cruciferae) have four petals (resembling a cross) and four sepals. In Lamiaceae the stems are square in cross section, aiding identification, whilst in Cyperaceae (the sedge family) they are triangular, which helps distinguish them from Poaceae (the grasses) which have cylindrical stems (Figure 2.12). Plant family names always end in -aceae.

Carl Linnaeus – the Father of Taxonomy (1707–1778)

Carl Linnaeus worked as an assistant and later as professor of botany at Uppsala University in Sweden. He brought together all recorded knowledge of the natural world known at that time and classified it into three kingdoms: minerals, plants and animals. He named some 7,700 plants and 4,400 animals in his lifetime and used an innovative classification system which was first set out in his *Systema Naturae* in 1735. Each kingdom was subdivided into a hierarchy of classes, orders, families, genera and species which replaced existing classification systems and is still largely in use today. For plants, he used the structure of flowers, in particular the male and female parts, as the basis of his classification. This 'sexual system' was a practical and easily learned approach which enabled him and his students to study a large number of species, although the florid way in which he described flowers caused an uproar at the time!

In addition, Linnaeus is credited with establishing the **binomial** system of nomenclature. Although the use of binomials for plants was not new, Linnaeus applied them consistently to all plant species alongside the cumbersome many-worded 'phrase names' which were current at the time and thus laid the foundation for a simple and universal naming system which has been adopted ever since. As such, Linnaeus's work *Species Plantarum*, first published in 1753, forms the basis for plant nomenclature right up to the present day.

Figure 2.11 Carl Linnaeus (Carl von Linné) dressed as a Laplander from a painting by Hendrik Hollander in 1853. He explored Lapland in 1732, collecting plants, birds and rocks and used the 600-mile trip to apply his ideas of classification and nomenclature. He described 100 newly identified plant species in his book *Flora Lapponica*, including twinflower (*Linnaea borealis*) which he named after himself. By Hendrik Hollander – University of Amsterdam, Public Domain, https://commons.wikimedia.org/w/index.php?curid=2612040

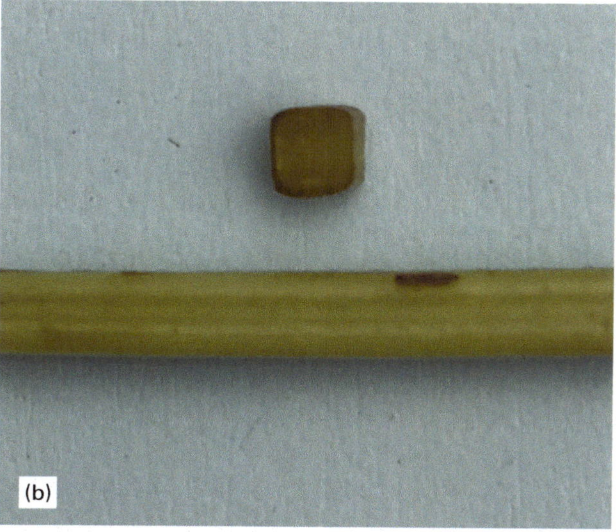

Figure 2.12 (a) A eudicotyledonous angiosperm: *Rosa rugosa* (rose) flower with five petals and succulent fruits (hips); (b) square stems of lemon balm (*Melissa officinalis*) characteristic of Lamiaceae

The family is an important grouping in horticulture as all plants within a family have certain characteristics in common, so predictions can often be made about other family members. For example, many members of the family Rosaceae are susceptible to the disease fireblight (see p. 301), whilst members of the Brassicaceae (the cabbage family) can be infected by club root (see p. 293). Thus, the gardener can anticipate these diseases when growing these groups of plants and prevent their spread.

Within a family, plants are organized into groups of similar plants called **genera** (sing. genus). A family may contain many genera, for example the Asteraceae (the daisy family) with 1,900 or more genera including *Lactuca* (lettuce), *Taraxacum* (dandelion) and *Dahlia*, or a few such as Hostaceae with just one, *Hosta*.

A genus is made up of groups of plants called **species**. A species is a group of individual plants which show the greatest degree of mutual resemblance and which, most importantly, are generally able to breed among themselves and produce viable offspring but cannot breed naturally with plants from another species.

A **genus** is a group of species within a family that have characteristics in common. A **species** is a group of individuals within a genus that have characteristics in common and are able to interbreed among themselves.

What is the APG?

Classification systems for flowering plants have been developed by individual botanists over time and this has led to a number of systems being used simultaneously in different parts of the world. These have largely been based on plant structure, especially flower structure (see Linnaeus), and plant chemistry. Since the early 1990s, a large group of botanists across the world known as the Angiosperm Phylogeny Group (APG) have collaborated to produce a new classification for flowering plants based on their evolutionary relationships or phylogeny. The groups in this system include all the descendants of a common ancestor, unlike previous classification systems. The development of increasingly sophisticated methods for studying plant genetics through DNA sequencing underlies their work. Their most recent update was published in 2016 (APG IV). Four groups are recognized: basal angiosperms which contain the most primitive flowering plant groups such as the Nymphaeales (which includes the waterlily family); the Magnoliidae (which includes the magnolia, laurel and pepper families); the monocots containing typical groups such as Liliales (which includes the lily family) and Poales (grasses); and the eudicots, the remaining group and the most advanced, which has replaced the previous group, the dicotyledons.

The APG has caused some controversial reclassifications of flowering plant species leading to many name changes (see p. 25). Nevertheless, eventually this should provide a more stable and objective classification. The work of the APG is steadily gaining acceptance and many botanic gardens, including Kew, have now adopted their system.

Naming plants – the binomial system

Classification and nomenclature (naming plants) are often confused but they are quite different. Classification is about putting plants together in groups based on their characteristics and relationships, whereas naming plants is what you do when you have decided where they go! Assigning names to plants was initially ungoverned, leading to much confusion, but a framework of rules (the **International Code of Nomenclature** for algae, fungi and plants – the **ICN**) must now be followed to allocate a botanical name.

Linnaeus utilized a naming system which included the name of the genus to which it belonged followed by its individual species name, both written in botanical Latin. This is called a binomial after the two parts and is often referred to as a plant's 'botanical name'. For example, the English oak is in the genus *Quercus* and is the species *robur*; note that the genus name begins with a capital letter, while the species has a small initial letter. Other examples are *Ilex aquifolium* (holly), *Magnolia stellata* (star magnolia) and *Ribes sanguineum* (redcurrant). The genus and species names should be written in *italics*, or underlined where this is not possible.

The name given to a plant species is very important and is the key to identification in the field or garden. Botanical plant names are stable and unambiguous, therefore their use avoids confusion. They are an international form of identity used by researchers and gardeners alike, in an internationally understood language. Armed with a botanical name, information on a specific plant can be reliably sourced from books and the internet.

Figure 2.13 Marsh marigold (*Caltha palustris*)

A botanical name is required before breeders can legally protect the new plants they have bred and also means that the correct plant can be selected and identified in planting schemes, preventing wastage. When dealing with medicinal plants and herbs, poisonous plants can be avoided.

Common names that we use for plants, such as daisy, potato and lettuce are, of course, acceptable in English but are not universally used. Common names may vary with location, for example, *Caltha palustris* has 140 names in Germany, 60 in France and 90 local names in Britain and Ireland including marsh marigold, kingcups and Mayblobs! (Figure 2.13).

Alternatively, the same common name can describe several different species.

Bluebell is the local name for *Campanula rotundifolia* in Scotland, *Hyacinthoides non-scripta* in England, *Wahlenbergia saxicola* in New Zealand, *Clitoria ternatea* in West Africa and *Phacelia minor* in the USA, none of which are related (Figure 2.14).

Figure 2.14 Two plants known as bluebells: (a) *Hyacinthoides non-scripta* (English bluebell) (source: Shutterstock, Olga S photography); (b) *Campanula rotundifolia* (known as bluebell in Scotland and harebell in England)

Common names may be in a variety of languages and scripts and often plants are introduced without a common name, for example *Camellia sinensis*, or with one invented by the seller. A scientific method of naming plants therefore enables every plant to be unambiguously identified with an accurate name that is universally recognized.

Specific names can also give information about the plant. Some are commemorative such as *willmottiae* after Ellen Willmott 1860–1934, the English gardener; some are geographical, either local such as *cantabridgensis* (of Cambridge) or national such as *lusitanicus* (from Portugal); and some are more general such as *occidentalis* (of the west). They may refer to a plant's original habitat such as *palustris* (of the meadows) and *sylvatica* (of the woods). Many specific names describe the plant's season such as *praecox* (early), life cycle such as *annuus* (annual) or use such as *edulis* (edible) and *utilis* (useful). The name *officinalis* refers to its being sold in apothecary shops and is therefore often applied to medicinal plants. By far the largest group are those that refer to some plant attribute: for example *alba* (white), *repens* (creeping), *mollis* (soft), *spinosus* (thorny), *cordatus* (heart shaped) and *nana* (small). The list is endless.

Plants within a species can vary genetically in the wild giving rise to a number of naturally occurring individuals with distinctive characteristics, much as people vary in their appearance. Where these differ significantly from the original species (but not so much as to be a different species) they may be given an additional name after the species name and are called a **subspecies** (subsp.), **varietas** (var.)

or **forma** (f.) depending on the degree of difference (forma being the least different and subspecies the most). These extra names are written in Latin and are italicized. They follow the species name, beginning with a small letter and with the category abbreviated and unitalicized in front of them, for example *Euphorbia characias* subsp. *wulfenii*, *Ceanothus thyrsiflorus* var. *repens* and *Primula sieboldii* f. *lactiflora*. Subspecies often arise in widely different locations and reflect a long period of isolation and subsequent evolution away from the original species, heading towards becoming a new species themselves. On the other hand, formas mostly arise within a population of the same species due to a small genetic change, often leading to a different flower colour for example.

Finally, botanical names must always include an **authority citation** after the species name, although these are often omitted in practice. It refers to the person who published the name of the plant (not the person who discovered it), for example, the 'L.' in *Fragaria vesca* L. stands for Linnaeus.

Hybrids

When cross-pollination occurs between two genetically different plants, either in the wild or through plant breeding, hybridization results and the offspring usually bear characteristics distinct from either parent. Sexual hybridization can occur between different cultivars within a species, sometimes resulting in a new and distinctive cultivar. Alternatively, it may occur between two different species within a genus, resulting in an **interspecific hybrid** (Figure 2.1), for example *Prunus* × *yedoensis* and *Erica* × *darleyensis*. A much rarer hybridization between species from two different genera results in an **intergeneric hybrid** such as ×*Hesperotropis leylandii* (leyland cypress, formerly ×*Cupressocyparis leylandii* or ×*Cuprocyparis leylandii*) and ×*Fatshedera lizei*. The names of the resulting hybrids include elements from the names of the parents, connected or preceded by a multiplication sign (×) depending on the type of hybrid. A **graft hybrid**, or chimaera, consists of tissue containing genetic material from two distinct parents which arises from a graft, not a sexual cross. It is indicated by a plus sign (+) in front of the name. The graft hybrid +*Laburnocytisus adamii* arose from a graft between *Cytisus purpureus* and *Laburnum anagyroides*.

Cultivars and trade designations

Selection and breeding involving intervention by humans has produced variations in species referred to as cultivated varieties or **cultivars**. These have not usually arisen in the wild and must be maintained, either by specific breeding programmes to produce seed or by vegetative propagation. The cultivar is given a name often chosen by the plant breeder who produced it, for example *Rhododendron arboreum* 'Tony Schilling' or *Cornus alba* 'Sibirica', and this is always a non-Latin, vernacular name, unitalicized and enclosed in single quotation marks. Cultivar names can also provide information about a plant's characteristics, for example, a dessert apple that is suitable for small gardens, 'Red Devil', produces bright red fruit; a thornless blackberry, 'Loch Ness', was raised in Scotland and shows considerable winter hardiness; *Penstemon* 'Apple Blossom' describes the pale pink and white flowers. This information is useful for gardeners. Cultivar names can also be written, where applicable, after a common name, often for fruits and vegetables, for example tomato 'Ailsa Craig' and apple 'Bramley's Seedling'.

The general term '**variety**' (which is not the same as 'varietas' described earlier) is often used by horticulturists to refer to any plant type which varies from the original species. As such, it is frequently used interchangeably with 'cultivar'. In this publication, however, the correct term 'cultivar' will be used throughout.

Examples of plant groupings and how they are named in the family Rosaceae are shown in Figure 2.15.

> A **cultivar** is a variation within a species that has usually arisen from plant breeding and has been maintained through propagation. A **horticultural variety** is a general, non-botanical term for plants that vary from the original species.

A **trade designation** is an additional 'selling name' which is considered more acceptable than the cultivar name and replaces it at the point of sale. It is always written without quotation marks and in a different font, for example *Choisya ternata* **SUNDANCE** is the selling name for *Choisya ternata* 'Lich' (Figure 2.16). Cultivar names should appear on the plant label along with the trade designation but rarely do!

Plant breeders' rights

Plant breeders can apply for **plant breeders' rights** (PBRs) to give legal protection to their investment in developing a new cultivar. This gives the producer exclusive rights over 25–30 years to propagate, sell, market, export, import or stock the cultivar or licence others to do so. In effect it is a form of patenting. PBRs are governed by the International Union for the Protection of New Varieties of Plants (UPOV).

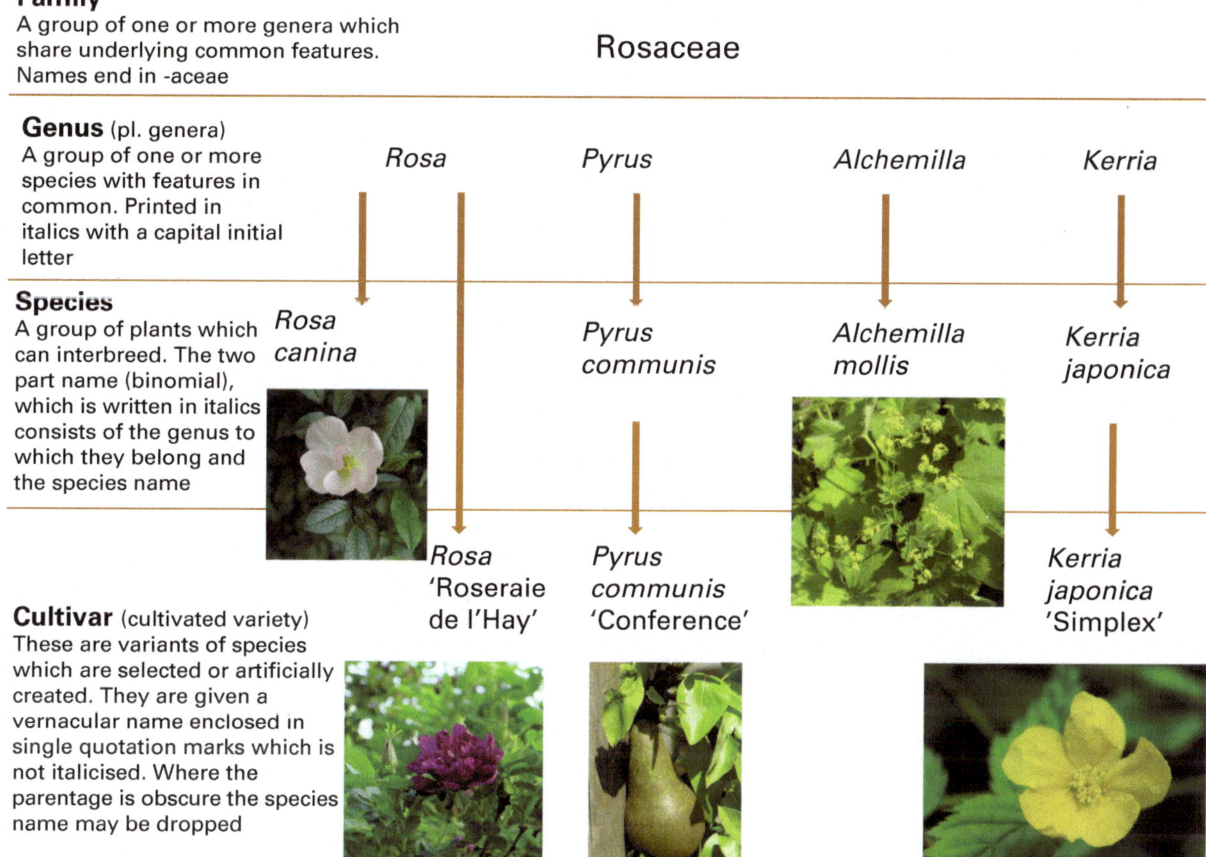

Figure 2.15 Some plant groups within the family Rosaceae

In UK, the Animal and Plant Health Agency (APHA) is the national body responsible and operates the scheme for 50 or more horticultural and agricultural crops, including trees, woody shrubs and climbers, herbaceous perennials, bulbs, cut flowers and pot plants, lawn grasses, top and soft fruit and some vegetables. PBRs can be very costly to obtain because tests must be carried out to check that the cultivar is genuinely **distinct** from previous cultivars, and that it is **uniform** and **stable** so individual plants do not vary any more than would be expected within a population or over time. The chosen name must also fulfil plant naming rules set by the ICNCP (the International Code of Nomenclature for Cultivated Plants). With a PBR, the breeder has the right to take someone to court if they are using or selling their cultivar without permission.

Figure 2.16 *Choisya ternata* 'Lich', otherwise known as *Choisya ternata* SUNDANCE (its trade designation or 'selling name')

Plant name changes

Most horticulturists yearn for stability in the naming of plants. However, the reasons for change are justifiable.

Firstly, new scientific findings may show that a genus or species needs reclassifying; it may belong in a different section of a plant family and in a new genus. For example, the two genera *Mahonia* and *Berberis* are no longer distinct enough to warrant being placed in separate genera so all species are now 'sunk' in *Berberis* (Figure 2.1). There may be differing views among scientists over whether a genus should be 'lumped' together with an existing genus, as with *Berberis*, or 'split' further into smaller units as happened to Liliaceae which originally had very many genera but currently only contains 15, or just left unchanged. At the end of the day this is a taxonomic judgement.

In addition, plant names may change to comply with naming rules. A common problem is that the same plant may have been 'discovered' independently by different plant collectors at different times. On each occasion a new name will have been given, whereas they are all in fact the same species. For example, over time, 29 different names have been given to the common daisy *Bellis perennis*. To rectify this, only the earliest published name is legitimate so all subsequent names (synonyms) are invalid. Sometimes this '**Law of Priority**' can be waived and the name 'conserved' if changing the name would have disastrous repercussions. This was the case for the houseplant *Fittonia* (published in 1865) which was allowed to keep its name despite the existence of an older generic name *Adelaster* (published in 1861).

Finally, plants can sometimes be introduced to the market simply with an incorrect name which then becomes established. An example is *Sutera cordata* (Figure 2.17), a popular trailing plant for hanging baskets and containers, introduced into the UK in 1992 and sold as either *Bacopa* 'Snowflake' or *Sutera diffusa*, neither of which is correct!

It does seem likely, though, that changes in plant names will continue to be a fact of horticultural life.

Checking plant names

The Royal Botanic Gardens, Kew, provide an accurate and current list of plant names for nearly 1.5 million plant species (Plants of the World Online) which is a very helpful resource. In addition, the names of plants and plant cultivars for sale in Britain and Ireland, from almost 400 nurseries, can be found in the RHS Plant Finder which is updated and published annually. Where plants are sourced directly from growers, plant names can be checked, but many garden centres buy plants in from elsewhere and their labels are not always reliable (Figure 2.18).

Figure 2.17 *Sutera cordata* sold as *Bacopa* 'Snowflake'

Further classifications of plants

Plants can also be grouped into other useful categories. A classification based on the length of their life cycles (ephemerals, annuals, biennials and perennials) has long been used, and these are important when considering plants in a garden. The classification reflects their life cycle strategies in response to environmental conditions.

Ephemerals have several life cycles in one season and frequently produce prolific amounts of seed to maximize their reproduction in as short a time as possible. They are opportunists, their seeds often germinating rapidly in disturbed ground when they are brought to the surface before they are outcompeted by other plants. **Annuals** have one life cycle in a season and the species continues in the form of seeds (see p. 32) protected by the soil during periods of inhospitable environmental conditions such as low winter temperatures or drought when the adult plant would not survive. **Biennials** usually grow vegetatively in their first season, growing quickly to compete with surrounding plants and storing up the energy

Figure 2.18 (a) A garden centre plant (b) The plant label – *Lithops* and *Pleiospilos* are two separate genera. *Lithops pleiospilos* does not exist neither does *Pleiospilos rubra*. *Lithops rubra* is a synonym of *Lithops optica*

required to reproduce in their roots, sometimes dying down over winter. They then regrow, flower and set seed in the second season when they complete their life cycle.

Perennials live for more than two years. Most herbaceous perennials, whose soft aerial growth would not survive adverse environmental conditions, exist below ground over winter as roots, often developing perennating organs (see pp. 71 and 72) which store energy in the form of starch for rapid growth in the spring. Most **herbaceous** perennials die back fully in winter and produce new foliage in the spring, but some 'evergreen' herbaceous perennials do retain some living structure above ground, for example *Bergenia cordifolia* and *Helleborus argutifolius*, although these never develop a woody structure. **Woody** perennials are the longest-lived plants having a stem and branch structure which can withstand tough environmental conditions and persist above ground. Some are **deciduous**; they will lose their leaves in the winter in response to those conditions whilst others remain evergreen with adapted leaves which can survive in a hostile environment.

Perennials, both herbaceous and woody, tend to be used to provide ornamental interest and also in the fruit garden and, being more permanent planting, they have a greater potential for providing food and shelter for wildlife. Annuals are often planted in containers, in formal bedding schemes and as fillers for spaces in borders and these, together with biennials, are particularly represented in the vegetable garden.

Many naturally perennial plants, such as *Pelargonium zonale*, *Lobelia erinus* and *Erysimum* × *cheiri* (wallflower) may be grown as annuals or biennials that are removed and replaced after their first or second season of growth. These may be referred to as 'annuals' and 'biennials' by horticulturists – an example of where botany and horticulture disagree!

A distinction can also be made between different types of woody plants such as **trees** and **shrubs** (single-stemmed or multistemmed). Another classification can be made based on a plant's temperature tolerance (**hardiness**), separating those plants that are able to withstand a frost (hardy), those that cannot (tender) and those which can withstand a few degrees of frost but may need some winter protection (half-hardy). The hardiness scheme in Table 2.2 is based on the classification produced by the Royal Horticultural Society (RHS). It must be remembered that, while withstanding cold conditions is the main factor in a plant's hardiness rating, other factors, such as wind and soil conditions (relating to the species origins), will play a part.

Table 2.2 brings together these useful terms, providing some definitions and some plant examples.

Table 2.2 Some commonly used terms that describe the life cycles, structure, leaf retention and hardiness of plants

	Description	Example(s)
Life cycles		
Ephemeral	A plant that has several life cycles in a growing season.	*Senecio vulgaris* (groundsel)
Annual	A plant that completes its life cycle in one growing season.	*Limnanthes douglasii* (poached egg flower)
Biennial	A plant with a life cycle that spans two growing seasons.	*Digitalis purpurea* (foxglove)
Perennial	A plant with a life cycle of more than two growing seasons.	*Quercus robur* (English oak), *Viburnum opulus*, *Acanthus spinosus*
Plant structure		
Herbaceous perennial	A perennial that usually loses its stems and foliage at the end of the growing season, surviving below ground. It produces new stems and foliage above ground the following season and never becomes woody.	*Papaver orientalis* (Oriental poppy), *Humulus lupulus* (hop)
Woody perennial	A perennial that maintains live woody stem growth at the end of the growing season through to the next season.	Bush fruit, shrubs, trees, climbers (e.g. *Vitis vinifera*, grapevine)
Shrub	A woody perennial plant having side branches emerging from near ground level appearing multistemmed. Up to 5 m tall.	*Syringa vulgaris* (lilac), *Osmanthus delavayi*
Tree	A large woody perennial unbranched for some distance above ground so on a single stem. Usually more than 5 m tall.	*Aesculus hippocastanum* (horse chestnut), *Liriodendron tulipifera* (tulip tree)
Leaf retention		
Deciduous	A plant that sheds all its leaves at once and remains leafless for a period of time.	*Philadelphus delavayi* (mock orange), *Fagus sylvatica* (beech)
Evergreen	A plant retaining leaves in all seasons.	*Aucuba japonica*, *Pinus sylvestris* (Scots pine)

(Continued)

Table 2.2 (Continued)

	Description	Example(s)
Semi-evergreen	A plant that retains some of its leaves well into winter, shedding some leaves under severe cold or extreme conditions.	*Lonicera nitida, Ligustrum ovalifolium* (privet)
Hardiness ratings		
H1a	Heated tropical glasshouse. Temperature > 15°C. Needs to be grown as houseplant or under glass all year round.	*Anthurium andraeanum, Maranta leuconeura Nepenthes × hookeriana*
H1b	Heated sub-tropical greenhouse. Temperature 10°C to 15°C. Can be grown outdoors in summer in sunny and sheltered locations but generally performs best as a houseplant or under glass all year round.	*Strelitzia reginae* (bird of paradise) *Monstera deliciosa* (Swiss cheese plant), *Ficus elastica* (rubber plant)
H1c	Heated greenhouse. Warm temperate. Temperature 5°C to 10°C. Can be grown outdoors in summer throughout most of Britain and Ireland while daytime temperatures are high enough to promote growth.	*Pelargonium* cultivars, *Coleus scutellarioides* cultivars, *Brugmansia* species
H2	Tender – cool or frost-free greenhouse. Temperature 1°C to 5°C. Tolerant of low temperatures but will not survive being frozen. Except in frost-free inner-city areas or coastal extremities requires glasshouse conditions in winter, but can be grown outdoors once risk of frost is over.	*Citrus x limon* 'Meyer', *Musa basjoo, Nicotiana sylvestris, Petunia* cultivars, *Cleistocactus* species
H3	Half-hardy. Unheated greenhouse/mild winter. Temperature –5°C to 1°C. Hardy in coastal/mild areas except in hard winters and at risk from sudden (early) frosts. May be hardy elsewhere with wall shelter or good microclimate. Can survive with artificial winter protection.	*Clianthus puniceus, Dahlia* cultivars, *Lathyrus odoratus* cultivars, *Cordyline australis, Dicksonia antarctica*
H4	Hardy – average winter. –10°C to –5°C. Hardy through most of Britain and Ireland apart from inland valleys, at altitude and central/northerly locations. May suffer foliage damage and stem dieback in harsh winters in cold gardens. Plants in pots are more vulnerable.	*Hesperocyparis macrocarpa* 'Goldcrest', *Daphne bholua* 'Jacqueline Postil', *Escallonia* cultivars
H5	Hardy – cold winter. –15°C to –10°C. Hardy through most of and Ireland even in severe winters. May not withstand open or exposed sites or central/northerly locations. Many evergreens suffer foliage damage and plants in pots will be at increased risk.	*Nymphaea* cultivars, *Meconopsis* cultivars, *Platycodon grandiflorus*
H6	Hardy – very cold winter. –20°C to –15°C. Hardy throughout Britain and Ireland and northern Europe. Many plants grown in containers will be damaged unless given protection.	*Laburnum × watereri, Hemerocallis* cultivars, *Miscanthus sinensis* cultivars
H7	Very hardy. < –20°C. Hardy in the severest European continental climates including exposed upland locations in Great Britain and Northern Ireland.	*Abies koreana, Achillea* 'Moonshine', *Betula* species and cultivars, *Phlox paniculata* cultivars
Half-hardy	A plant able to survive temperatures between 1°C to –5°C. Tolerant of a few degrees of frost. Will survive a mild winter but generally requires an unheated glasshouse over winter.	See H3 earlier
Tender	A plant able to survive temperatures1°C to 5°C. Tolerant of low temperatures but will not survive frost. Requires a cool or frost-free glasshouse in winter. Can be grown outside after danger of frost is over.	See H2 earlier
Heated glasshouse	A plant requiring temperatures above 5°C to survive. Some can be grown outside in summer when daytime temperatures are high enough or in a sheltered position. Some must be grown under glass or as houseplants all year. Can be divided into further categories depending on the plants' requirements for temperatures above 5°C.	See H1a–c earlier

The following terms are derived from the use of plants:

- **Bedding**: fast-growing species, often flowering annuals, used to make a temporary, usually formal display (see p. 26). They are grown from seed and discarded after a few months to provide separate summer and winter bedding displays often on the same site for example *Petunia*, *Pelargonium*, *Sedum*, *Viola*, *Bellis*. More recently there has been a trend away from annual bedding for more informal perennial planting which has a lower carbon footprint and is less wasteful of plants.
- **Tropical**: a non-native, usually exotic, tender plant, often perennial, used for indoor or seasonal outdoor display in herbaceous borders and bedding schemes, for example *Canna*, castor oil plant (*Ricinus communis*)
- **Edging**: low and often slow-growing species, grown in similar groups to create an edge to a path or boundary between planted areas, often in bedding schemes, for example *Thymus*, *Viola*, non-spreading hardy geraniums, and many low-growing, compact perennials
- **Dot plant**: a single plant, usually tall, planted to create a focal point within a bedding scheme, for example foxtail lily (*Eremurus*), standard roses, *Phormium*, *Acer palmatum*

Figure 2.19 Shaggy ink cap, a fungus showing its fruiting bodies

- **Ground cover**: low growing, usually evergreen plants designed to completely cover the soil, for example ivy (*Hedera helix*), periwinkle (*Vinca major*), *Epimedium x rubrum*

Fungi

The kingdom Fungi is a very diverse group about which much remains to be studied. It has been estimated that there are 1.5 to 5 million species of which only 5% have been classified to date! Some fungi are single celled (such as yeasts), but others are multicellular, such as moulds and the more familiar mushrooms and toadstools (Figure 2.19).

Most are made up of a mycelium, which is a mass of thread-like filaments (hyphae) which generally remains hidden from view. The mushrooms we see at certain times of the year are the spore-producing part of the life cycle (see p. 288).

Fungi obtain their food directly from other living organisms (heterotrophic nutrition), sometimes causing disease (see Chapter 18), or from dead organic matter, so contributing to its beneficial breakdown in the soil (see Chapter 12). They achieve this by secreting digestive enzymes onto their food source and absorbing the soluble products. Fungi (mushrooms) are also an important horticultural food crop.

Oomycetes, which are not true fungi, have simple spore forms and cell walls made of cellulose, not chitin. They may also cause disease in horticultural crops (see p. 291).

Lichens (Figure 2.20) are an algae or a photosynthetic bacterium (cyanobacterium) combined with one or more fungi, whilst mycorrhiza are fungi which associate with plant roots. These are both examples of mutualistic relationships (see p. 141) in which the association benefits both partners.

Figure 2.20 Lichens – a combination of fungi and cyanobacterium: (a) *Xanthoria* sp.; (b) *Parmotrema* sp.; (c) *Usnea* sp.

Further reading

Adams, C., Early, M., Brook, J. and Bamford, K. (2015) *Principles of Horticulture Level 3*. 7th ed. Routledge.

Allaby, M. (1992) *The Concise Oxford Dictionary of Botany*. Oxford University Press.

Angiosperm Phylogeny Group. (*Missouri Botanical Garden*). www.mobot.org/mobot/research/apweb/

Blunt, W. (1971) *Linnaeus: The Compleat Naturalist*. Frances Lincoln.

Cubey, J. (Ed.) (2024) *RHS Plant Finder 2025*. Royal Horticultural Society.

Dorling Kindersley. (2022) *The Science of Plants: Inside Their Secret World* (DK Secret World Encyclopaedias). Dorling Kindersley.

Edwards, D., Konyves, K., Lancaster, N. and Dee, R. (eds.) (updated annually). *The RHS Plant Finder*. Summerfield Books.

Harrison, L. (2018) *RHS Latin for Gardeners*. Octopus.

Johnson, A.T. and Smith, H.A. (2008) *In Plant Names Simplified: Their Derivation and Meaning*. Old Pond Publishing.

Royal Botanic Gardens, Kew. *APG – Classification by Consensus*. www.kew.org/read-and-watch/apg-classification-consensus

Royal Botanic Gardens, Kew. *Plants of the World Online*. https://powo.science.kew.org/

Royal Horticultural Society. (2023) *AGM Plants April 2023 © RHS – ORNAMENTAL*. chrome-extension://efaidnbmnnnibpcajpcglclefindmkaj/https://www.rhs.org.uk/plants/pdfs/agm-lists/agm-ornamentals.pdf

Sterndale-Bennet, J. (2005) *Plant Names Explained*. David and Charles.

Zona, S. (2022) *A Gardener's Guide to Botany: The Biology Behind the Plants You Love, How They Grow, and What They Need*. Cool Springs Press.

The online material is accessible via the QR code and includes further information on many of the topics in the book, such as

- Plant Varieties and Seeds Act
- Non-plant kingdoms
- Plants of the world
- Plant name meanings
- Plant identification tools
- The ICN & IUCN

The plant life cycle

Figure 3.1 Autumn colour in the leaves of Japanese maple (*Acer palmatum*) indicating leaf senescence

This chapter includes the following topics:

- Growth and development
- The seed – viability, germination requirements and dormancy
- The seedling – hypogeal and epigeal germination, tropisms
- Juvenile growth
- The adult plant
- Senescence and death

DOI: 10.4324/9781003581260-3

Growth and Development

Like people, plants follow a series of distinct phases in their lives, described as the plant's 'life cycle'. Unlike humans, the plant life cycle can be either very short, a matter of weeks in ephemerals such as groundsel (*Senecio vulgaris*), or many thousands of years as in the bristlecone pine (*Pinus aristata*), some individuals of which are thought to be more than 5,000 years old. In most cases, a plant's life begins with fertilization (see p. 103) and the development of the embryo within a **seed**. Germination of the seed gives rise to a seedling, which undergoes a **juvenile** period of growth and development. On reaching **adulthood** the plant is able to reproduce, then having produced fruits and seeds, a period of **senescence** ending in **death** of the plant ensues (see Figure 3.2).

The changes that take place in the structure, form and behaviour of a plant through its life cycle can usefully described as '**plant development**'. This is in contrast to the term '**plant growth**' which refers to the increase in a plant's weight and size. Plant growth is brought about in two ways, by new cells being produced in the meristems (see p. 49) and the new cells expanding due to turgor pressure (see p. 124).

The production of new cells is fuelled by the processes of photosynthesis, respiration and mineral uptake (see Chapters 7 and 8). The typical life cycle of plants, from seed to death and some of the horticultural implications for each phase, are described in this chapter.

Plant growth is the increase in the size of cells, organs and the whole plant due to cell division and cell expansion. **Plant development** describes the changes in structure, form and behaviour throughout its life cycle.

The seed

Following fertilization, the seed represents the first stage of the plant's life cycle. Seeds contain and protect the plant embryo, which will grow into the new plant, and act as a food store which supports growth until the plant is able to photosynthesize and manufacture its own food supply. A seed enables the plant to withstand periods when the environment is not suitable for growth and may have mechanisms which bring about seed dormancy. It is also the means by which many plants spread away from the parent plant and colonize new areas, thereby reducing competition for water and nutrients and increasing the success of survival of the species. Seed structure is described in Chapter 6.

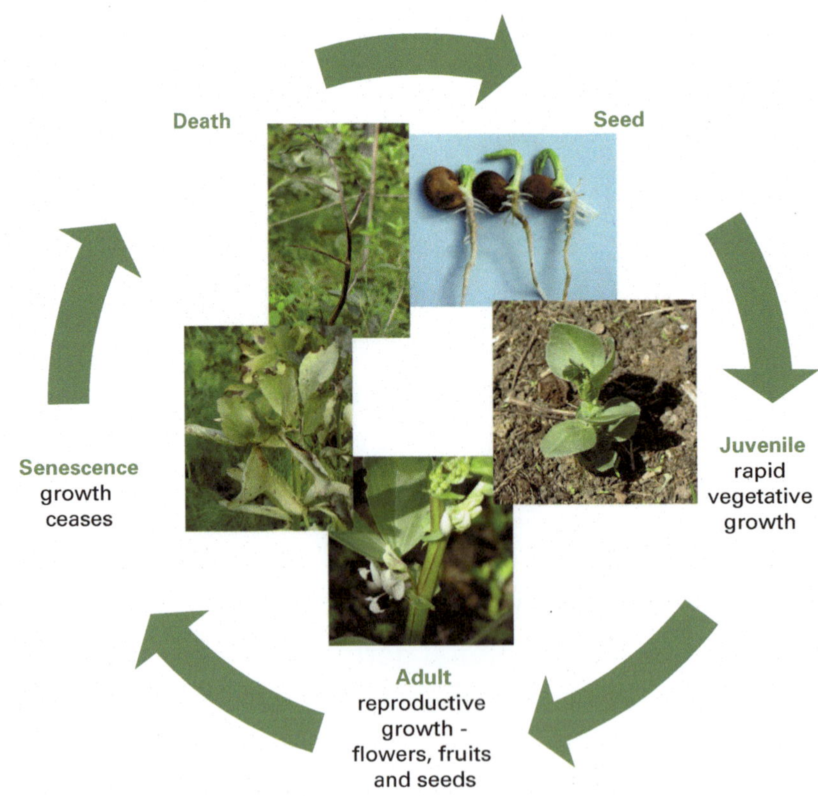

Figure 3.2 Stages in the life cycle of a *Vicia faba* (broad bean) plant

Figure 3.3 (a) Flower of sacred lotus (*Nelumbo nucifera*) with developing seed head; (b) seed head showing embedded seeds

Seed viability

A **viable seed** has the potential for germination when the required external conditions are supplied. Its viability is, therefore, an indication of whether the seed is 'alive'. Most seeds remain viable until the next growing season, but many can remain so for a number of years until conditions are favourable for germination. Some seeds are amazingly long-lived, especially if stored in very dry conditions. For example, seeds of the sacred lotus (*Nelumbo nucifera*) survived in a dry lakebed in China for 1,300 years (Figure 3.3), whilst a date palm seed germinated after being buried for 2,000 years! In general, viability of a batch of seed diminishes with time, its maximum viability period depending largely on the species. For example, celery seed quickly loses viability after the first season, whereas wheat has been reported to germinate after scores of years. The germination potential of any seed batch will depend on the storage conditions of the seed, which should generally be cool (but not freezing) and dry, thereby slowing down respiration (see p. 119).

The majority of seeds, described as **'orthodox'** seeds, can be stored successfully for long periods in these conditions, for example French bean (*Phaseolus vulgaris*), carrot (*Daucus carota*), lobelia (*Lobelia erinus*), *Lolium perenne* and love-in-a-mist (*Nigella damascena*), while other **'recalcitrant'** seeds lose viability quickly if dried out so have to be stored in moist, cool conditions for only a short period of time. These tend to be large fleshy seeds such as horse chestnut (*Aesculus hippocastanum*), oak (*Quercus robur*), sweet chestnut (*Castanea sativa*), willow (*Salix* spp.), avocado (*Persea americana*) and mango (*Mangifera indica*). Cool, dry conditions are achieved in commercial seed stores by means of sensitive control equipment. Packaging of seed for sale (Figure 3.4) takes account of these requirements and often includes a waterproof lining of the packet, which maintains a constant, low water content together with low oxygen levels which reduce respiration in the seeds (see respiration p. 119).

The ability of seeds to germinate depends greatly on their storage conditions particularly the initial moisture content, the temperature in store and the length of time they are stored, as viability will decrease with time.

Figure 3.4 A seed packet with a waterproof and airtight insert containing the seed

Seed germination

For seeds to germinate successfully, a number of environmental conditions must be supplied:

- water
- oxygen
- a suitable temperature
- light (in some seeds).

Water is needed to trigger germination. The water content must be increased from around 10% in the dry seed to 70% or more. Seed structure is shown on p. 106. Water is initially absorbed (imbibed) by the seed entering the structure of the seed coat (testa) in a way similar to a sponge and softening it. It is followed by uptake through a pore in the testa called the micropyle. The cells of the seed take up water by osmosis (see p. 124), often assuming twice the size of the dry seed, and this is a passive process, that is, it does not require energy. A continuous water supply is now required if germination is to proceed at a consistent rate. Eventually the new root (radicle) breaks through the seed coat and emerges followed by the new shoot (plumule).

Oxygen is essential for respiration (see p. 119). On imbibition of water, respiration rates increase dramatically as the seed's food stores, the cotyledons or endosperm, are broken down to produce energy and building materials for rapid cell division enabling the new root and shoot to develop. The growing medium, whether it is outdoor soil or compost in a seed tray, must not be waterlogged because oxygen would be withheld from the growing embryo.

Correct temperature is a very important germination requirement and is usually specific to a given species or even a cultivar. It acts by fundamentally influencing the activity of the enzymes (see p. 48) involved in the biochemical processes of respiration which occur between 0°C and 40°C. However, species adapted to specialized environments may respond to a narrower range of germination temperatures.

In general terms temperate plants need temperatures above 5°C, whereas those from tropical and sub-tropical areas need 10°C or more. For example, cucumbers (*Cucumis sativus*) require a minimum temperature of 15°C and tomatoes (*Solanum lycopersicum*) 10°C, whereas lettuce (*Latuca sativa*) germination may be inhibited by temperatures higher than 30°C, and in some cultivars a period of induced dormancy occurs at 25°C. Some species, such as mustard (*Sinapis alba*), will germinate in temperatures just above freezing and up to 40°C, provided they are not allowed to dry out. Typical temperature ranges for various plants are given in Table 3.1.

Light is a factor that may influence germination in some species, but most species are indifferent. Seed of *Rhododendron*, *Veronica* and *Phlox* is inhibited in its germination by exposure to light, while that of celery, lettuce, most grasses, conifers and many herbaceous flowering plants is slowed down when light is excluded. On the other hand, some species such as birch (*Betula pendula*) and foxglove (*Digitalis purpurea*) need light to germinate, both being woodland plants that take advantage of the light penetrating to the woodland floor when clearings appear. This is also a common feature of many weed seeds which lie buried in the soil and germinate only when they are brought to the surface by cultivation.

When a viable seed fails to germinate because any one of these three factors is not suitable, for example if water is not present or the temperature is too low, the seed is said to be '**quiescent**'.

Seed germination is defined as the emergence of the young root or radicle through the testa, usually at the micropyle.

Table 3.1 Optimum germination temperature ranges

Seed	Plant type	Temperature range °C
tulip (*Tulipa*)	Mediterranean bulb	10–15
busy lizzie (*Impatiens*)	Summer bedding	20–25
pansy (*Viola*)	Winter bedding	13–16
sea holly (*Eryngium*)	Sun-tolerant herbaceous	20–25
solomon's seal (*Polygonatum*)	Shade-tolerant herbaceous	10–15
grasses	Turf grasses	10–12
bottlebrush (*Callistemon*)	Sun-tolerant shrub	20–25
spotted laurel (*Aucuba*)	Shade-tolerant shrub	15–18
carrot (*Daucus carota*)	Temperate vegetable	10–30
tomatoes (*Solanum lycopersicum*)	Tropical vegetable	15–30
aubergine (*Solanum melongena*)	Tropical vegetable	25–30

Seed dormancy

As soon as the seed germinates, the plant is vulnerable to damage from cold or drought. Seed dormancy prevents germination occurring when poor growing conditions prevail. It also brings about staggered germination so that even if some seedlings perish when the weather turns, others will survive. Dormant seeds are unable to germinate even though water, oxygen and the correct temperature are given to them (unlike quiescent seeds), and they use a variety of mechanisms to delay germination until conditions become more favourable. For example, the seed coat may prevent water and oxygen entering, or may be so hard that the embryo is unable to penetrate it, as in many leguminous plants such as sweet pea (*Lathyrus odoratus*). Water penetration may also be impeded by the surrounding tissue in fleshy fruits which is high in sugar so that the osmotic movement of water (see p. 124) is out of the seed and into the fruit flesh, for example in tomato (*Solanum lycopersicum*), rather than in the opposite direction. Many seeds, for example in apple (*Malus* spp.), have chemical inhibitors in their seed coat and embryo which prevent germination whilst others are shed with immature embryos which need a period of time to develop fully before germination can commence, such as in ash (*Fraxinus excelsior*). The balance between the natural plant growth regulators within the seed (**abscisic acid** which promotes dormancy and **gibberellic acid** which promotes germination) changes with time.

In the wild, such dormancy mechanisms are gradually overcome through the abrasive action of the soil, the action of soil microorganisms, ingestion and excretion of seeds by birds and exposure to cold temperature cycles together with freezing and thawing which soften and break down the seed coat and surrounding fruit structure. Inhibitors are washed out by rain or are broken down chemically, and embryos can mature during a period of dormancy. In some seeds, *Protea* spp. for example, chemicals in smoke overcome dormancy. Horticulturists have a range of techniques to mimic nature and reduce the period of dormancy or 'break' dormancy such as:

- **Scarification** – physically damaging the seed coat by nicking or scratching it or rubbing with sandpaper, for example in *Lathyrus* and *Paeonia* seeds. Commercially, seeds can be rotated in drums with an abrasive lining and for *Gleditsia triacanthos* soaking in sulphuric acid can be used.
- **Soaking** – in hot or cold water, for example in *Beta vulgaris* seeds, to soften the seed coat and remove inhibitors. Tomato (*Solanum lycopersicum*) seeds are soaked and fermented in water for three to seven days to completely remove the flesh around the seed.
- **Stratification** – soaked seeds are stored in warm or cold temperatures in a moist environment such as sand or compost, for example in *Lupinus*, *Aconitum* and *Euonymus*. The compost must be well aerated and the seed can be stored in bags, or pots or pits in the ground for large quantities. Cold storage may be outside over winter or in a fridge between 1°C and 5°C. Some seeds require cycles of warm and cold stratification.

The potential effects of **climate change** on seed dormancy and germination are a subject of research. Because plants differ in their requirements for breaking dormancy, it is difficult to predict. Dormancy is affected by temperature and water stress as the seed matures on the plant, even before it is shed, and once shed, the length and depth of dormancy is influenced by seasonal changes affecting soil temperature and moisture levels. Once released from dormancy, germination itself is triggered by water availability and temperature. Climate change will affect all these factors and therefore potentially the dormancy and germination behaviour of seeds. Some effects may be:

- higher temperatures and lower rainfall in summer when the seed is on the plant will reduce seed dormancy after it is shed
- where cold temperatures are needed to break dormancy, shorter periods of low temperature or warmer temperatures may not overcome it
- an earlier spring may trigger earlier germination
- a longer and warmer autumn may result in premature germination.

The provenance of the seed (see p. 149) is important when selecting seed which will be able to combat climate change, as plants of the same species producing seed under different climatic conditions and altitudes vary in their dormancy requirements and germination behaviour.

The seedling

The emergence of the plumule above the growing medium is usually the first occasion that the seedling is subjected to light. This stimulus prevents rapid extension of the stem so that it becomes thicker and stronger. The leaves unfold and become green (produce chlorophyll) in response to light, which enables the seedling to photosynthesize and so support itself. Seedlings which are deprived of light show **etiolated growth** with elongated internodes, pale reduced leaves, few branches and no chlorophyll (Figure 3.5). At this stage the seedling is still very susceptible to attack from pests and damping off diseases.

The cotyledons are often the first part of the seed to develop and they may emerge from the testa and remain in the soil (**hypogeal germination**), as in broad bean (*Vicia faba*), Oriental lilies (e.g. *Lilium martagon*) and pea (*Pisum sativum*), or be carried with the testa into the air, above the soil where the cotyledons expand (**epigeal germination**) as in tomato (*Solanum lycopersicum*), French bean (*Phaseolus vulgaris*) and Asiatic lilies (e.g. *Lilium cernuum*) (Figure 3.6). In epigeal germination the delicate new shoot is protected by the hooking of the hypocotyl as it emerges above the soil and also by the cotyledons which enclose it.

Figure 3.5 Etiolation in broad bean (*Vicia faba*) seedlings germinated and grown in the dark

Hypogeal germination occurs when the cotyledon/s remain below the ground. **Epigeal** germination occurs when the cotyledon/s emerge above the ground.

The cotyledons in epigeal germination are sometimes called 'seed leaves' when they emerge and they usually look quite different from 'true leaves'. They contribute initially to photosynthesis in the seedling, but the true leaves very quickly unfold and take over this function (Figure 3.7).

Once the food store in the cotyledons and/or endosperm has been exhausted, the seedling must rapidly produce its own food supply and begin to photosynthesize. It must therefore respond to its environment to establish the correct direction of growth. Such a response is termed a **tropism** and is very important in the early survival of the seedling.

A **tropism** is a directional growth response to an environmental stimulus.

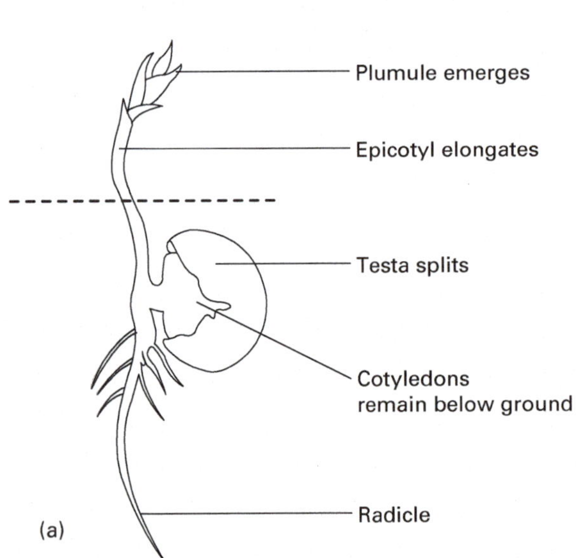

Figure 3.6 (a) Hypogeal germination in broad bean (*Vicia faba*); (b) broad bean on the left, French bean on the right; (c) epigeal germination in French bean (*Phaseolus vulgaris*)

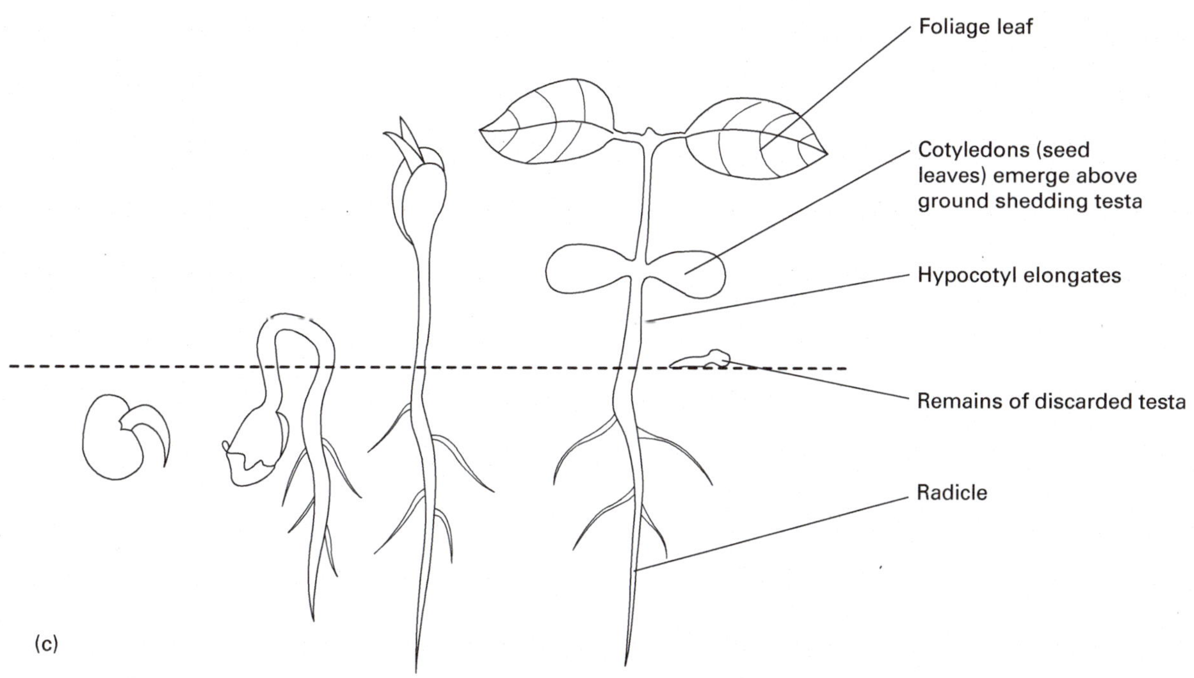

Figure 3.6 (*Continued*)

Figure 3.7 Germinating blackthorn (*Prunus spinosa*) seeds showing the simple 'seed leaves' or cotyledons which are produced first, followed by the true leaves

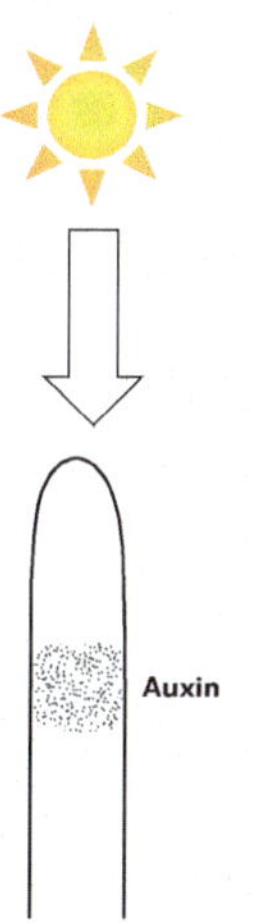

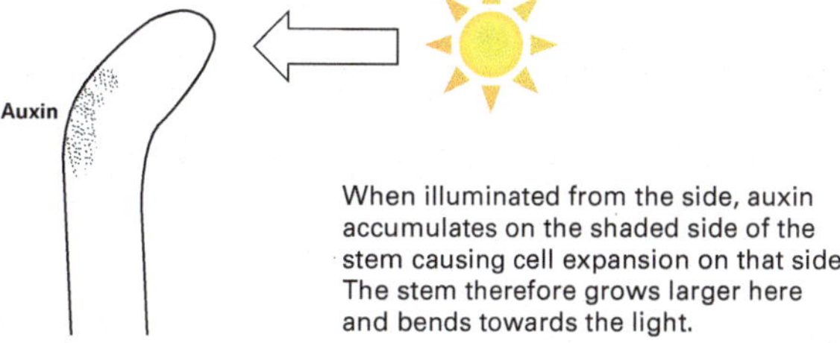

Figure 3.8 The mechanism of positive phototropism in a shoot

Geotropism (also called **gravitropism**) is a directional response to gravity. The emergence of the radicle from the testa is followed by growth of the root system, which must quickly take up water and minerals to enable the shoot system to develop. A seed germinating near the surface of a growing medium must not put out roots that grow on to the surface and dry out but must establish roots that grow downwards to tap water supplies. Conversely, **phototropism** enables the shoot to grow towards a light source that provides the energy for photosynthesis. If the light source is to the side of the plant, a bend takes place in the stem just below the tip as cells in the stem away from the light grow larger than those nearer to the light source. A greater concentration of a substance called **auxin** (a plant growth regulator) in the shaded part of the stem causes the extended growth (Figure 3.8). If the light source is directly above the shoot, auxin is distributed equally across the stem so it grows straight up. Roots display

Figure 3.9 Leaf shapes in ivy (*Hedera helix*) (a) juvenile growth with lobed leaf; (b) adult growth showing fruits and leaf without lobes

positive geotropism (they grow in the direction of the gravitational pull) and negative phototropism (they grow away from the light) whilst shoots display the reverse, they are positively phototropic and negatively geotropic.

Juvenile growth

The early growth stage or **juvenile stage** is the period after germination that is capable of **rapid vegetative growth** and is **non-reproductive sexually**. By putting its energies into growth rather than producing seeds, the plant can establish itself more effectively in competition with others.

Juvenile growth can be characterized by certain physical appearances and activities that are different from those found in adult growth. Plant **growth habit** may differ; the juvenile stem of ivy (*Hedera helix*), for example, tends to grow horizontally, without terminal buds, and is vegetative in nature with adventitious roots for climbing (Figure 5.14), and its energy is put into internode extension rather than leaves. Adult growth is vertical, with shortened internodes, large leaves and terminal shoots which bear flowers and fruit.

In some *Citrus* such as lemon (*Citrus × limon*), juvenile non-fruiting stems grow vigorously and bear thorns for defence against herbivores unlike adult mature stems which bear fruit. Often these can be found on the same plant. Often **leaf shapes** vary, for example, the juvenile leaf of ivy (*Hedera helix*) is three-lobed while the adult leaf is very different, more oval in shape, as shown in Figure 3.9.

The juvenile leaves on many *Eucalyptus* spp. can vary greatly from the adult plant. In *Eucalyptus gunnii*, they are round and often without petioles (sessile) compared with the adult plant and juvenile stems are much used by florists (Figure 3.10).

Other leaf characteristics such as colour and arrangement on the stem may differ. Differences in juvenile growth are also common in conifer species, where the complete appearance of the plant is altered by the change in leaf form, for example in *Juniperus* species such as *Juniperus chinensis*. In the genera *Chamaecyparis* and *Thuja*, the juvenile condition can be achieved permanently by repeated vegetative propagation producing plants called **retinospores**, which display their juvenile leaf shape and are used as decorative features in the garden.

Figure 3.10 *Eucalyptus gunnii* leaves: left: juvenile leaves; right: adult leaves

Figure 3.11 *Pseudopanax crassifolius*: (a) juvenile form; (b) adult form (source: Shutterstock, Steve Todd)

An interesting example is the endemic New Zealand genus *Pseudopanax* (Figure 3.11). Juvenile plants of lancewood trees (*Pseudopanax crassifolius*) have very narrow, downward-pointing leaves up to a metre long but when adult, after 10 to 20 years, they take on a very different form which is more like a conventional tree. It is suggested that the juvenile form protected the plant from a giant flightless bird called a moa (found in the fossil record) which became extinct nearly 600 years ago and was the dominant herbivore. Once it had grown above the height of the moa, the tree assumed its adult form.

Leaf retention (marcescence) is sometimes a characteristic of juvenility and can be significant in species such as beech (*Fagus sylvatica*) and hornbeam (*Carpinus betulus*). Annual pruning in the dormant season keeps the plants in juvenile growth and provides an attractive hedge with colourful leaves retained throughout the winter (Figure 3.12). This can create additional protection in windbreaks, although the barrier created tends to be too solid to provide ideal wind protection.

Juvenile growth is non-reproductive (vegetative) growth whereas **adult growth** is sexually reproductive (flowering) growth.

In **propagation**, juvenility is related to rooting success. Softwood cuttings are non-flowering and root easily but as the season progresses and shoots switch from juvenile to flowering growth, rooting becomes more difficult. Rooting hormones may need to be applied to stimulate rooting in semi-ripe and hardwood cuttings for this reason. Since flowering growth often roots less easily, flowers and flower buds must be removed if juvenile material is not available. Adult growth should be removed from stock plants to leave the more successful juvenile growth for cutting.

Juvenility may also affect **pest and disease resistance**. In Dutch elm disease, juvenile elm trees are resistant but they succumb at around 15–20 years old when they are mature.

Sometimes juvenile leaf characteristics are desired rather than flowers, as in *Eucalyptus* spp., Indian bean tree (*Catalpa bignonioides*), smoke bush (*Cotinus* spp.) and tulip tree (*Liriodendron tulipifera*). The juvenile leaves greatly increase in size giving foliage interest in the garden but the flowers are sacrificed. Juvenility may be maintained by **coppicing or stooling**, that is, cutting the plants to a low framework or 'stool' each year in their dormant seasons, and **pollarding**, where trees are cut to a single stem a short distance above the ground (Figure 3.13). These are traditional pruning techniques which have been used for many

Figure 3.12 Leaf retention (marcescence) in (a) a young beech tree (*Fagus sylvatica*); (b) a formal beech hedge in winter; (c) a single hornbeam (*Carpinus betulus*) pruned as an ornamental column

centuries in woodlands and elsewhere to produce wood for many purposes, such as fencing or leaves for fodder. When carried out periodically, typically every 15 years in the case of hazel (*Corylus avellana*) for example, trees are maintained in a juvenile or partially juvenile state and this can lengthen their lives considerably. Many veteran trees are ancient pollards (see p. 148).

The adult plant

The adult stage is defined by the ability of the plant to **reproduce sexually** and therefore produce flowers, fruit and seed. The progression from a vegetative to a flowering plant involves profound physical and chemical changes. This change may simply be genetically programmed with a plant switching to adult growth after a certain number of leaves are produced or when it has reached a certain size. Often, however, an environmental stimulus is required, such as a period of low temperature and/or a specific daylength which links flowering to an appropriate season. In this way many plants flower characteristically at particular times of year, for example lilac (*Syringa vulgaris*) in the spring, *Buddleja alternifolia* in early summer, *Hypericum calycinum* in late summer to autumn and *Viburnum× bodnantense* in the winter.

In the reproductive phase of the life cycle, flowers are potent 'sinks' for a plant's resources, drawing the sugars made in photosynthesis towards them at the expense of vegetative growth. Plant growth therefore slows and all but ceases as all the plant's energies are redirected to producing flowers, fruits and seeds. Adult growth often shows different growth patterns to juvenile growth as in ivy (*Hedera helix*) and changes in rootability of cuttings (see p. 39).

In many plants, adult growth is also linked to producing stores of food towards the end of the growing season for overwintering and growth the following spring (**perennation**). Once flowering is over, sugars may be redirected and stored as starch in the stems of woody plants or in modified roots, leaves and stems such as **bulbs, rhizomes and tubers** in herbaceous plants These may also be a means of vegetative spread (**asexual reproduction**) and are useful to the horticulturist as propagating material (see Chapters 4 and 5).

Pruning in adult plants

Plant pruning is often carried out to reduce the competition within the plant for the available resources. The plant is encouraged to grow, flower or fruit in the way the horticulturist requires. A reduction in the number of flower buds, for example, in chrysanthemum (*Chrysanthemum* × *morifolium*) will cause the remaining buds to develop into larger flowers; a reduction in fruiting buds of apple trees will produce bigger apples, and a reduction in the branches of soft fruit and ornamental shrubs will allow the plants to grow stronger when planted densely.

To encourage flowering and improve the quantity and quality of blooms, species that flower on the previous year's growth of wood such as *Forsythia* should be pruned soon after flowering has stopped. Conversely, species that flower later in the year on the present year's wood, such as *Buddleja davidii*, should be pruned the following spring, to maximize the growth period for flower production.

Figure 3.13 (a) Juvenile leaves in Indian bean tree (*Catalpa bignonioides* 'Aurea') which has been pollarded; (b) the same tree pollarded in winter

As flowers age, they begin to use up a considerable amount of the plant's energy in the production of fruits. In addition, chemicals (plant growth regulators) produced by the fruit switch off flower development because they have done their job. By removing dead flowers (**deadheading**), plants may continue to flower for many weeks longer than those allowed to retain their dead flowers and the appearance of a garden border can be maintained. Examples of species benefitting from this procedure are seen in bedding plants which flower over several months including African marigold (*Tagetes erecta*); in herbaceous perennials *Delphinium* and *Lupin*; in small shrubs such as *Penstemon fruticosus*; and in climbers, sweet pea (*Lathyrus odoratus*) and many *Rosa* spp.

Many species such as begonia (*Begonia semperflorens*) and busy lizzie (*Impatiens walleriana*) used as bedding plants have been specially bred so that flowers do not produce fruits and continue flowering over a longer period. In such cases, there is not such a great need to deadhead, but this activity will help to prevent unsightly rotting brown petals from spoiling the appearance of foliage and newly produced flowers. On the other hand, such flowers will not benefit wildlife in the garden, as they will not produce seeds.

Pruning – some general principles

Pruning affects the shape of the plant, through a property of plants called **apical dominance**. This is where the apical bud inhibits growth of buds further down the stem (see p. 54). By removing the apical bud, lateral shoots are released and develop. The success of such pruning depends very much on the skill of the operator, so a good knowledge of the species habit and response to pruning is required together with an appreciation of the purpose of pruning.

- **Young plants** should be trained in a way that will reflect the eventual shape of the more mature plant (formative pruning). For example, a young apple tree (called a 'maiden') can be pruned to have one dominant 'leader' shoot, which will give rise to a taller, more slender shape. Alternatively, selecting a few branches at the desired height and cutting back the leader in the first few years forms a bush apple which is not too tall and has an open centre. A cordon is a plant in which there is a leader shoot, often trained at 45° to the ground, and where all side shoots are pruned back to one or two buds. Cordon fruit bushes are usually grown against walls or fences. Similarly, fans and espalier forms can be developed. However, regular trimming of hedges produces a mass of laterals with no single leader making a dense, well-shaped hedge.
- **The pruning cut** should be made just above a bud that points in the required direction (usually to the outside of the plant). In this way, the plant is less likely to acquire too dense growth in its centre. Some plants such as fruit trees, roses and gooseberries are made less susceptible to disease by the creation of an open centre to reduce humidity.
- **Pruning should remove any shoots that are diseased or are rubbing together (crossing),** as they may have damaged bark which could be a point of entry for disease.
- **Weak shoots should be pruned the hardest** where growth within the plant is uneven, and strong shoots pruned less, since pruning causes a stimulation of growth.

Root pruning is used in the growing of bonsai plants to keep the plant small. By reducing the root area the plant will reduce its leaf area to compensate, keeping water uptake and water loss in balance (see p. 130). Similarly, if large shrubs or trees are transplanted the shoots may be cut back prior to moving them as the roots will inevitably be damaged.

The senescent plant

The term '**senescence**' in the plant life cycle refers to the period between adulthood and death of the plant. It is the stage after flowering and fruiting where growth has ceased and a gradual deterioration occurs. This is most obvious in ephemeral, annual and biennial plants which flower and fruit only once before senescence and death.

In perennial plants, the same cycle of seed, juvenile growth, adult growth, senescence and death occurs through the plant's lifetime, as in all plants, but the term 'senescence' is also used to describe the changes that take place through the year and are repeated each season in leaves and fruits. In deciduous trees and shrubs, changes in leaf colour associated with the autumn are due to **pigments** in the leaves which are revealed as the chlorophyll (the green pigment) is broken down and absorbed by the plant prior to leaf fall (Figures 3.1 and 3.14) and waste products accumulate.

Pigments are substances that are capable of absorbing light; they also reflect certain wavelengths of light, which determine the colour of the pigment. In the actively growing plant, chlorophyll, which reflects mainly green light, is produced in considerable amounts, and therefore the plant, especially the leaves, appear predominantly green. Other pigments are present; for example, the carotenoids (yellow) and xanthophylls (red), but usually the quantities are so small as to be masked by the chlorophyll. In some species, such as copper beech (*Fagus sylvatica* Purpurea Group), other pigments predominate, masking chlorophyll. Many colours are displayed in the leaves at this time in such species as *Acer platanoides*, turning gold and red, *Prunus cerasifera* 'Pissardii' with light purple leaves, European larch (*Larix decidua*) with yellow leaves, Virginia creeper (*Parthenocissus quinquefolia*) and *Vitis* spp. with red leaves. Colourful autumn fruits are also prized in the garden as they mature and senesce such as the berries of *Cotoneaster* and *Pyracantha* which, together with coloured stems which can become more intense as in dogwoods (*Cornus* spp.), provide autumn displays at a time when fewer flowering plants are seen outdoors.

In deciduous woody species, as leaf senescence continues, the leaves drop in the process of **abscission**, which can be triggered by shortening daylengths. An abscission layer of loose cells forms at the base of the leaf stalk (petiole) which enables the leaf to be shed in the wind. To reduce risk of water loss from the remaining leaf scar, a corky layer is also formed behind the abscission layer before the leaf falls.

Figure 3.14 Autumn colour in (a) blueberry (*Vaccinium corymbosum*), (b) *Viburnum* and (c) *Photinia* showing loss of chlorophyll and emergence of xanthophylls

In many **fruits**, ripening is associated with an increase in respiration and changes in colour, sweetness, flavour, texture and scent. This is soon followed by natural progression to a senescent stage in which the fruit deteriorates and the seed ripens before it is shed, although growers aim to harvest fruit before this stage is reached and arrest further development through a range of postharvest techniques such as the use of reduced temperatures and control of the gases surrounding the fruit. Ripening, senescence and abscission are all processes which are triggered by a plant growth regulator called **ethylene (ethene)**.

Eventually, senescence is followed by **death** at the end of the season in annuals or at the end of the plant's life in biennials and perennials. All metabolic processes cease and the plant matter is returned to the soil to be taken up by, and to sustain, future plants.

Further reading

Adams, C., Early, M., Brook, J. and Bamford, K. (2015) *Principles of Horticulture Level 3*. 7th ed. Routledge.

Brickell, C. and Joyce, D. (2017) *RHS Pruning and Training*. Dorling Kindersley.

Capon, B. (2022) *Botany for Gardeners: An Introduction to the Science of Plants*. 4th ed. Timber Press.

Chalker-Scott, L. (2015) *How Plants Work*. Timber Press.

Collins, S. and Mabbit, M. (2021) *RHS How to Grow Plants from Seed*. Mitchell Beazley.

Cushnie, J. (2007) *How to Prune*. Kyle Cathie.

Farrimond, S. (2023) *The Science of Gardening: Discover How Your Garden Really Grows*. Dorling Kindersley.

Hartman, K. (2010) *Plant Propagation, Principles and Practice*. Prentice Hall.

Hodge, G. (2013) *Practical Botany for Gardeners*. University of Chicago Press.

Royal Horticultural Society. (2013) *RHS Botany for Gardeners: The Art and Science of Gardening Explained & Explored*. Mitchell Beazley.

The online material is accessible via the QR code and includes further information on many of the topics in the book, such as

- List of ephemerals
- List of annuals
- List of biennials
- List of perennials
- Seed storage
- Endospermic/non-endospermic seeds
- Seed dormancy

CHAPTER 4

Plant cells and tissues

Figure 4.1 Freshly cut oak (*Quercus robur*) logs showing the anatomy of woody stems brought about by the process of secondary thickening

This chapter includes the following topics:

- Plant cells and their contents
- Plant tissues
- Stem and root anatomy
- Growth and differentiation of the stem and root
- Woody stems and secondary thickening

DOI: 10.4324/9781003581260-4

Plant tissues, and the cells that form them, work together to enable plants to carry out the functions essential for life. As in humans, specialized tissues support the plant body, enable movement of substances round it, manufacture the substances it requires to live and to reproduce and protect it from its environment. An understanding of how these components work and interact with each other enables gardeners to provide successfully for their plants' needs, thus optimizing their growth.

The internal structure (anatomy) of the plant is made up of different **tissues**. Each tissue is a collection of specialized **cells** carrying out one function, such as xylem tissue conducting water and nutrients. An **organ** is made up of a group of tissues carrying out a specific function such as a leaf producing sugars for the plant.

A **tissue** is a collection of cells carrying out a specific function. Tissues can be **simple**, containing only one cell type, or **complex**, containing several cell types.

Plant cells and their contents

Without the use of a microscope, the horticulturist will not be able to see cells, since they are very small, about a twentieth of a millimetre in size.

Cells and the microscope

The original inventor of the microscope is hard to establish but it seems likely that the earliest instruments were probably developed in the Netherlands in the early 1600s. In 1665 Robert Hooke, scientist, inventor, architect (and experimenter with flying machines), worked on improving the instrument and examined many natural objects under his microscopes (Figure 4.2). He published *Micrographia*, a book of observations made using microscopes and telescopes which included many illustrations. He coined the term 'cell', referring to the resemblance between the cork cells he studied and the cells which monks lived in. He estimated that a cubic inch of cork would contain about 1,259 million cells! These first microscopes were the familiar optical microscopes which use light to view the subject; these are still widely used today in various forms. They can produce a magnification up to about 2,000×. Modern electron microscopes, however, use beams of electrons to illuminate a subject and produce an image with much greater magnification up to 10,000,000×. Scanning electron microscopes produce images of the surfaces of objects such as the leaf shown in Figure 4.3, revealing incredible detail.

Figure 4.2 A microscope manufactured by Christopher Cock of London for Robert Hooke (source: Billings Microscope Collection, National Museum, USA)

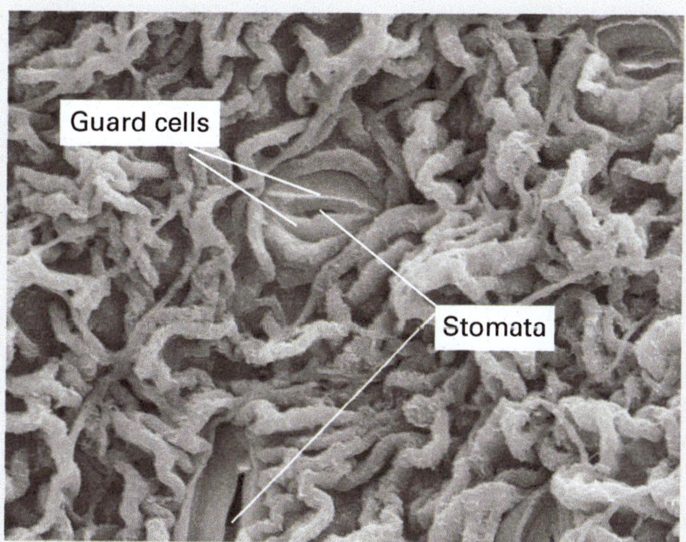

Figure 4.3 Scanning electron microscope image showing stomata on the surface of a birch (*Betula pendula*) leaf. Each small pore (stoma) is surrounded by a pair of guard cells (source: Vivian Vislap and Tõnu Järveots)

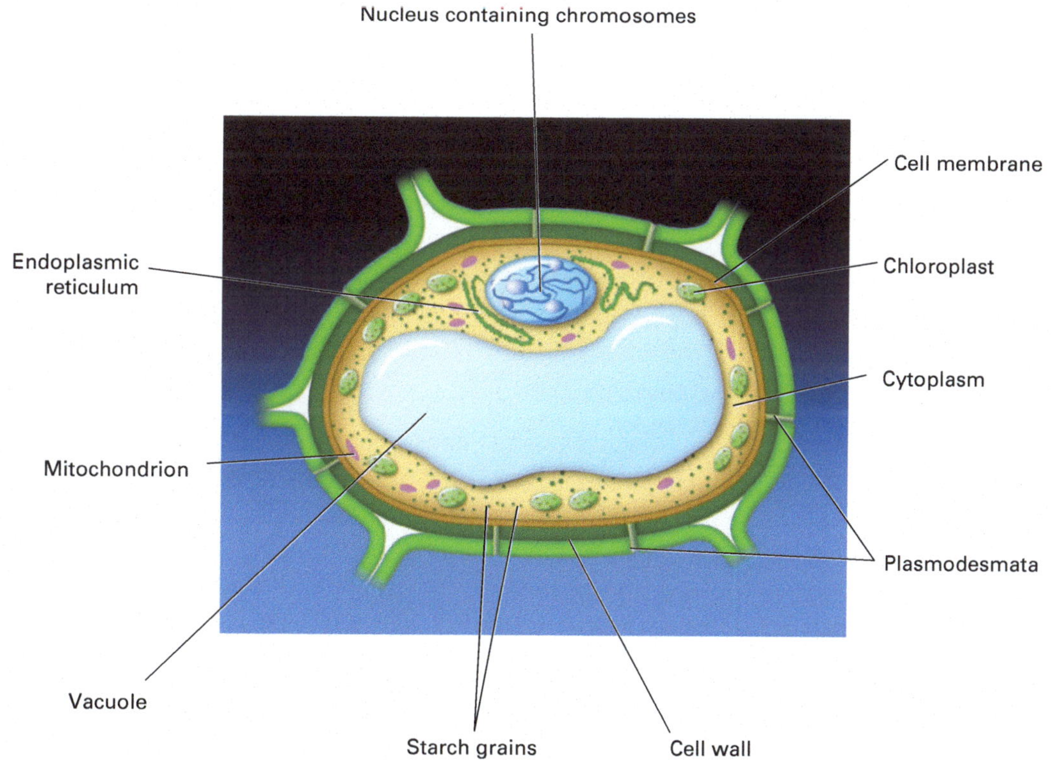

Figure 4.4 An unspecialized plant cell (source: Shutterstock, Alexonline)

A simple, unspecialized plant cell (Figure 4.4) consists of an outer **cellulose** cell wall with a cell membrane inside it which contains the cell contents suspended in the jelly-like **cytoplasm**.

The **cell wall** is formed of fibres made of cellulose laid down in a mesh which allows the wall to stretch as the cell expands. Once the cell is mature, the cell wall loses its elasticity and the permanent shape of the cell is set. In living cells, dissolved substances can pass freely through the open mesh of the cell wall. Larger items such as fungal spores, however, can be excluded. Within the mesh framework are many apertures which, in living cells such as parenchyma, allow cell membranes and strands of cytoplasm (called **plasmodesmata**) to interconnect between adjacent cells. These strands allow the passage of substances such as nutrients and plant growth regulators between cells. When a plant lacks water, it wilts and its cell contents shrink, but the plasmodesmata normally retain their links with adjacent cells and the plant can recover. In the situation of 'plasmolysis', however (see p. 125), the plasmodesmata break, wilting is irreversible and cell death ensues. The cell walls of adjoining cells are held together by a layer of calcium pectate (pectin), a glue-like substance which is an important setting ingredient in jam making. Some types of cell, for example xylem vessels, lose their cell contents and die in order to carry out their function efficiently, leaving a central cavity (lumen). The cell wall becomes thickened by additional cellulose layers and **lignin**, which is a strengthening and waterproofing substance. Cell walls, especially those strengthened by lignin, give support and mechanical strength to the cell and ultimately the plant. Grey mould (*Botrytis cinerea*) damages plant cells by releasing enzymes such as cellulases and hemicellulases which decompose cell walls to provide nutrition for the pathogen (see p. 296).

The **cell membrane** not only encloses the cell contents but also controls the movement of substances in and out of the cell. It is selective. Water crosses the membrane through the process of osmosis (see p. 124) but other substances such as sugars are excluded. Mineral nutrients required by the cell are bound to specific protein carriers embedded in the membrane. This process of mineral uptake into cells often requires energy and is then called 'active transport' (see p. 130). It allows the plant to take in useful substances from the water in the soil and reject others depending on which protein carriers are found in the membrane.

The **cytoplasm**, which is largely water, enables dissolved substances to move around the cell and take part in chemical reactions within it. It surrounds small membrane-bound structures (organelles) which have specific functions, described later, and has a network of protein strands which hold these in place to prevent them sinking to the bottom of the cell. The strands can also move them around as required. For example, chloroplasts move up and down cells according to light levels.

The **nucleus**, which is enclosed in its own membrane, coordinates the activities of the cell. The long chromosome strands that fill the nucleus contain the complex chemical DNA (deoxyribonucleic acid). In addition to its ability to produce more of itself for the process of cell division, DNA is also constantly manufacturing smaller but similar RNA (ribonucleic acid) units, which are able to pass through the nucleus membrane and attach themselves to other organelles. In this way, the nucleus is able to transmit instructions for the assembly, or destruction, of proteins and especially enzymes which speed up the chemical reactions within the cell. The coded information for these processes is found in the genes within the chromosome. Crown gall (*Rhizobium radiobacter*) is a bacterial disease of herbaceous plants such as *Dahlia* spp. and woody plants such as *Salix* spp. The bacterium combines its DNA with the host's DNA and triggers cell division causing galls, areas of abnormal tissue which form in roots and stems.

The **mitochondria** release energy, in a controlled way, by the process of respiration (see p. 119). The energy is transferred via a chemical called ATP (adenosine triphosphate). The meristem areas of the stem, root and flower, where rapid cell division and growth take place, have cells with the highest number of mitochondria as they require the most energy.

The **chloroplasts**, also bound by their own membrane, containing the green pigment chlorophyll, are involved in the production of sugar through the process of photosynthesis (see p. 112). Excess sugars are often stored as starch, and starch grains can be found in the chloroplasts and throughout the cytoplasm in living cells as well as in storage organs such as bulbs and tubers.

Both mitochondria and chloroplasts contain their own DNA and replicate within the cell independent of the cell's nucleus. This suggests they evolved from bacteria which were incorporated into cells early on during the evolution of plants!

The **endoplasmic reticulum** is a complex membrane structure that enables transport of chemicals within the cell and links with the cell membrane. Small structures called ribosomes, which are made up of RNA and protein, are commonly attached to the endoplasmic reticulum and make proteins, including enzymes.

The **vacuole** is a sac within the cell, bound by a membrane. It contains a dilute sugar solution, nutrients, pigments and waste materials. It may occupy the major volume of the cell and its main functions are storage of waste products and maintaining the cell shape through controlling cell turgor (see p. 124).

The whole of the living matter of a cell, its membrane, nucleus and cytoplasm, is collectively called the **protoplasm**. Plant cells differ from animal cells by having a cell wall, a vacuole and chloroplasts.

Proteins and enzymes

Proteins are molecules made up of simple molecules called amino acids. A protein will often contain one or more chains of amino acids enabling it to fold up into a specific three-dimensional structure. **Enzymes** are proteins which act as catalysts, that is, they speed up chemical reactions within the cell but are unchanged themselves. Their ability to do this depends crucially on maintaining their three-dimensional structure which can be destroyed by high temperatures and unsuitable pH values, causing the chains to unravel. Enzyme names end in -ase, for example, amylase is the enzyme which breaks down starch in germinating seeds. Ribulose 1,5 bis-phosphate decarboxylase (RuBisCo for short) is a key enzyme in photosynthesis (see p. 112) and is probably the most abundant protein on earth!

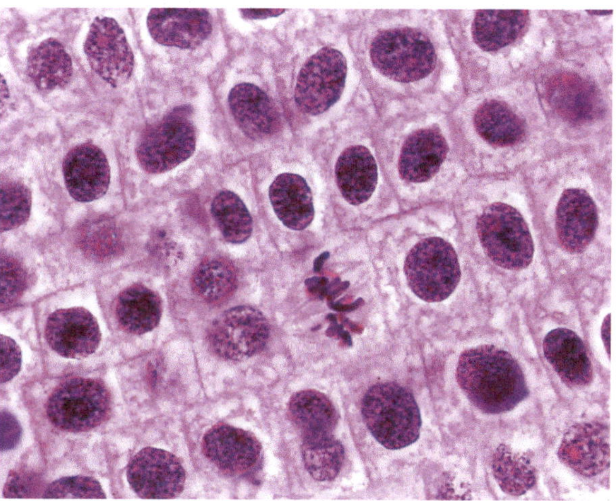

Figure 4.5 Meristematic cell in the tip of an onion root. Chromosomes are visible in the cell at the centre which is about to undergo cell division. The nucleus occupies most of the cell space

Plant tissues

Plant tissues are collections of cells performing a specific function. They may be grouped into six categories according to the functions they perform:

- **Meristematic** tissues are where new cells are produced by cell division. The cells in this tissue are **totipotent**, that is, they are able capable of developing into any cell type in the plant and are **undifferentiated**, that is, they are as yet unspecialized for any particular function. They continually divide, each cell forming two new cells which can in turn themselves divide. **Apical meristems** are responsible for lengthwise growth (**primary growth**) and are found at the tips of roots and shoots. **Lateral meristems** produce widthways growth (**secondary growth**) as the plant increases in size and needs to support itself. Two important lateral meristems in woody stems are the **vascular cambium** which gives rise to xylem and phloem tissue and the **cork cambium** which produces the outer cork layer of the bark. Meristematic cells (Figure 4.5) are cuboid in shape, have a small vacuole and a large nucleus and contain many mitochondria to provide the energy required for creation of new cells through cellular respiration (see p. 119).
- **Protective** tissues, for example, the epidermis, cover the entire plant surface in young plants and in herbaceous plants. The cells of the epidermis hold the plant together and are flattened and tightly connected to exclude air and prevent water loss. They may be modified to enclose a pore called a stoma (pl.stomata) which allows oxygen to pass to the living cells below and carbon dioxide to move out (see p. 128). They can also carry out specialized functions such as prickles on rose stems and the hairs of the insectivorous plant sundew (*Drosera anglica*) (Figure 4.6). The epidermis is

Figure 4.6 Modified epidermal outgrowths: (a) sticky hairs of sundew (*Drosera anglica*) (source: Shutterstock, Max Suduk); (b) prickles in *Rosa* spp.

replaced by a corky layer in woody stems which is the outer surface of bark.

- **Transport** or **vascular** tissues are the plumbing system of the plant and comprise two tissues. **Xylem** tissue transports water and minerals (see p. 127). It contains long, wide, open-ended cells called **xylem vessels** which lose their end walls and are joined end to end forming long tubes. These have very thick walls containing a strengthening and waterproofing substance called lignin which enables the vessels to withstand the high pressures of the water and minerals which they carry. The cells are dead, and the cell contents disappear to leave a central cavity or lumen. They are connected by **pits**, gaps in the wall where plasmodesmata were in the living cell. **Tracheids** are similar to xylem vessels but retain their end walls and are also found in xylem tissue. They are the only water transport cells in conifers. **Phloem**, again, consists of long, tube-like cells (**sieve tubes**) connected end to end and is responsible for transporting food (sucrose) manufactured in the leaves to the roots, stems or flowers (see p. 131). The sieve tubes, in contrast to xylem, are living cells with cellulose cell walls and cytoplasm. They are unusual in not containing a nucleus. The end walls are only partially broken down to leave sieve-like structures (sieve plates) at intervals along the sieve tubes. Alongside every sieve tube cell there is a small **companion cell** with a nucleus, which regulates the flow of sucrose through the sieve tube. Both xylem and phloem also transport plant growth regulators around the plant.
- **Packing** tissue, such as parenchyma, makes up the bulk of the plant. **Parenchyma cells** (Figure 4.8) are differentiated cells but are often

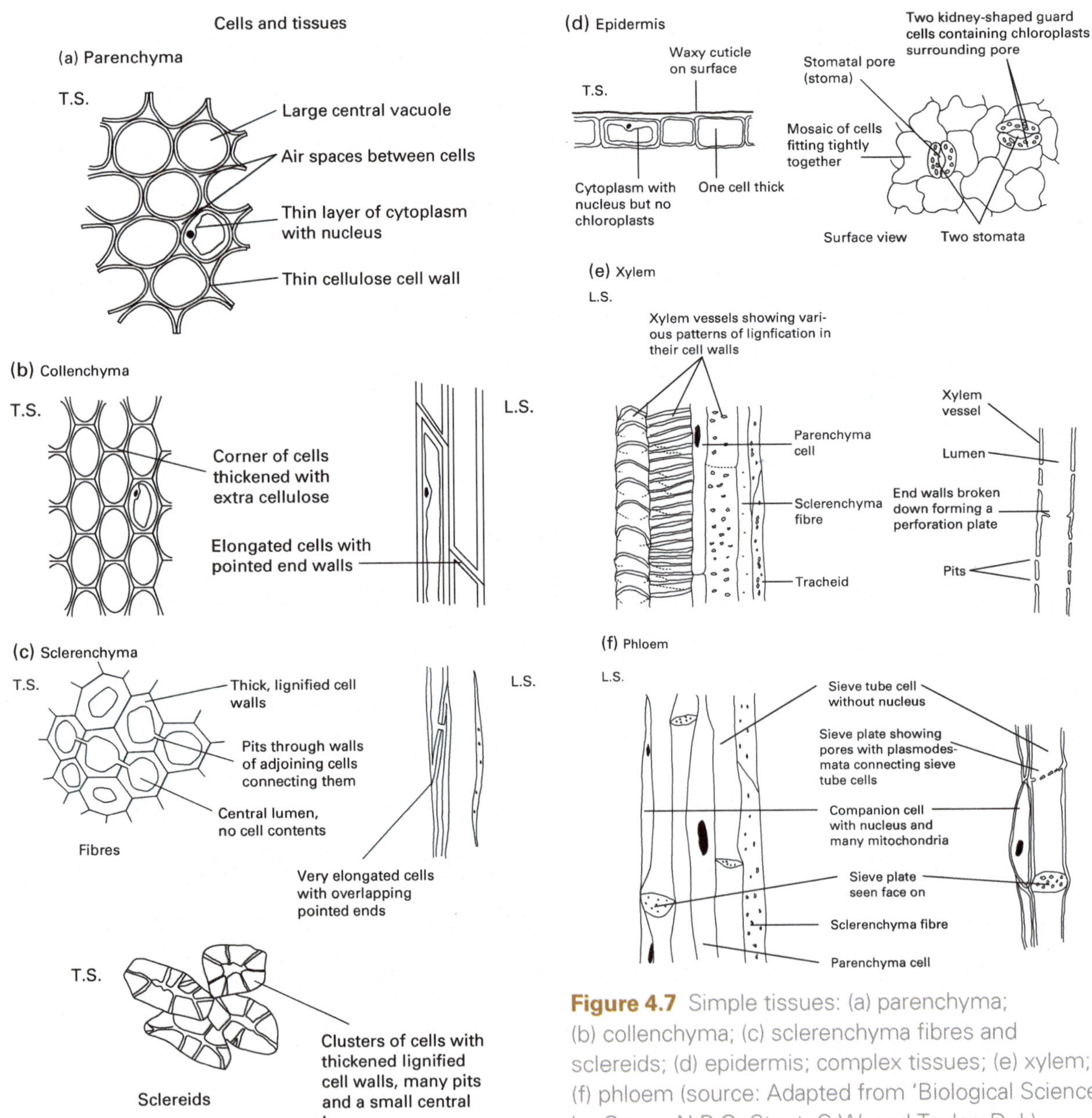

Figure 4.7 Simple tissues: (a) parenchyma; (b) collenchyma; (c) sclerenchyma fibres and sclereids; (d) epidermis; complex tissues; (e) xylem; (f) phloem (source: Adapted from 'Biological Science' by Green. N.P.O, Stout, G.W. and Taylor, D.J.)

unspecialized and may sometimes be adapted for specific functions (see box). The cells are thin walled and maintained in an approximately spherical shape by osmotic pressure (see p. 124) with many air spaces between them.

- **Supporting** tissues reinforce leaves and stems. The cell walls in these tissues have extra thickenings of cellulose in **collenchyma** tissue or lignin in **sclerenchyma** tissue. Like xylem vessels, sclerenchyma cells are dead cells connected by pits with only the thickened, lignified cell wall remaining. They are often elongated in shape with pointed ends (sclerenchyma fibres) which interlock for strength, or they may appear in a variety of shapes such as the stone cells which give pears their rough surface and gritty texture.

Some examples of plant cells and tissues are shown in Figure 4.7.

The versatile parenchyma cell

Parenchyma cells are multipurpose cells which underpin the structure and function of all plant organs. Although relatively unspecialized they can fulfil a range of roles in the plant. **Storage** is a key function. They are the main cells of the pith and cortex in stems and roots, as well as in perennating organs (see pp. 71 and 72) such as the potato tuber where they store starch. Starch is an energy source made up of units of glucose which can be released for respiration as and when needed (see p. 120). Parenchyma can also store proteins, oils and pigments and are particularly important for **water storage** in succulent plants adapted to arid regions (see p. 79). They are also involved in secretion of resins and volatile oils, for example, in *Arnica* spp. and *Centaurea cyanus* (cornflower). The large vacuole in parenchyma cells is filled with water and is responsible for **turgor pressure** which supports young plants and herbaceous plants (see p. 124). In the parenchyma rays of woody stems, living parenchyma cells **transport** sugars, nutrients and waste products across the dead xylem tissue. Parenchyma cells also form the mesophyll layers of leaves and green stems where they contain chloroplasts and are the **main photosynthetic tissue** (chlorenchyma) (see p. 64). In aquatic plants and plants adapted to waterlogged soils the parenchyma tissue (**aerenchyma**) (see p. 83) has large, interconnected air spaces between cells to **aid buoyancy**, for example in *Nymphaea* spp. (waterlily) leaves and to **store oxygen and transport it** from the aerial parts of the plant to the submerged organs as in *Juncus effusus* (soft rush).

Parenchyma cells in the pith and cortex can sometimes be triggered to undergo cell division, a useful property when a plant has been damaged or infected with a pathogen. The resulting mass of parenchyma cells is called **callus** and helps cover the wound and heal it and in extreme instances can be seen as a gall or tumour on the plant. Some of the callus cells can also differentiate into roots and stems, a useful property when cuttings are taken. Similarly, when plants are grafted, callus formation is the first step in joining the rootstock and the scion together (see p. 60). Callus is also produced when plants are regenerated by micropropagation from fragments of leaves, roots or buds and can be persuaded to develop into roots or shoots artificially by manipulating plant growth regulators in the growing medium. Parenchyma cells can be associated with other types of cells in complex tissues too. In phloem tissue, **companion cells and transfer cells** are specialized parenchyma cells involved in loading and unloading sucrose into the phloem sieve tubes and controlling their transport (see p. 131) whilst in xylem tissue, parenchyma cells can create balloon-like **tyloses** which block xylem vessels to prevent air bubbles disrupting the flow of water (see p. 127) and disease spread. In a few cases, parenchyma cells can have thick walls, for example in the seed endosperm of date palm and coffee where the sugars present in these thick walls become the nutrients for the germinating embryo.

Truly the parenchyma cell is the work horse of plant tissues!

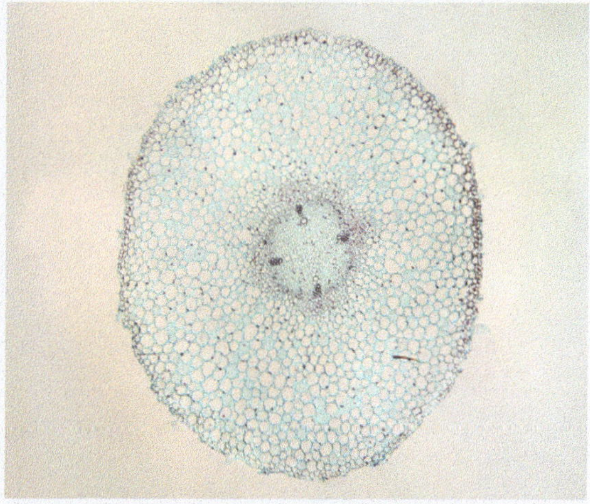

Figure 4.8 Parenchyma cells in the cortex of a broad bean (*Vicia faba*) root with thin cell walls and air spaces between cells

Science

A tissue is a collection of cells performing a specific function. It may contain one type of cell (**simple tissue**), such as parenchyma tissue or several cell types (**complex tissue**), such as xylem or phloem tissues. The name of a simple tissue is also used to describe the cell type of which it is made. The term 'tissue' may also be used to describe the different areas within plant organs such as the cortex of roots and stems or the mesophyll of plant leaves.

Primary stem and root structure

When roots and stems are first formed (primary growth), their tissues are arranged in distinctive ways. In young roots and stems, herbaceous plants and monocotyledons, and in the area behind the root and shoot tips, this structure persists. In shrubs and trees, however, which become woody, the structure eventually changes through a process called secondary thickening or secondary growth which enables sideways growth of the root and stem.

Eudicot stems

The internal structure of a eudicot stem, seen in cross section, is shown in Figure 4.9a and Figure 4.10. The stem is broadly divided into four areas of tissue: the epidermis, the cortex, the vascular tissues and the pith.

The protective **epidermis** consists of a single, transparent layer of cells on the outside of the stem which produces a waterproofing waxy layer of cutin on its surface called the **cuticle**. Pores called **stomata** punctuate the epidermal layer to allow oxygen to enter and carbon dioxide to be released from the living tissues below (see p. 128). Opening and closing each pore is controlled by a pair of epidermal cells called guard cells (Figure 4.3).

The **cortex** of the stem lies beneath the epidermis and is largely made up of **parenchyma tissue** composed of relatively unspecialized parenchyma cells. The mass of parenchyma cells maintains the plant shape. They store and release energy from starch, through respiration (see p. 119), for use in the surrounding tissues.

Pith refers to the central zone of the stem, which is also mainly made up of parenchyma cells. It may sometimes break down to give a hollow stem, for example in elder (*Sambucus nigra*). In bramble (*Rubus fruticosus*) hollow older stems are used by bees for nesting, and overwintering sites for other invertebrates are also provided by the hollow stems of hogweed (*Heracleum sphondylium*).

Inside the cortex in eudicot stems is a ring of **vascular bundles**, so named because they contain the two vascular tissues that are responsible for transport. The phloem is found to the outside and the xylem to the inside of the stem in most species. Also contained within the vascular bundles of eudicots is the **vascular cambium**. This is a **lateral meristem** which contains actively dividing cells producing more xylem and phloem tissue to increase the girth of the stem in woody plants as it grows **(secondary thickening)**. In monocots and herbaceous plants which do not become woody, this will be absent. The vascular bundles of the stem in eudicots are arranged in a ring which gives strength and support to the stem in much the same way as steel rods do in reinforced concrete.

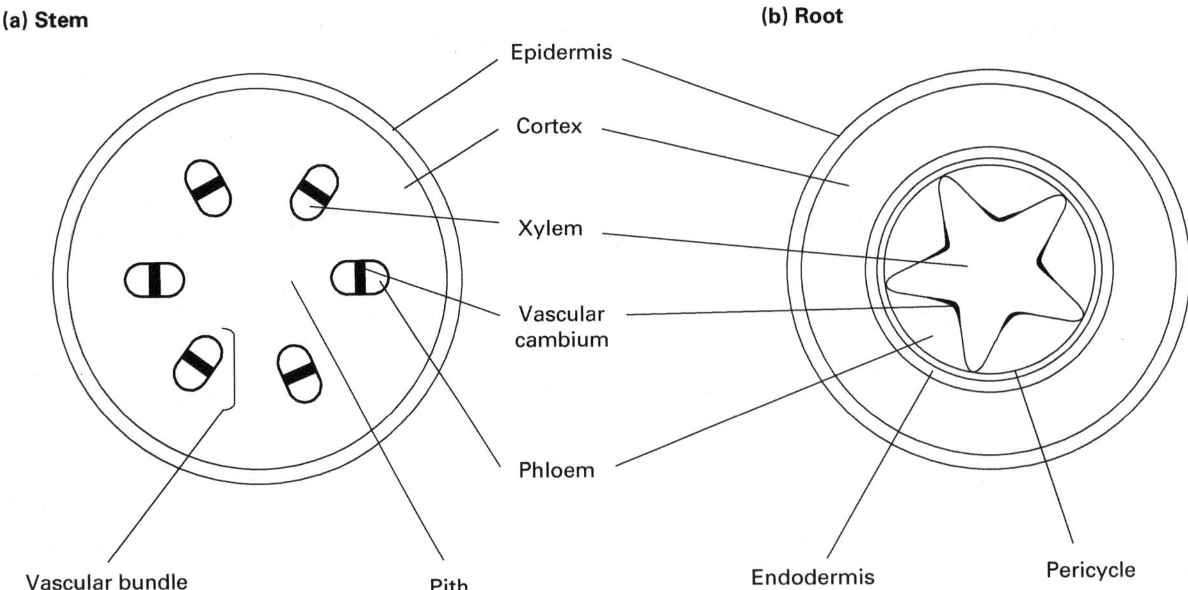

Figure 4.9 Transverse section of a young (primary) eudicot (a) stem; (b) root

Collenchyma and **sclerenchyma** tissues are often found just inside of the epidermis and around the vascular bundles and are responsible for **support** in the young plant. Both tissues have cells with specially thickened walls and, characteristically, there are no air spaces between the cells. In collenchyma, cell walls have extra cellulose, whilst in sclerenchyma cells, the thickness of the wall is increased by the addition of lignin (see p. 48). These sclerenchyma cells, which are long and tapering and interlock for additional strength, consist only of cell walls with a central cavity (Figure 4.17).

Eudicot roots

The internal structure of a eudicot root, seen in cross section, is shown in Figures 4.9b and 4.11. In eudicot roots the **epidermis** is comparable with the epidermis of the stem; it is a single layer of cells which has a protective as well as an absorptive function. Unlike the stem, it lacks a cuticle since reducing water loss is unnecessary in the root. Inside the epidermis is the parenchymatous **cortex**. The main function of this tissue is respiration to produce energy for growth of the root and for the absorption of mineral nutrients. The cortex can also be used for the storage of starch where the root is an overwintering organ (see p. 27).

The central region of the root (called the **stele**) is separated from the cortex by a single layer of cells, the **endodermis** which is not found in the stem. It has the function of controlling the passage of water and nutrients into the central xylem (see p. 126). Water and dissolved minerals pass through the endodermis to the **xylem** tissue of the stele, which transports them up to the stem and leaves (see p. 000). The arrangement of the xylem tissue varies between species but often appears in transverse section as a star with up to seven 'arms'. Since support is unnecessary in roots surrounded by soil, this arrangement can maximize water uptake. The root also has a **vascular cambium** in eudicots that undergo secondary thickening to increase its girth. As in the stem, **phloem** tissue is present for transporting sugars from the leaves to provide energy for the living cells of the root. A distinct area in the root forming a single layer of cells just inside the endodermis, the **pericycle**, is also not found in the stem. It has cells which are able to divide and produce lateral roots, which push through to the main root surface from deep within the structure.

Growth and differentiation of the stem and root

Plant growth, that is elongation of the plant stem and root, takes place in two stages (Figure 4.12). The first phase is cell division, in which new cells are formed at the apical meristems in the **zone of cell division** and are as yet undifferentiated (Figure 4.12). The second phase is cell expansion, which occurs at the base of the meristem in the **zone of cell elongation**. Here, the tiny unspecialized cells begin to take in water and nutrients and form a cell vacuole. As a result, each cell elongates and the stem or root rapidly grows.

Once expanded, cells begin to specialize and adopt their final role. They create cell walls and the connections between cells (plasmodesmata). The exact shape and chemical composition of the

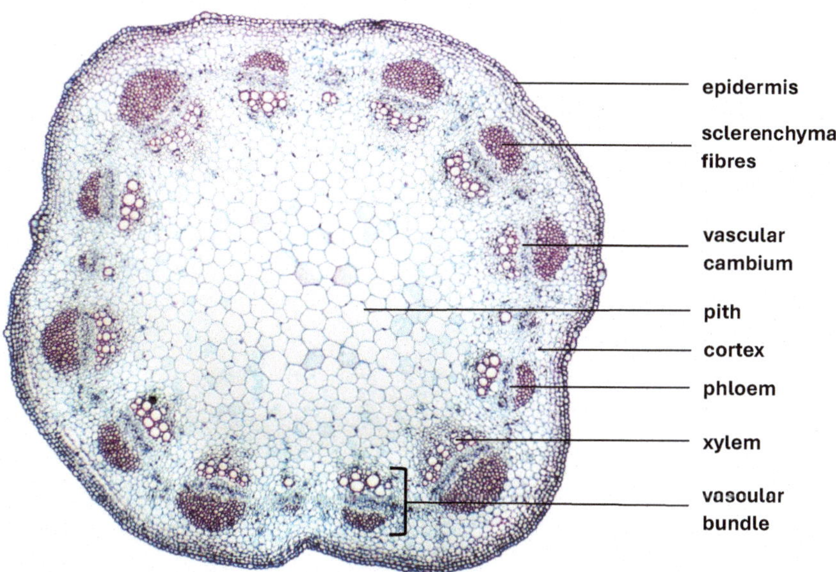

Figure 4.10 A eudicot stem: transverse section of a young (primary) stem in sunflower (*Helianthus annus*) (source: Shutterstock/D. Kucharski K. Kucharska)

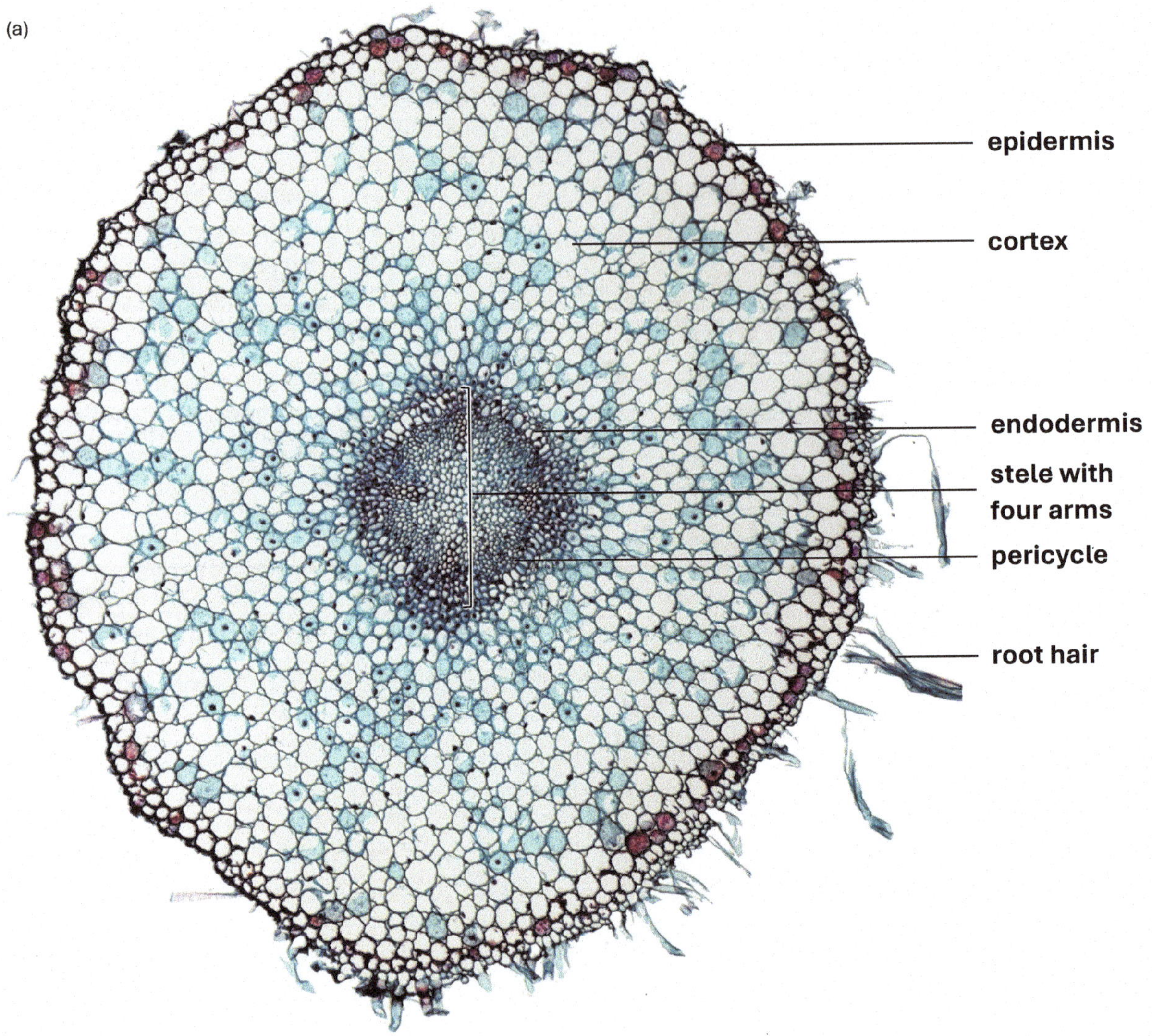

(a)

epidermis

cortex

endodermis

stele with four arms

pericycle

root hair

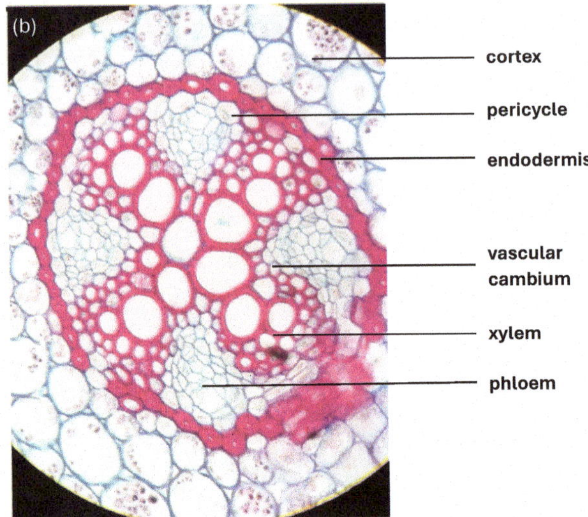

Figure 4.11 Eudicot roots: (a) transverse section of a young (primary) root of broad bean (*Vicia faba*) (source: Shutterstock/D. Kucharski K. Kucharska); (b) central stele in a buttercup (*Ranunculus acris*) root showing vascular tissue (source: Shutterstock/ Kallayanee Naloka)

wall are different for each type of cell, depending on its function. This takes place in the **zone of differentiation** where the epidermis, cortex, vascular tissues and pith become distinct. In the stem tip the apical meristem gives rise to small leaves (bud scales) which protect the meristem. These and the meristem collectively form the **apical bud**. Buds located lower down the stem in the angle of the leaf are called **axillary buds** which have their own meristems and often give rise to side branches and sometimes flowers (see p. 63). Leaf tissues similarly develop from the apex and form specialized tissues to carry out the process of photosynthesis.

Behind the root tip, single epidermal cells become hugely elongated to form many thousands of **root**

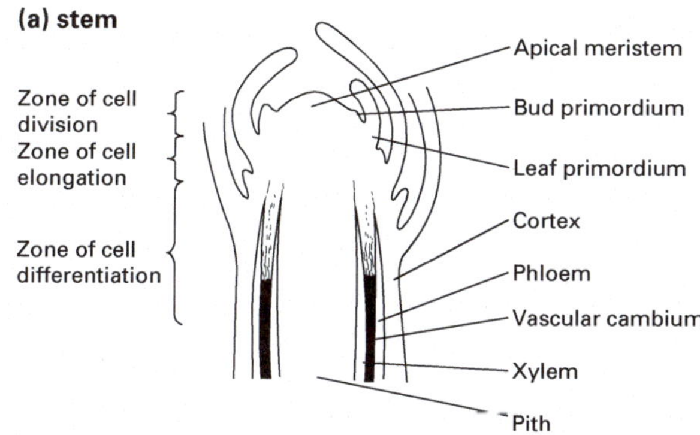

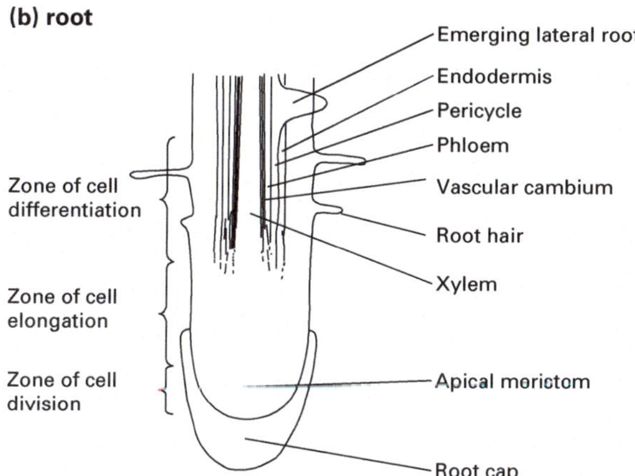

Figure 4.12 Longitudinal sections through a eudicot stem and root tip showing zones of cell division, elongation and differentiation: (a) stem; (b) root

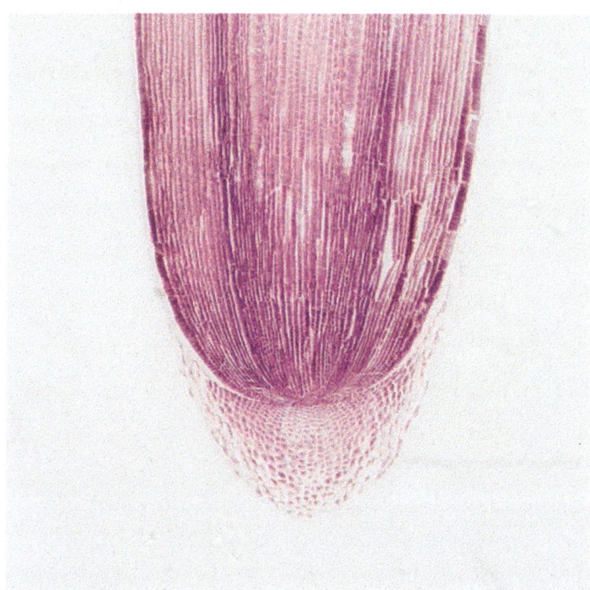

Figure 4.13 A eudicot root tip showing the protective root cap

hairs which increase the surface area for water uptake. The root tip itself is protected by a **root cap** (Figure 4.13) which exudes a gel enabling the root tip to grow through the soil more easily and whose cells are continuously worn away and replaced. Farther back from the root tip, **lateral roots** develop from the pericycle.

In some plant families like Poaceae (the grasses), the primary meristem remains at the base of the leaves just above a node **(intercalary meristem)** which protects them against some herbicides, such as 2,4-D (see p. 230). This also means that grasses regrow from their base after animals have grazed them and that grasses can be mown, which enables us to create lawns. Mowing would kill eudicot plants that have their stems cut off at the base and therefore lose their meristems. However, many eudicot species, for example daisy (*Bellis perennis*), overcome this by having very short internodes, creating a **rosette** of leaves where the growing point stays below the cutting height of the mower (or grazing animal). The process of cutting back the grass also leads to it sending up several shoots from the base instead of just one (**tillering**), which helps to thicken up the turf sward to make it a useful surface for sport as well as decoration. Intercalary meristems are also found in the nettle family Lamiaceae, for example in mint (*Mentha* spp.), just below stem nodes (see p. 62).

Tissues in monocotyledonous stems and roots

These have the same functions as those of a eudicot, therefore the cell types and tissues are similar. However, the internal arrangement of the tissues does differ. Monocot stems (Figure 4.14a) have their vascular bundles scattered throughout the stem and do not have a clearly defined cortex and pith, whereas in eudicots they are arranged in a ring between the cortex and the pith tissues. Monocot roots generally have stele with multiple arms rather than the relatively few found in eudicots. A major distinction is the absence of a vascular cambium in monocots, which therefore do not undergo secondary thickening and show limited increase in stem diameter, generally not becoming woody. In monocots such as palms and bamboos a 'woody' stem does develop but this comes about through a different mechanism to that in eudicots. Mostly, the stem relies on extensive sclerenchyma tissue for support, which is often found as a band below the epidermis or a sheath around each of the scattered vascular bundles. The orchid root shown in Figure 4.14b has a multilayered epidermis or velamen layer, which is commonly found in the aerial roots of epiphytic orchids, and an exodermis (similar to the endodermis) directly beneath it (see p. 69).

Secondary thickening

In eudicots as the stem length increases, the stem girth must also increase to support the taller

(a)

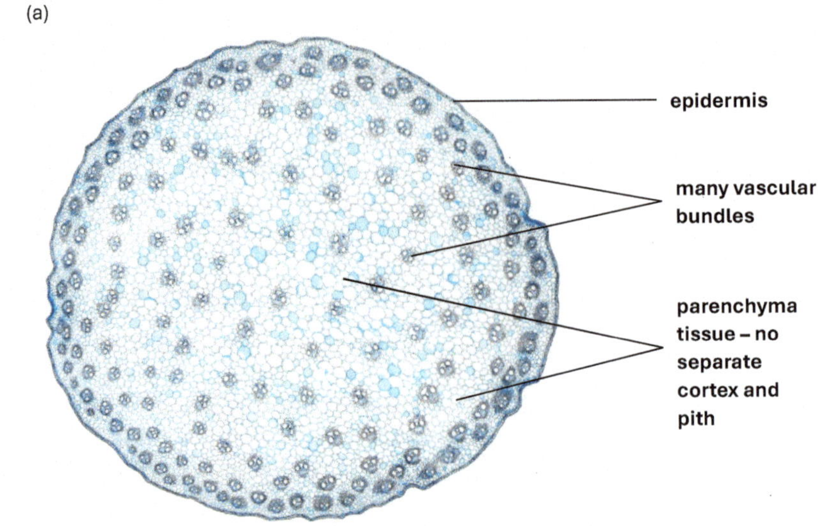

(b)

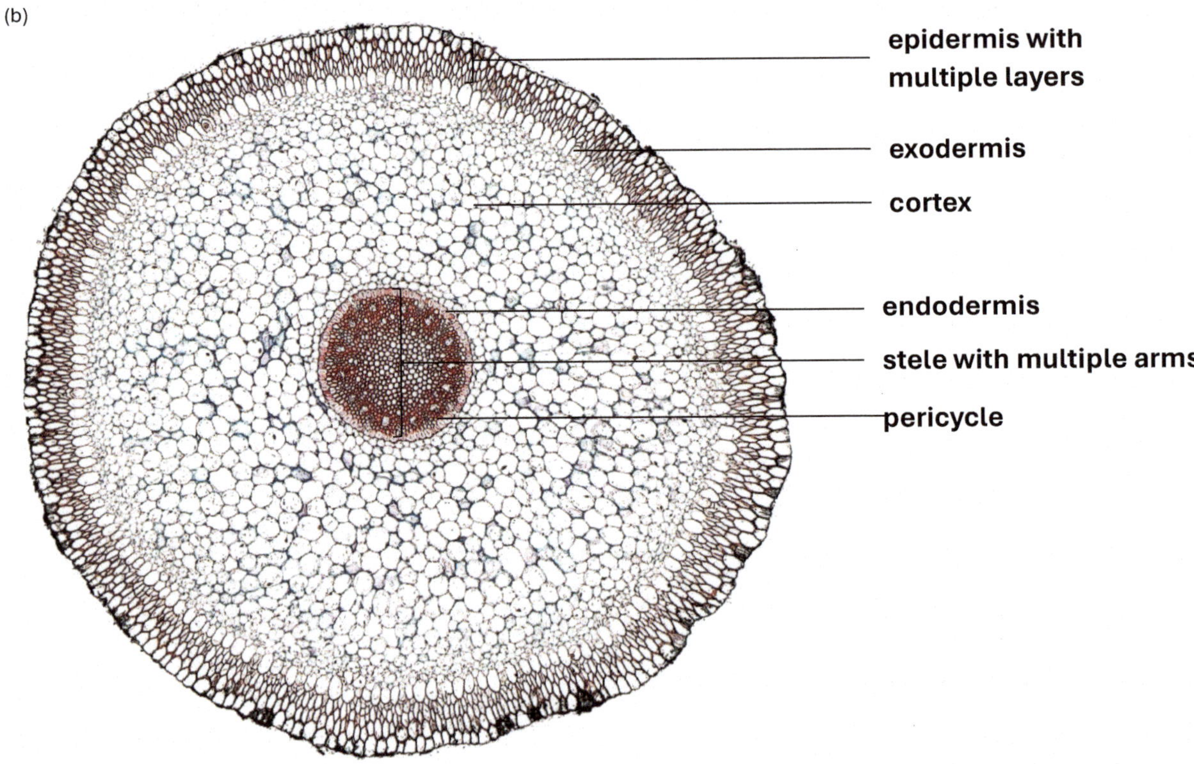

Figure 4.14 Monocots: (a) transverse section of a maize (*Zea mays*) stem (source: Shutterstock/Peter Hermes Furian); (b) transverse section of an orchid root (source: Shutterstock/Mike Rosecope)

plant and supply the greater amount of water and minerals required. This process is called **secondary thickening** or **secondary growth** and, in many cases, it results in the production of wood and woody stems.

At the end of primary growth vascular bundles containing primary xylem and phloem are arranged in a circle within the plant stem (Figure 4.9). Secondary growth commences with the parenchyma cells of the cortex between the vascular bundles, becoming meristematic and joining up with the vascular cambium within the vascular bundles. The **ring of vascular cambium** now starts to produce new secondary phloem to the outside of the stem and new secondary xylem to the inside of the stem. As more secondary growth takes place, so more phloem and xylem tissues are added. Eventually, the primary xylem and phloem disappears as it is crushed by the new secondary xylem and phloem. More xylem than phloem is produced so that the majority of the stem consists of secondary xylem (the **wood**) with only a thin layer of secondary phloem just beneath the surface of the stem (Figure 4.15).

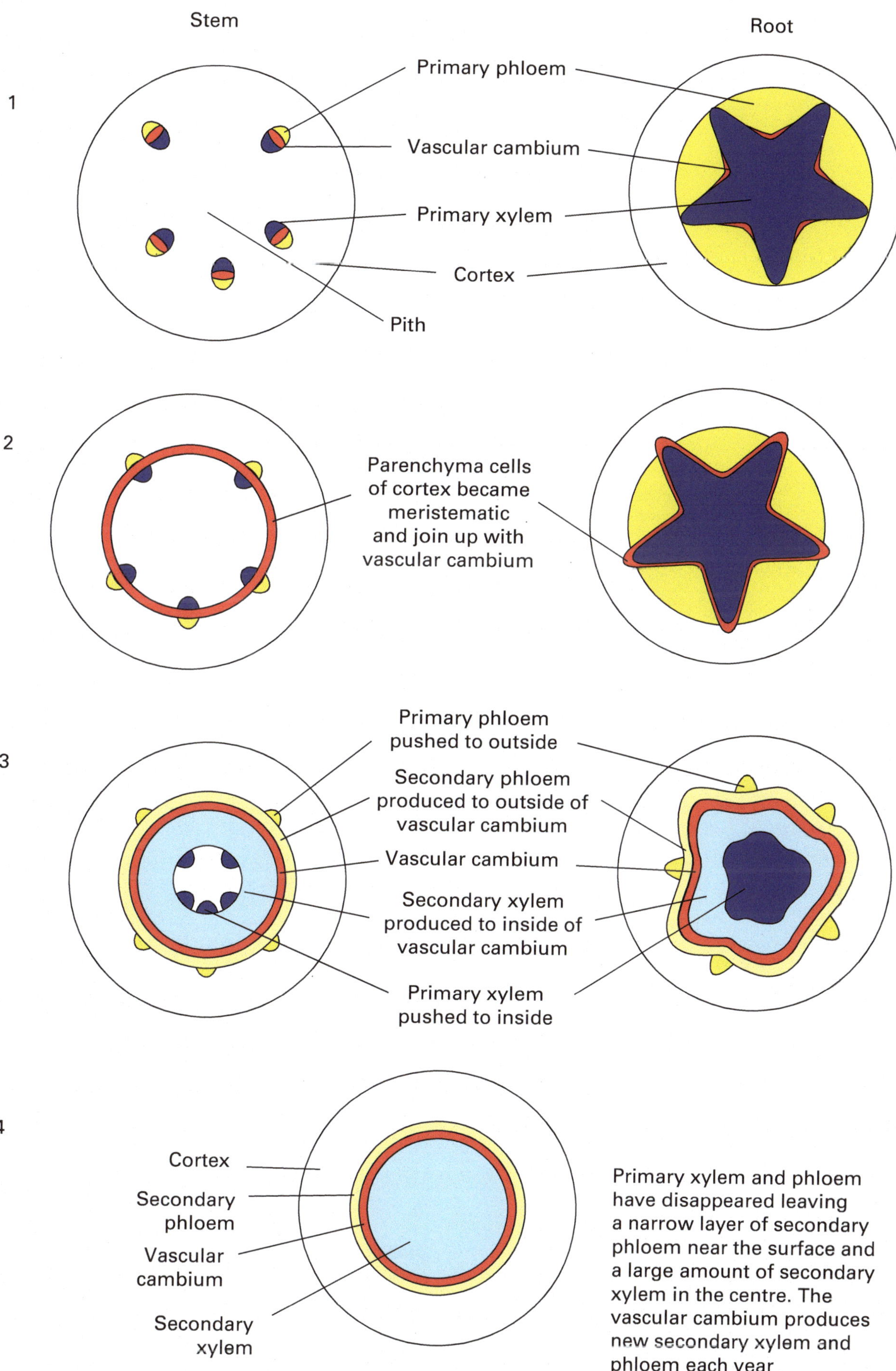

Figure 4.15 Secondary thickening in eudicot stems and roots (cork production not shown)

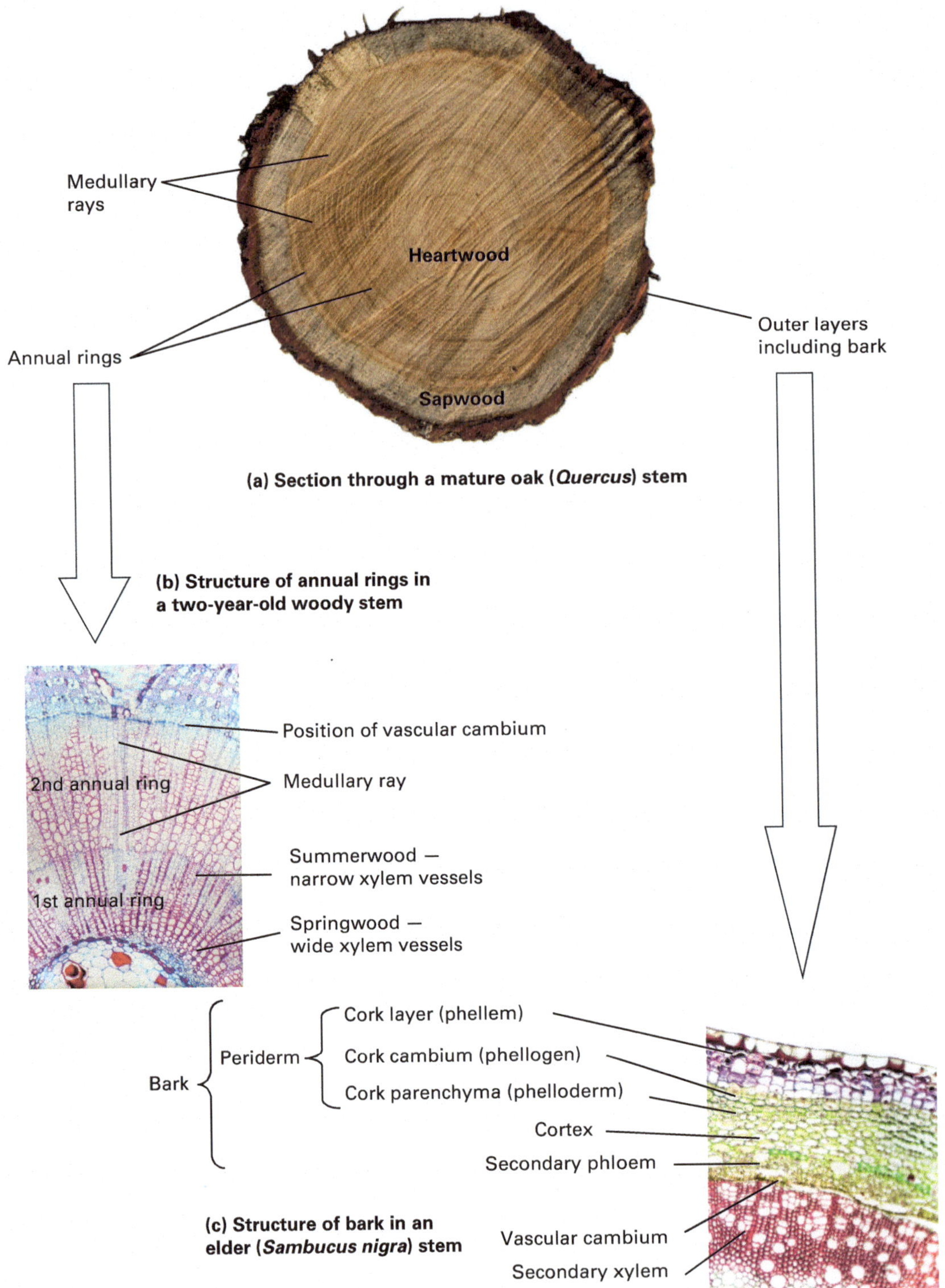

Figure 4.16 Internal features of a woody stem: (a) section through a mature oak (*Quercus robur*) stem; (b) structure of annual rings in a two-year-old woody stem; (c) structure of bark in an elder (*Sambucus nigra*) stem

As these secondary tissues increase, the circumference of the stem also increases and the stem epidermis stretches and breaks away. Therefore, another lateral meristem, the **cork cambium**, is produced in a ring just to the inside of the epidermis (Figure 4.16c).

The cork cambium (**phellogen**) cells divide to produce a layer of corky cells containing suberin, an impermeable wax, on the outside of the stem (**phellem**) and more cortex (**phelloderm**) to the inside of the stem. The three layers together are called the **periderm**, and they prevent water loss and damage by pests and diseases if cracks should occur. Because the cork layer is also airtight, and oxygen is required by the living tissues within the stem, **lenticels** replace the stomata of the epidermis. These are small areas of loosely packed cork cells with air spaces between them, which

Figure 4.17 Lenticels in the bark of a cherry (*Prunus*) tree

develop at the surface of the cork layer. In some species such as *Cornus* and *Forsythia* the lenticels can be seen clearly as specks on the surface of the stem whilst in *Betula* and *Prunus* they appear as lines (Figure 4.17).

The cortex and secondary phloem tissue are eventually pushed against the cork layers by the increasing volume of xylem so that a woody stem appears to have two distinct layers, the wood in the centre and the **bark** on the outside which is formed of all the layers between the vascular cambium and the outside of the stem (Figure 4.16c). If the bark is removed, the secondary phloem, which is only a postcard's thickness even in a large tree trunk, will also be lost, leaving the vascular cambium exposed. The stem's food transport system from leaves to the roots is thus removed and, if a trunk is completely **ringed** (or 'girdled'), the plant will die. Rabbits, squirrels or deer may cause this sort of damage (Figure 4.18), and young trees should always be protected with a barrier if this is likely to be a problem. 'Partial ringing', that is removing the bark from almost the whole of the circumference, is an old method to create a deliberate reduction in growth rate of vigorous tree fruit cultivars and woody ornamental species.

Initially, bark is smooth and shiny but with age it thickens and the outer cork layer starts to peel or flake off. This is replaced from below and the cork gradually takes on its characteristic colours and textures. Many trees have attractive bark and are particularly valued for winter interest (see Figure 5.8). Bark can also play a protective role against heat as in cork oaks (*Quercus suber*) when fire passes through grassland underneath. Many gum trees (*Eucalyptus* spp.) shed their bark in strips, possibly to maintain a thin bark layer to allow oxygen to enter and also to prevent accumulation of parasites and their eggs.

In a mature woody stem the central region of xylem sometimes becomes darkly stained with gums and resins (**heartwood**) (Figure 4.16a) and performs the long-term function of support for

Figure 4.18 Damage to bark caused by a deer

a heavy trunk or branch. Beech heart rot caused by a bracket fungus (*Ganoderma applanatum*) causes decay of heartwood in beech (*Fagus* spp.) leading to lack of structural integrity. The xylem in this region is no longer functioning as the vessels become blocked with air and other substances (see p. 127). The outer xylem, the **sapwood** (Figure 4.16a), is still functional in transporting water and nutrients and is often lighter in colour. The xylem tissue produced in the spring has larger diameter vessels than summer-produced xylem, owing to the greater volume of water that must be transported, and a distinct ring is therefore produced where the two contrasting types of xylem meet. As these rings will be formed each season, their number can indicate the age of the branch or trunk; they are called **annual rings** (Figure 4.16b). Annual rings are often absent in trees growing in tropical areas because the lack of seasonal variation means that the xylem vessels are the same diameter all year round. Dutch elm disease (*Ophiostoma novo-ulmi*) and Chalara ash dieback (*Hymenoscyphus fraxineus*) are pathogens which cause the xylem vessels to block and prevent water and nutrient transport up the infected tree (see p. 298).

A further feature of a woody stem is the mass of lines radiating outwards from the centre, most obvious in the xylem tissues. These are

Figure 4.19 A grafted tree (apple 'Topaz' scion on MM106 rootstock)

medullary or parenchyma rays (Figure 4.16a and b) consisting of strands of living parenchyma tissue which traverse the dead xylem. These allow air, water, sugars and nutrients to move into the stem and across it from cell to cell. The oxygen in the air is needed for the process of **respiration** (see p. 119), which provides energy for the movement of substances into the centre of the stem where they are deposited in the heartwood. Rays are often linked to a lenticel at the stem's surface which can be a means of entry of some diseases, for example fireblight (see p. 301).

Since cells in the vascular cambium are dividing, this can be taken advantage of in grafting. When the **scion** (a cut stem) is joined to the **rootstock** (a root from another tree), the vascular cambium tissues of both components must be positioned as close to each other as possible. The success of a graft depends very much on the rapid **callus** (see p. 51) growth derived from the vascular cambium, from which new cambial cells form and give rise to new xylem and phloem. These need to connect to complete the union and allow water and sugars to flow between the scion and the rootstock (Figure 4.19).

Roots also undergo secondary thickening in a similar fashion to stems (Figure 4.15) and also form bark on their outer surface. Because the cork layer of the bark is impermeable to water the root hairs at the root tip are particularly important as the main site of water uptake. However some water is still taken up over the root surface since water loss is not such a problem as in the aerial parts of the plant.

Further reading

Adams, C., Early, M., Brook, J. and Bamford, K. (2015) *Principles of Horticulture Level 3*. 7th ed. Routledge.

Allaby, M. (2006) *A Dictionary of Plant Sciences*. Oxford University Press.

Bowes, B.G. (1996) *A Colour Atlas of Plant Structure*. Manson Publishing.

Capon, B. (2022) *Botany for Gardeners: An Introduction to the Science of Plants*. 4th ed. Timber Press.

Chalker-Scott, L. (2015) *How Plants Work*. Timber Press.

Clegg, C.J. (2003) *Green Plants: The Inside Story*. Hodder Murray.

Royal Horticultural Society. (2013) *RHS Botany for Gardeners: The Art and Science of Gardening Explained & Explored*. Mitchell Beazley.

Rudall, P.J. (2020) *Anatomy of Flowering Plants: An Introduction to Plant Structure and Development*. 4th ed. CUP.

The online material is accessible via the QR code and includes further information on many of the topics in the book, such as

- Tree rings
- The size of things

CHAPTER 5

Roots, stems, leaves and their adaptations

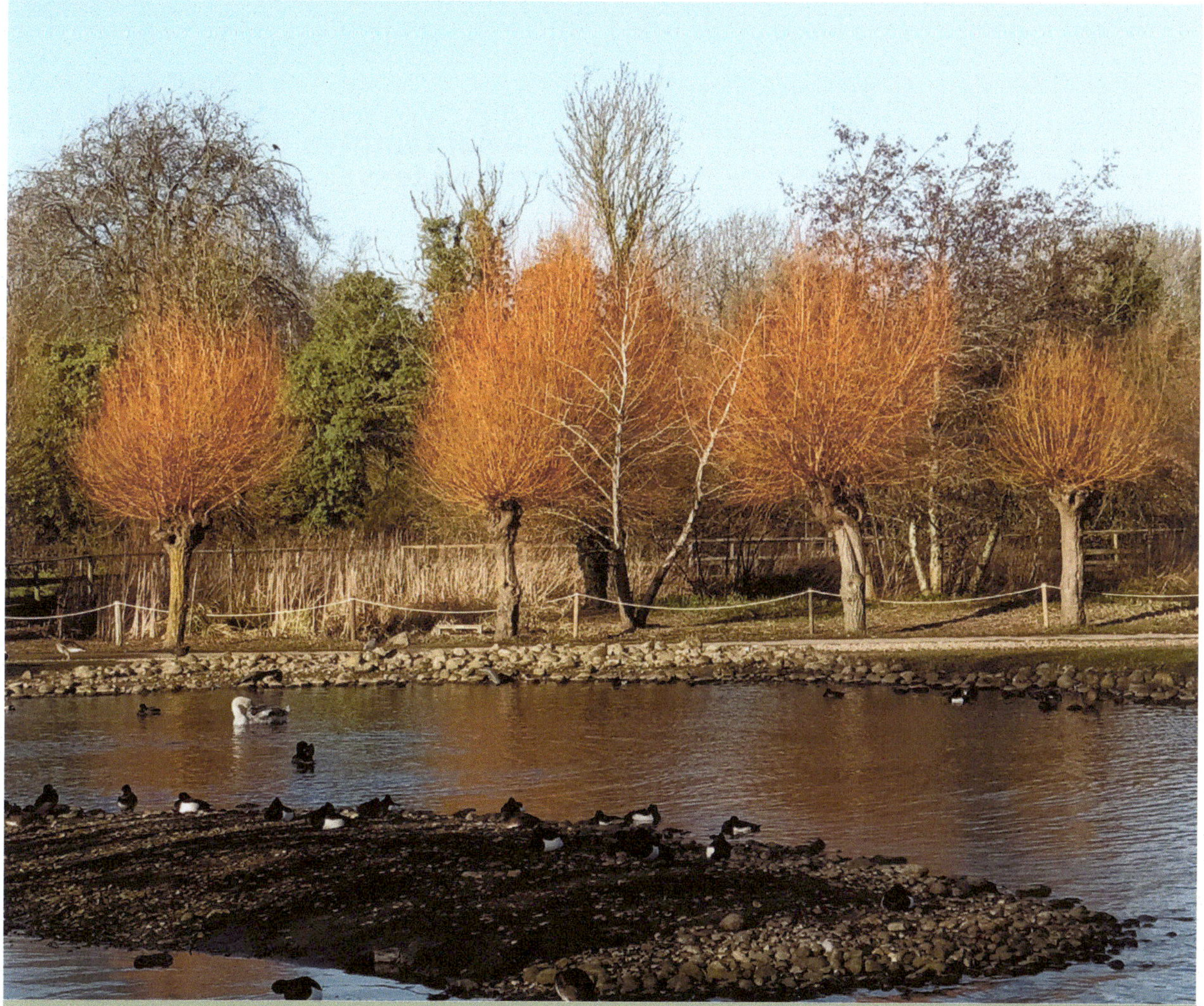

Figure 5.1 Winter stems of pollarded willow (*Salix* spp.) trees make a colourful display

This chapter includes the following topics:

- External features of roots, stems, buds and leaves
- Adaptations of roots, stems and leaves

DOI: 10.4324/9781003581260-5

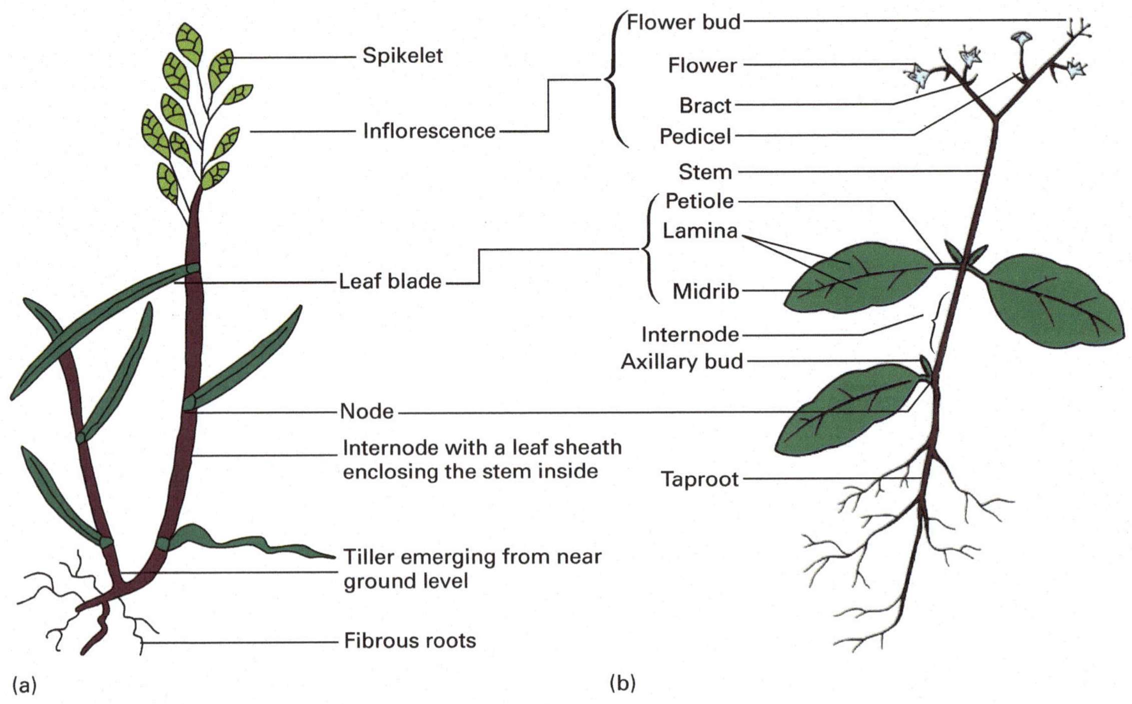

Figure 5.2 Generalized structure of (a) a monocot and (b) a eudicot showing plant parts

While the internal structure of a plant organ can give us an idea of its function and how it is designed to carry out that function, it is the external appearance of plant organs, the plant's **morphology,** with which we are most familiar. Most plant species at first sight appear very similar, since all four organs, the root, stem, leaf and flower, are present in approximately the same form and have the same major functions. However, it is the differences in appearance of these plant parts that enable us to distinguish between them. An appreciation of how plants and plant organs vary in their size, shape, colour, arrangement or other characteristics helps us to correctly identify them. Knowledge of how to accurately describe them also helps to ensure that the plants we are buying are correctly chosen. In addition, studying plant form and appearance gives us a better understanding of how they are adapted to their particular habitats.

Flower structure is the most reliable feature used to identify a plant because sexual reproduction is the most important function, so the structure of the reproductive parts hardly changes. Nevertheless, leaf, root and stem features can also be helpful, especially when flowers are absent. Some basic terms used to describe the various plant organs are shown for a generalized eudicot and for a grass as an example of a monocot in Figure 5.2.

More detailed descriptions of the parts of flowers, fruits and seeds are given in Chapter 7. The internal structures of roots and stems are described in Chapter 5, and leaf structure, as an organ of photosynthesis, is described in Chapter 8.

External features of roots, stems, buds and leaves

Roots

The main functions of the root system are to:

- take up water from the growing medium or soil
- take up mineral nutrients dissolved in water from the growing medium or soil
- anchor the plant in that medium or in the soil.

To achieve maximum water and mineral uptake, roots must have as large a surface area as possible. The **root hairs**, which can be seen just behind the root tip (see Figure 4.12b), greatly increase the root surface area with as many 200 – 400 hairs per square millimetre. The loss of root hairs during transplanting can check plant growth considerably, and the hairs can be points of entry for diseases such as clubroot (see p. 293). Root hairs are replaced frequently as the root grows, a single rye plant producing more than a million a day!

Two main types of root system are produced. A **taproot** (primary root) is a single large root which grows directly from the radicle (see p. 106) in the embryo (Figure 5.3a). It has many smaller **lateral roots** (**secondary roots**) growing out from it at intervals. Taproot systems are a distinctive feature of eudicots and are often storage organs as in carrot (*Daucus carota*), for example. In contrast, a **fibrous** root system consists of many roots with no dominant root. It is characteristic of monocot plants such as grasses but can also be found in many annuals and herbaceous perennial eudicots such as groundsel (*Senecio vulgaris*), *Epimedium* spp. and *Heuchera*

Figure 5.3 (a) A taproot system in a dandelion (*Taraxacum*), a eudicot; (b) a fibrous root system in a grass, a monocot

spp. Most fibrous root systems are made up of adventitious roots which grow from the bottom of the stem as the primary root fails to develop or dies away.

Adventitious roots do not derive from the radicle of the plant embryo. They often grow in unusual places such as on the stem or on other organs.

A **taproot (primary root)** is a single large root which will have many **lateral (secondary)** roots growing out from it at intervals. **Primary** roots originate from the radicle of the embryo. A **fibrous root system** consists of many roots growing from the base of the stem with no dominant root. **Adventitious** roots grow in unusual places and do not originate from the radicle of the embryo.

Stems

The stem's main functions are to:

- physically support the leaves in the optimum position for photosynthesis
- physically support the flowers in the optimum position for pollination and seed dispersal
- transport water, minerals and food (sugars) between roots, buds, leaves and flowers.

The main features of a woody stem are shown in Figure 5.4.

The leaf joins the stem at a **node** and the space between one node and the next is termed the **internode**. Young stems are enclosed by the epidermis with stomatal pores allowing gas exchange between the air and the living tissues within the stem. Stems may be green and carry out photosynthesis (Figure 5.5), and sometimes this persists. When a stem becomes woody the epidermis is replaced with bark and the stomata with lenticels (see p. 58) which can be useful for identification (Figure 4.17).

A **bud** is a condensed stem which is very short and has small leaves enclosing and protecting the meristem (Figure 5.6). A **terminal (apical) bud** is present at the tip of a main stem or branch and contains a meristem from which lengthwise vegetative growth or, less commonly, a flower will emerge. Further down the stem, **axillary buds** are found in the angle where the leaf joins the stem. Axillary buds can grow out to produce lateral shoots (vegetative buds) or flowers (flowering or fruiting buds). On the outside of the bud, the leaves are often thicker and darker forming **bud scales** to resist drying and damage from animals and disease. They may also contain chemical inhibitors which delay bud break until the spring. The horse chestnut buds (*Aesculus hippocastanum*) shown in Figure 5.4, for example, exude a resinous substance to deter insects, hence the name 'sticky buds'.

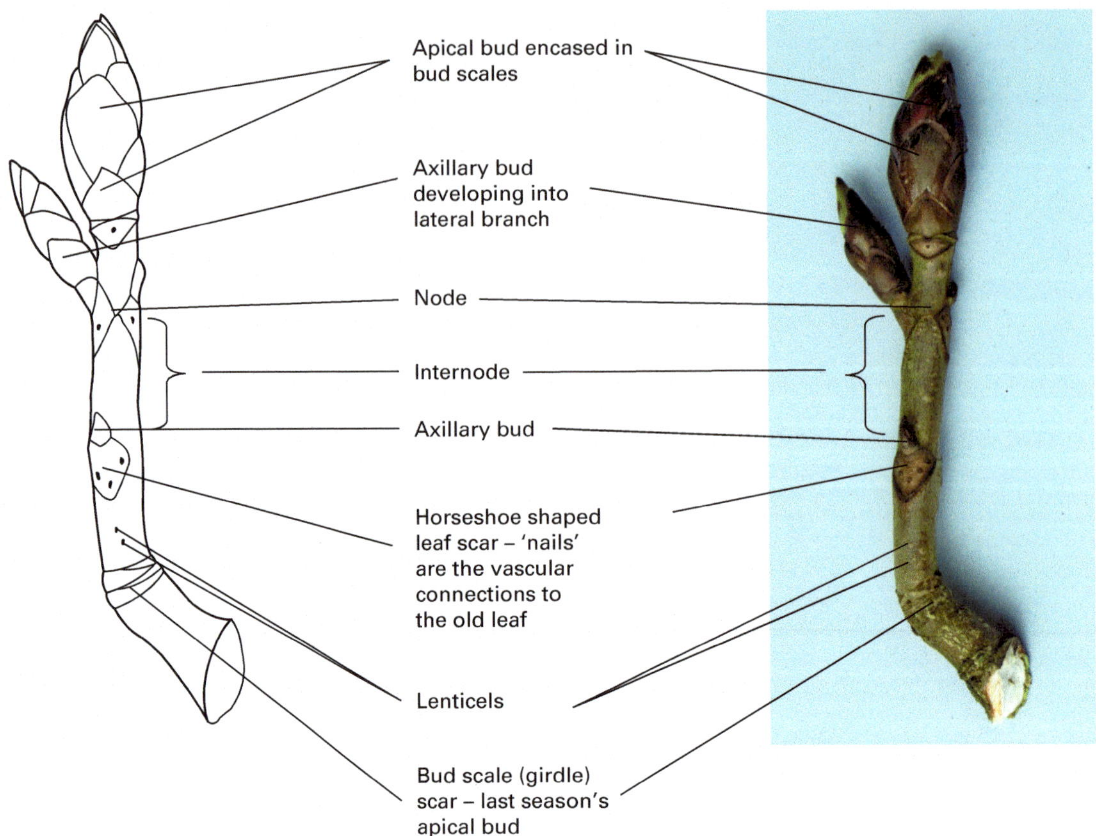

Figure 5.4 External features of a woody stem in horse chestnut (*Aesculus hippocastanum*)

Winter stems in deciduous woody plants will show a **leaf scar** at the node where a leaf was attached and a **bud scale scar** (girdle scar) where last year's **apical bud** was positioned, and this can be useful in determining which part of the stem is current, one-year-old or two-year-old wood when pruning.

Bud characteristics can be useful in identifying plants – for example, the native ash (*Fraxinus excelsior*) has black buds (Figure 5.7a), those of beech (*Fagus sylvatica*) are long and pointed (Figure 5.7b) and *Magnolia* buds are hairy (Figure 5.7c). Flower buds tend to be much larger and plumper than vegetative buds (Figure 5.7d), which give rise to new shoots and leaves, and this is useful when pruning. In spur pruning of apples, for example, the vegetative buds on the ends of lateral shoots are easily identified and removed, encouraging development of flower buds whilst the flower buds should be retained.

The colour and texture of plant stems, especially the bark of trees and shrubs, is a useful aid to identification (Figure 5.8) and a decorative feature, particularly in winter.

Shrubs such as *Cornus alba* 'Sibirica' (a dogwood) with bright red stems, *Salix alba* var. *vitellina* with yellow-orange stems and *Rubus cockburnianus*

Figure 5.5 Photosynthetic stems

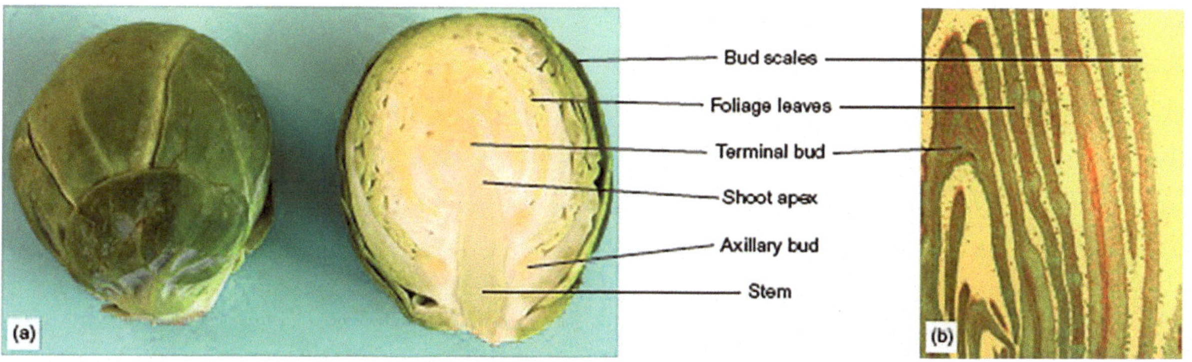

Figure 5.6 Structure of a bud; (a) section through a Brussels sprout; (b) magnified image

Figure 5.7 Bud characteristics: (a) ash (*Fraxinus excelsior*); (b) beech (*Fagus sylvatica*); (c) *Magnolia* spp.; (d) *Camellia japonica*, a large flower bud next to a narrow vegetative bud

with white stems make stunning displays. These shrubs are cut back hard in late winter (coppiced or pollarded) as the young stems provide the brightest colour (Figures 5.1 and 5.9).

Leaves

The main function of the leaf is to carry out **photosynthesis** (see Chapter 8).

The **leaf** consists of the leaf bladc (**lamina**) and stalk (**petiole**) (Figure 5.2b), its shape and arrangement on the stem often depending on the water and light energy supply in the species' habitat. Sessile leaves lack a petiole, while in peltate leaves such as *Nasturtium*, the petiole attaches to the centre of the lamina rather than the base (Figure 5.10d).

Leaf shape and structure (Figure 5.10) are useful indicators in plant identification, especially when flowers are not available, although the shape can be influenced by plant age and the environment. Some plants have distinctly different juvenile leaves (see p. 38) and some leaves vary even within a tree depending on whether the leaves are on the outside of the canopy in sun or within it, in shade.

Some descriptive terms are described as follows but many more are used by botanists and horticulturists alike. Of course, plants are not aware of these categories and can show a range

Figure 5.8 Trees with ornamental bark: (a) moosewood (*Acer pensylvanicum* 'Erythrocladum'); (b) Tibetan cherry (*Prunus serrula*); (c) Tasmanian snow gum (*Eucalyptus coccifera*)

of shapes, venations and margins which are sometimes hard to match to a particular term!

- **Simple** leaves have a continuous leaf blade, with an axillary bud at the base of the petiole (Figure 5.10b–f). There is a multitude of leaf shapes in eudicots, whereas monocots often have strap-like, linear leaves which lack a petiole (Figure 5.10a and i), for example grasses and bulbs such as daffodils (*Narcissus* spp.).
- **Compound** leaves have separate **leaflets**, each with an individual base attached to one leaf stalk. The single axillary bud is at the base of the main leaf stalk where it attaches to the stem, which is useful for distinguishing between simple and compound leaves. Compound leaves may be palmate (hand-shaped) with the leaflets attached at a single point (Figure 5.10g) or pinnate with the leaflets arranged along the petiole (Figure 5.10h). Trifoliate compound leaves have three leaflets, as in clover (*Trifolium repens*).

Simple and compound leaves tend to be unchanging characteristics so are useful for identification purposes.

- **Shapes** of leaves vary hugely, and there are an equally large number of words describing them. A few are shown in Figure 5.10a, b, c and e. Leaves may be lobed, palmately in the *Geranium* example shown (Figure 5.10f) or pinnately as in oak (*Quercus robur*) leaves.
- **Margins** (edges) of leaves also vary. Commonly they may be smooth (entire) (Figure 5.10a–e) or toothed (serrate) (Figure 5.10h)
- **Leaf vein arrangement** (venation) is another important feature of leaves. Parallel venation is a characteristic of monocots (Figure 5.10i),

Figure 5.9 Winter stem colour: (a) red stem colour of dogwood (*Cornus alba* 'Sibirica'); (b) white stems of *Rubus cockburnianus*; (c) hairy bronze stems of *Rubus phoenicolasius*

whereas eudicots have a wide variety of venations such as the pinnate arrangement shown in Figure 5.10h and j and the palmate venation of *Geranium* leaves (Figure 5.10f). A netted venation (reticulate) is seen in the hellebore leaf in Figure 5.10k.

In cultivated plants, leaf size and shape, colour (Figure 5.10k) and variegation (see Figure 7.5) are also important identifying features since many cultivars have been bred to differ from the original species in these respects. In some plants, extra leafy structures called **stipules** may be found at the base of, or attached to, the petiole as in *Rosa* spp. (Figure 5.11).

The arrangement of leaves on the stem (a plant's **phyllotaxy**) maximizes light interception for photosynthesis and is also a very important identifying feature (Figure 5.12). For example, *Salix alba* has leaves arranged alternately along the stem whereas *Cornus alba* leaves are attached opposite each other. Similarly, *Liquidamber styraciflua* and *Acer* spp. trees have very similar palmate leaf shapes, but the former has alternate leaves while the latter has opposite leaves. Leaves may also be attached in whorls around the stem as in lilies (*Lilium* spp.) and the weed cleavers (*Galium aparine*) (see p. 242).

The major differences between monocots and eudicots are summarized in Chapter 2.

Leaves in the garden

We may easily overlook the contribution that the shape, texture, venation, colour and size of leaves can make to the general appearance of a garden. Flowers are the most striking feature, but they are often shortlived and it is foliage which gives more permanent interest.

Considering leaf shape and size, the long linear leaves of New Zealand flax (*Phormium tenax*) contrast with the large palmate leaves of *Gunnera manicata*, while on a smaller scale, the shade-loving hostas, with their lanceolate leaves, mix well with the pinnate-leaved *Dryopteris filix-mas* (male fern). Leaf texture is also important. Most species have quite smooth-textured leaves but *Verbascum olympicum*, lamb's tongue (*Stachys byzantina*), and the alpine, edelweiss (*Leontopodium alpinum*), all have woolly textures. In contrast holly (*Ilex aquifolium*) and *Pieris japonica* have striking glossy leaves.

A wide variety of leaf tones and colours are available to the gardener. The conifer Chinese juniper (*Juniperus chinensis*), shrubs of the *Ceanothus* genus and Christmas rose (*Helleborus niger*) are examples of dark green-leaved plants. Plants with yellow leaves, often more intense when young, include the false acacia tree (*Robinia pseudoacacia* 'Frisia'), the climber, golden hop (*Humulus lupulus* 'Aureus'), and the prostrate herbaceous perennial, creeping jenny (*Lysimachia nummularia* 'Aurea'). Unusually coloured foliage can be found in the small tree *Prunus* 'Shirofugen' (bronze-red) and the shade perennial bugle (*Ajuga reptans* 'Atropurpurea') (bronze-purple). In autumn, the leaves of several tree, shrub and climber species change from green to a striking orange-red colour. Japanese maple (*Acer japonicum*), winged spindle (*Euonymus alatus*), and Boston ivy (*Parthenocissus tricuspidata*) are examples.

Variegation gives a novel appearance to the plant. Spotted laurel (*Aucuba japonica*), *Euonymus fortunei* and ivy (*Hedera helix*) all have good variegated forms (see p. 117).

Plant adaptations

A knowledge of plant adaptations is very useful. It helps gardeners to select plants which are suitable for a given environment and therefore increases the chances of successful establishment. In a changing climate it also

Figure 5.10 Leaf shape: (a) linear, e.g. *Agapanthus*; (b) lanceolate, e.g. *Viburnum tinus*; (c) oval, e.g. *Garrya elliptica*; (d) peltate, e.g. *Nasturtium*; (e) hastate, e.g. *Zantedeschia*; (f) lobed, e.g. *Geranium*; (g) palmately compound, e.g. *Lupinus*; (h) pinnately compound, e.g. *Rosa*. Leaf venation: (i) parallel veins in a monocot leaf; (j) pinnate veins in a eudicot leaf. Leaf colour: (k) green *Helleborus*, yellow *Berberis* and purple *Ajuga* leaves.

Figure 5.11 Stipules of *Rosa* spp. along the petiole

helps inform plant selection in the future too. Adaptations allow plants to occupy challenging environments such as high temperatures and a reduced water supply (xerophytic adaptations), waterlogged or aquatic habitats and low or high light levels. Some of the ways in which roots, stems and leaves are adapted to meet these challenges and exploit their environment successfully are explored next. Note that bulbs, tubers, rhizomes and corms are often incorrectly referred to as 'bulbs' collectively in catalogues and garden centres.

A **xerophyte** is a plant that has adapted to survive in an environment where there is only a small amount of liquid water available. **Xeromorphic adaptations** may be structural or physiological

A **perennating organ** is a plant organ that stores food (usually as starch), enabling the plant to survive unfavourable conditions. Examples are bulbs, corms, tubers and rhizomes.

Root adaptations

▶ Support and climbing

Adapted **adventitious roots** are found in many tropical plants for support. Some examples are **buttress roots** (Figure 5.13a) in a strangler fig (*Ficus* spp.) which are plank-like outgrowths of the root at the base of the stem providing support on shallow soils; **stilt roots** dropping down from branches to the waterlogged ground, for example in mangroves such as *Rhizophora mangle* (Figure 5.13b), and **prop roots** at the base of tall stems as in maize (*Zea mays*) and the tropical tree *Pandanus utilis* (Figure 6.11c and d) which help prevent the plant being blown over.

Some trees such as alders (*Alnus* spp.) and swamp cypress (*Taxodium distichum*) are particularly adapted to grow in waterlogged soils and produce 'breathing' roots or 'knees' (**pneumatophores**) covered in many lenticels, which may act as snorkels enabling them to obtain oxygen for respiration from the air above through aerenchyma tissue (see p. 83) but more likely, they improve the anchorage of the tree (Figure 5.13e).

In climbing hydrangea (*Hydrangea petiolaris*) and ivy (*Hedera helix*), **adventitious climbing roots** develop along the stem attaching it to vertical surfaces such as tree trunks and walls to increase light for photosynthesis and raise the flowers up for better pollination opportunities (Figure 5.14). Ivy is not parasitic and obtains all its water and nutrients from its own roots in the ground; its adventitious climbing roots are there just for support. Contrary to popular belief it does not kill trees or damage their bark and, where it grows heavily into the crown, this is most likely where the tree is already in decline or is diseased. The ivy leaves can tolerate shade whereas the tree's leaves will be on the outside of the crown so they do not compete with each other for light. Sometimes it can overcome young trees, however, in which case it can be removed. Ivy rarely damages walls if the brickwork is sound, and on buildings it can insulate in the winter and cool the building in the summer. Furthermore, it is an effective trap for polluting airborne particulates. Ivy is also a valuable plant for wildlife in the garden (see p. 150).

Epiphytes are plants which are not rooted in the ground but are physically attached to aerial parts of other plants for support, enabling them to reach more light for photosynthesis, such as some ferns, bromeliads and orchids. Some epiphytic ferns are shown in Figure 9.10. Epiphytic orchids, which are mainly tropical, have specialized aerial roots called **velamen roots**. These have a multilayered epidermis called the velamen layer composed of dead cells giving the root a silvery appearance. The velamen protects the tissues beneath from harmful ultraviolet rays and may also absorb water and nutrients dissolved in moisture from the air. Beneath it there is an exodermis layer which is similar in structure to the endodermis and helps reduce water loss from the root tissue (see p. 56). Unusually in velamen roots, the underlying tissue is green and photosynthesizes (Figure 5.13f).

▶ Nutrition

Mistletoe (*Viscum album*) (Figure 9.12) is a hemiparasite (see p. 141) as it obtains some

5

Figure 5.12 Leaf arrangements: (a) alternate in hazel (*Corylus avellana*); (b) opposite in *Acer* spp.

Figure 5.13 Some root adaptations: (a) buttress roots; (b) stilt roots of mangrove; (c) prop roots of maize (*Zea mays*); (d) prop roots of *Pandanus utilis*; (e) emerging pneumatophores of swamp cypress (*Taxodium distichum*); (f) velamen roots in an orchid; (g) root tuber of a *Dahlia* (source: Shutterstock, OlgaSolo)

(but not all) of its nutrition from a host tree, commonly apple, poplar or lime, to supplement its own photosynthesis. It produces **haustorial roots** which penetrate the vascular system of the host and tap into water and sugars. Some other plant roots associate with organisms such as bacteria, for example in **nitrogen-fixing roots** (see p. 187 and 241) and fungi in **mycorrhizal roots** (see p. 184) to improve their uptake of water and nutrients.

- **Storage**
  Root tubers develop near the base of the plant, often from adventitious roots, as in *Dahlia* or bee orchids (*Ophrys* spp.). Root tubers such as the *Dahlia* tuber in Figure 5.13g can be distinguished from stem tubers in having lateral roots and no nodes. These, together with **swollen taproots** in carrot (*Daucus carota*) and dandelion (*Taraxacum*) are all **perennating organs** storing starch over winter which is used to provide for new growth in the spring before leaves are produced and start photosynthesizing. Roots may also store water, a strategy in many Cactaceae and some members of the Cucurbitaceae which grow in arid regions. Swollen taproots such as in carrot (*Daucus carota*), sweet potato (*Ipomea batatas*) and beets (*Beta vulgaris*) (Figure 5.15) may store water as well as starch.

Figure 5.14 Adventitious roots along the stem of ivy (*Hedera helix*)

Figure 5.15 Swollen taproot of sugar beet (*Beta vulgaris*)

- **Drought**
  In environments that have a low water supply, one way of dealing with water shortage is to **avoid** it. Many annuals of arid regions do not show specific adaptations but survive periods of drought as seeds, germinating only when water is available. They have shallow roots and many more root hairs which can absorb water as quickly and efficiently as possible so that they can germinate, grow, flower and set seed rapidly. Shallow, spreading roots are also a feature of many cacti and succulents so that they can capitalize on rain when it does fall. Some trees and shrubs take the opposite approach. Taproots connect with the water table deep below ground as in tamarisk (*Tamarix* spp.), oleander (*Nerium oleander*), *Acacia* spp. and the mesquite species of the Americas. These include *Netumis juliflora*, a small, thorny tree which has been planted in arid regions of Africa to combat soil erosion and is now an invasive weed. It has extremely deep roots; one specimen's roots were recorded at a depth of 58 metres in Arizona!
- **Protection**
  Roots may also exude chemicals which reduce growth in other species therefore giving them a competitive advantage. For example, capsaicin in *Capsicum* (peppers) spp. leaves has been shown to deter herbivores such as caterpillars and prevent seed germination of other plants.

Stem adaptations

- **Support and climbing**
  Many stems are adapted for climbing. In runner bean (*Phaseolus coccineus*), honeysuckle (*Lonicera* spp.) and *Wisteria* spp., **twining stems** wind around other upright structures for support either clockwise, for example in *Wisteria sinensis*, or anticlockwise as in *Wisteria floribunda*. Such twining stems can become large and woody, which are then called **lianes** (Figure 5.16a and b). Bramble (*Rubus fruticosus*) produces rapidly growing stems or **stolons** which, together with their stem prickles, enable them to scramble over other plants (Figure 5.16g).
- **Protection**
  Thorns, which are modified branches growing from axillary buds (and hence have a vascular connection to the stem), can have a protective function, for example in hawthorn (*Crataegus* spp.) and *Pyracantha* spp., discouraging herbivores (Figure 5.16d). **Prickles** are specialized outgrowths of the stem epidermis (so are easily rubbed off), which not only protect but also assist the plant in scrambling over other vegetation, as in many roses (Figure 5.16e). Both

Figure 5.16 Some stem adaptations: (a) twining stems of a mature *Wisteria sinensis*; (b) strangler fig (*Ficus* spp.) uses a tree for support which it may eventually kill, leaving a hollow core (source: Shutterstock, Torsten Pursche); (c) cladodes of butcher's broom (*Ruscus aculeatus*) showing flowers and fruits attached to stem; (d) thorn of *Pyracantha*; (e) prickles on the stem of *Rosa sericea* subsp. *omeiensis* f. *pteracantha*, grown as an ornamental for its large red prickles; (f) stem tuber of potato (*Solanum tuberosum*) with new shoots growing from buds on the stem; (g) stolon of bramble (*Rubus fruticosus*)

provide shelter and protection for nesting birds and invertebrates. Butcher's broom (*Ruscus aculeatus*) is a native plant which is adapted to dry, shady woodlands. Its stems are leaf-like **cladodes** carrying out the functions of the leaves which are themselves reduced to a protective spine at the cladode tip. The small buds, flowers and berries are borne in the centre of the cladode indicating that it is a stem (Figure 5.16c).

- **Storage**
  Many stem adaptations important to horticulture are also **perennating organs** storing starch such as some **rhizomes**, which are stems growing horizontally usually underneath, and sometimes just above, the ground as in many *Iris* species and cultivars. Nodes and internodes can be seen clearly along the stem together with adventitious roots which anchor it in the soil and take up water and mineral nutrients. As the stem branches, new rhizomes are formed from lateral buds, each with a shoot at its tip. This eventually enables the plant to spread. Rhizomes are found in many *Iris* spp. such as *I. germanica* and its cultivars, including the bearded *Iris* in Figure 5.17.

Corms, found in *Gladiolus*, *Crocosmia* and *Crocus*, are compressed underground shoots in which the stem is swollen with starch (Figure 5.18).

They have dry scale leaves on their outer surface underneath which nodes, internodes and axillary buds can be seen. Each year a new corm forms on top of the old one. As well as small adventitious roots, rhizomes and corms have specialized, thickened contractile roots which help to pull them to an appropriate level in the soil by corkscrewing like a spring. **Stem tubers** are typified by potato (*Solanum tuberosum*) (Figure 5.16f). The tubers of potato can be distinguished from root tubers by having vestigial nodes and axillary buds (the 'eyes') which grow into shoots. They also turn green (and poisonous) on exposure to light since stems can photosynthesize but roots cannot, hence the need to 'earth up' potatoes to exclude light. *Anemone blanda*, *A. coronaria* De Caen Group, *Begonia* × *tuberhybrida* (tuberous begonia), *Gloxinia* and *Cyclamen* tubers develop from the hypocotyl (see p. 106) but are often classed as stem tubers.

Vegetative spread

Rhizomes and **stem tubers** can also be methods of vegetative (asexual) spread (see p. 87). Others include the **runners** of *Fragaria* spp. (strawberry), horizontal stems which grow just above the ground and root at nodes along the stem or at stem tips producing plantlets. These are often found in rosette plants and grow from buds at the base of the plant. The term **stolon** is often used interchangeably with runner but it also includes plants such as bramble (*Rubus fruticosus*) (Figure 5.16g) which have long arching branches with few leaves. They grow very quickly and root where their tips touch the ground, forming new plants. Non-perennating rhizomes can be a nuisance when trying to eradicate perennial weeds such as hedge bindweed (*Calystegia sepium*), field bindweed (*Convolvulus arvensis*) and couch grass (*Elymus repens*) (Figure 5.19) as their rhizomes can penetrate up to 5 m deep and they can regenerate a new plant from buds on even the tiniest section (see p. 242). **Suckers** are stems that grow from adventitious buds (buds which do not derive from the plumule in the seed (see p. 106)), which are found on the roots of many plants, such as stag's-horn sumach (*Rhus typhina*) (Figure 5.20). These can be useful in propagating plants vegetatively but can also be a nuisance in borders if their spread is difficult to control and can penetrate paved areas, causing problems.

Drought

As well as storing water in their **swollen stems**, **pleated stems** in many cacti enable them to survive in arid environments. The pleating casts a shadow which reduces exposure to

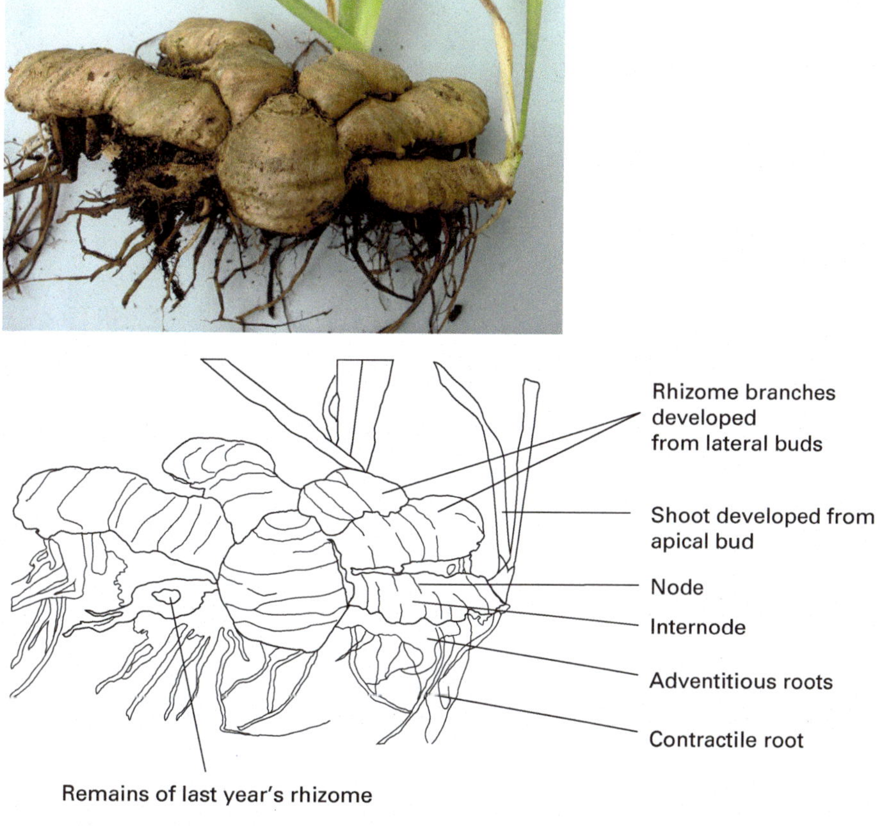

Figure 5.17 Structure of a bearded *Iris* rhizome

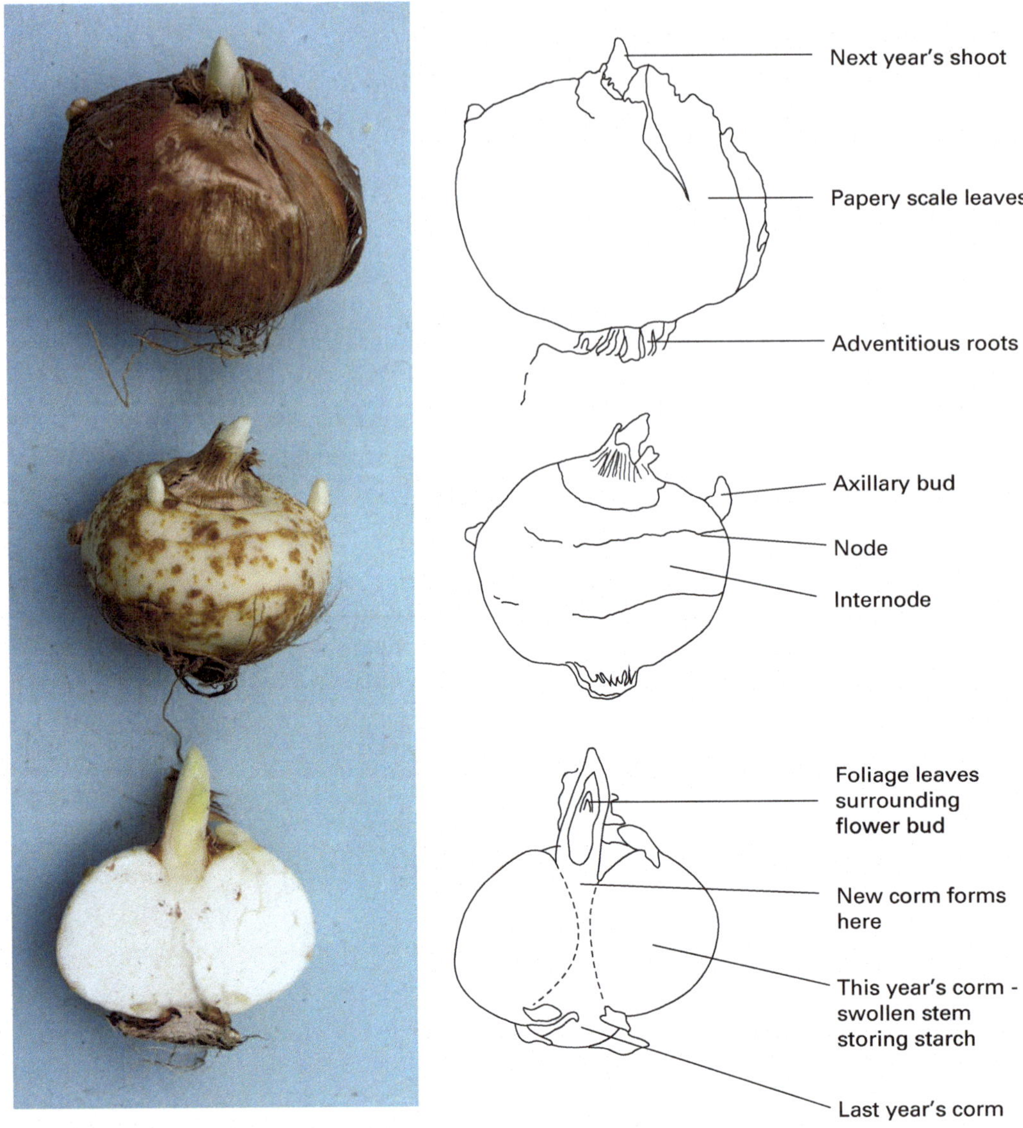

Figure 5.18 Structure of a crocus corm

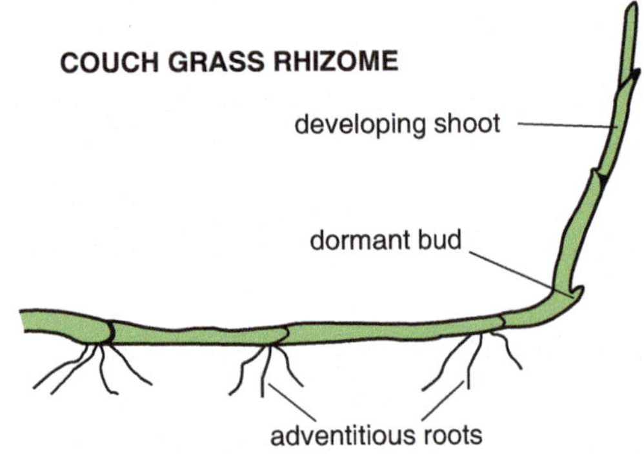

Figure 5.19 Diagram of a couch grass (*Elymus repens*) rhizome

direct sunlight as the sun moves round through the day, lowering the tissue temperature and therefore reducing transpiration (see p. 129) water loss (Figure 5.21a). Baobab trees such as the bottle tree (*Adansonia gregorii*) store water in their massive trunks and have much reduced leaves and branches (Figure 5.21b), and the spherical shape of some cacti maximizes the water storage capacity of the stem (Figure 5.21c). In some plants, shortened stems create a rosette habit (Figure 5.32a) as in *Aloe* spp. and *Agave* spp., in which leaves shade each other and trap humid air close to the ground.

Leaf adaptations

▶ Climbing and support

Modifications for climbing enable the plant to compete effectively with other plants for light, therefore increasing its photosynthetic capacity and also improving its chances of pollination and seed dispersal. Many plants use modified plant parts called **tendrils** which wrap themselves around a support to climb upwards. Sometimes the tendril itself will coil like a spring, pulling the plant even closer to the support. Leaf tendrils may be formed from slender extensions of the leaf and are of three types.

Figure 5.20 Stag's horn sumach (*Rhus typhina*) with multiple stems derived from suckers (source: Shutterstock/Ritvars)

In *Clematis* spp., the leaf petiole curls round the stems of other plants or garden structures to support the climber (Figure 5.22a), while sweet pea (*Lathyrus odoratus*) holds on with tendrils modified from the end-leaflets of the compound leaf (Figure 5.22b). The monocotyledonous climber *Smilax china* has tendrils provided by modified stipules (found at the base of the petiole).

Tendrils can be formed from various plant organs and often it is difficult to establish their origin. In members of the grape family (Vitaceae), for example, which includes *Parthenocissus* represented by Boston ivy (*P. tricuspidata*) and Virginia creeper (*P. quinqefolia*), grapevine (*Vitis vinifera*) and grape ivy (*Cissus alata* formerly *C. rhombifolia*), there is much debate about whether the tendrils are formed from a leaf, a stem tip or even a group of flower buds. *Parthenocissus* spp. have an additional mechanism in the form of sucker pads on the ends of the tendrils which secrete a sticky substance enabling them to attach tightly to smooth surfaces such as walls. In passion flowers (*Passiflora* spp.), a group of axillary buds is found at the base of each petiole, one of which forms a tendril, one or two form flower buds and another a vegetative shoot making it difficult to decide what plant part the tendril actually develops from (Figure 5.23).

In cleavers (*Galium aparine*), both the leaf and stipules, borne in a whorl, bear small **epidermal prickles** acting as grappling hooks that allow the weed to sprawl over other plant species. Roses also have epidermal prickles which can have the same function as well as being protective.

- **Protection**

  Spines, which are modified leaves, and leaf hairs both act as a deterrent to herbivores. These modifications also reduce transpiration (see p. 129) and are often found in plants adapted to drought (see later). Plants may also modify the chemistry of their leaves. In *Cistus* and lavender (*Lavandula* spp.) the evaporation of **volatile oils** from glands in the leaves is thought to have a cooling effect and is also a deterrent to herbivores. Plant leaves can also release **volatile chemicals** which are carried in the air in response to disease or pest damage. This alerts other plants downwind to the potential hazard (an example of allelopathy) so they can raise the level of unpalatable chemicals such as tannins in their leaf tissue as a deterrent.

- **Nutrition**

  In **insectivorous plants** leaves are adapted to trap insects which are digested to supply nutrients not available in the waterlogged soils in which they grow. Such soils do not have sufficient oxygen to sustain organisms that

Figure 5.21 Water storage stems: (a) *Ferocactus pilosus*; (b) a baobab tree (*Adansonia gregorii*); (c) *Echinocactus grusonii*

break down organic matter and release nutrients (see p. 185). These include the native sundew (*Drosera* spp.) and butterwort (*Pinguicula* spp.), which trap their prey with glands on their leaves which exude a sticky substance (Figure 5.24a and b). Tropical pitcher plants (*Sarracenia* spp. and *Nepenthes* spp.) have leaves that form containers into which insects topple and are digested. In *Nepenthes* spp. (a type of pitcher plant), a tendril develops from the leaf and then the tip of the tendril forms the pitcher. The insects are attracted to the colour, scent and nectar of the plants but are prevented from escaping by a combination of slippery inner surfaces, barriers of stiff hairs, elastic threads in the liquid which wrap round them and even narcotic substances in the nectar (Figure 5.24c and d). Venus fly trap (*Dionaea muscipula*) actively traps insects between two leaves which snap shut suddenly as the insects crawl across the leaf surface. The rapid response is triggered by three hairs on the inner surface of the leaves which must be touched in turn by the insect, ensuring that energy is not wasted in trapping non-living objects (Figure 5.24e).

- **Storage**

  In a **bulb,** for example daffodil (*Narcissus* spp.), the outer papery scale leaves enclose succulent, light-coloured scale leaves containing all the food and moisture necessary for the bulb's emergence, making this a perennating organ, storing starch (Figure 5.25).

  The scales are packed densely together around the terminal bud towards the base of the shortened stem, minimizing the risks from extremes of climate or pests such as eelworms and mice. Bulbs have both adventitious and contractile roots (see corms earlier) attached to a basal plate.

- **Vegetative spread**

  As well as perennation, bulbs can bring about vegetative spread (see p. 87). Daughter bulbs form in the axils of the scale leaves

Figure 5.22 Some leaf adaptations: (a) petiole tendrils of *Clematis* spp.; (b) leaflet tendrils of sweet pea (*Lathyrus odoratus*); (c) pitcher developed from a leaf tendril in pitcher plant (*Nepenthe* spp.); (d) spines on an aubergine leaf (source: Shutterstock, Alison Taylor); (e) spines of *Berberis* spp.; (f) spines of a cactus *Ferocactus emoryi* subsp. *emoryi*

Figure 5.23 Tendrils of passion flower

which eventually detach, and these can be used by gardeners to propagate bulbs vegetatively (asexually). A novel method of asexual reproduction is seen in the houseplant *Kalanchoe daigremontiana* (formerly *Bryophyllum daigremontianum*) where the leaf margin bears **adventitious buds** which are able to drop to the ground below and develop into young plantlets.

- **Pollination**

  In many plants such as *Hydrangea* spp., the houseplant poinsettia (*Euphorbia pulcherrima*) and flowering dogwoods (*Cornus* spp.), colourful modified leaves called **bracts** take the place of petals in attracting pollinating insects (Figure 5.26).

- **Drought**

  Many perennial plants show a wide range of behavioural and structural leaf adaptations to combat low water supply. By **reducing leaf area** the overall quantity of water lost through transpiration is reduced, so many drought-tolerant plants have small leaves, for example, *Acacia* spp., *Ephedra* spp., rosemary (*Rosmarinus officinalis*), thymes (*Thymus* spp.) and conifers such as *Pinus* spp. and *Thuja* spp. (Figure 2.8). Often the stems remain green to supplement photosynthesis (Figure 5.5). Some trees and shrubs **shed their leaves** entirely and become dormant during periods of drought. **Wilting** is, in itself, a way in which plants reduce the leaf area exposed to the sun and trap humid air under the leaf, so reducing transpiration. **Phyllodes**, found in *Acacia* spp., are flattened leaf-like petioles which replace and carry out the functions of the mature leaves. The leaves of many monocots are believed to have originated from phyllodes.

  In many desert species, the leaves are transformed into **spines** (Figure 5.27) which minimize water loss by reducing leaf area, trapping moist air close to the stem and shading the stem, which takes over the function of photosynthesis, as well as deterring predators. **Rolled leaves** are found in some plants, for example cross-leaved heath (*Erica tetralix*),

Figure 5.24 Insectivorous plants: (a) butterwort (*Pinguicula* spp.) (source: Shutterstock, Vankich1); (b) sundew (*Drosera* spp.); (c) pitcher plant (*Nepenthes* spp.); (d) pitcher plant (*Sarracenia* spp.) (source: Shutterstock, Lakeview Images); (e) Venus fly trap (*Dionaea* spp.) (source: Shutterstock, Leah-Anne Thompson)

whilst others roll them in response to drought as in Marram grass (*Ammophila arenaria*), thereby reducing their surface area (Figure 5.28).

Other leaf modifications which reduce transpiration include **leaf and stem hairs** (Figure 5.29), which trap still air close to the leaf and stem surface, increasing humidity and also reflecting radiation, shading the leaf surface and lowering surface temperatures.

These hairs can often make the leaves appear silvery (Figure 5.31a). **Thick waxy cuticles** in many plants such as *Eucalyptus* spp. reduce water loss from leaves and stems and **sunken stomata**, which are found at the bottom of pits in the leaf in marram grass (Figure 5.28) and *Pinus sylvestris* (Figure 5.30), also trap humid air and reduce transpiration. *Pinus sylvestris* also has an internal layer of cells beneath the leaf epidermis (endodermis) which is similar to the endodermis of the root with cell walls thickened with a waterproofing wax (suberin). This reduces water loss from the xylem tissue in the centre of the leaf.

Another strategy in arid habitats is water storage in **succulent swollen leaves**, common in *Kalanchoe* spp., *Aloe* spp., *Euphorbia* spp., *Sedum* spp. and *Crassula* spp. (Figure 5.31b). An extreme example is found in living stone plants (*Lithops* spp.) where the aerial parts are reduced pairs of succulent leaves which sit on the soil surface mimicking stones as a camouflage to avoid being browsed (Figure 2.18a). Many drought-tolerant plants, particularly those with succulent stems and leaves, reverse their stomatal opening, closing their stomata during the day when temperatures are high and opening them at night. Carbon dioxide is taken up at night then stored in a chemical form until light is available to drive photosynthesis in the daytime. Plants with this behaviour are called **crassulation**

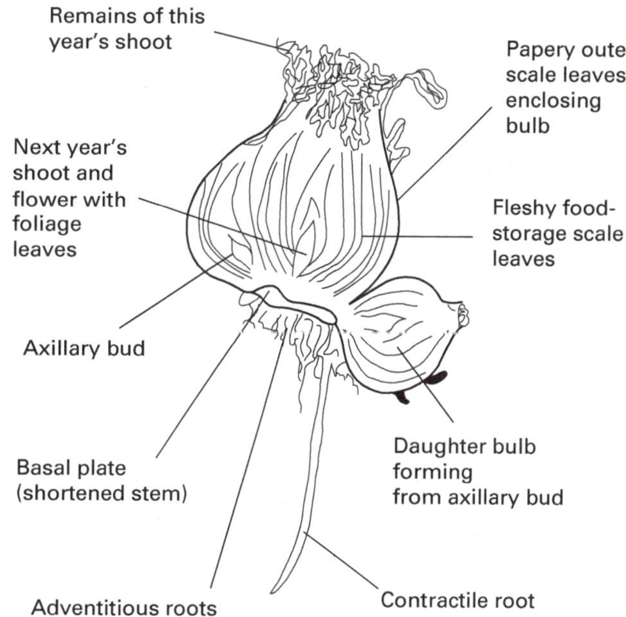

Figure 5.25 Structure of a *Narcissus* bulb

Figure 5.26 White bracts of flowering dogwood. The insignificant flowers are in the centre of the bracts

acid metabolism (CAM) plants, named after the family Crassulaceae (Figure 5.31b). Other plants are able to **osmoregulate** in response to drought or a high level of salts in the soil water, for example marram grass (Figure 5.28). They increase their cell solute concentrations to enable osmosis to continue in the right direction and prevent plasmolysis (see p. 125.)

High and low light levels

Many of the adaptations discussed earlier which enable plants to tolerate drought, such as leaf hairs, spines, pleated stems, sunken stomata and a rosette habit, also protect them from the **high light levels** found in arid environments

Figure 5.27 Leaves reduced to spines on unbranched, photosynthetic stems in (a) *Cleistocactus* spp. with multiple spines trapping humidity and (b) *Pachypodium lamerei* showing spines shading the stem

Ammophila arenaria (marram grass)

Marram is a native grass in Britain and Ireland found growing on coastal sand dunes. Its leaves are adapted to water loss in a number of ways. The upper (outer) surface of the leaf has a thick cuticle and few stomata. Hairs on the lower (inner) surface increase the width of the **boundary layer** (the layer of still air next to the leaf surface) reducing transpiration. Large epidermal cells (hinge cells) at the base of furrows on the lower leaf surface shrink quickly when the leaves are water stressed, causing the leaf to fold in on itself. When folded, the stomata are positioned towards the bottom of the furrows inside the leaf, humid air is trapped within and the leaf surface area is minimized, which leads to a reduction in the rate of transpiration. Marram grass is also very deep rooting and is often planted to stabilize sand dunes.

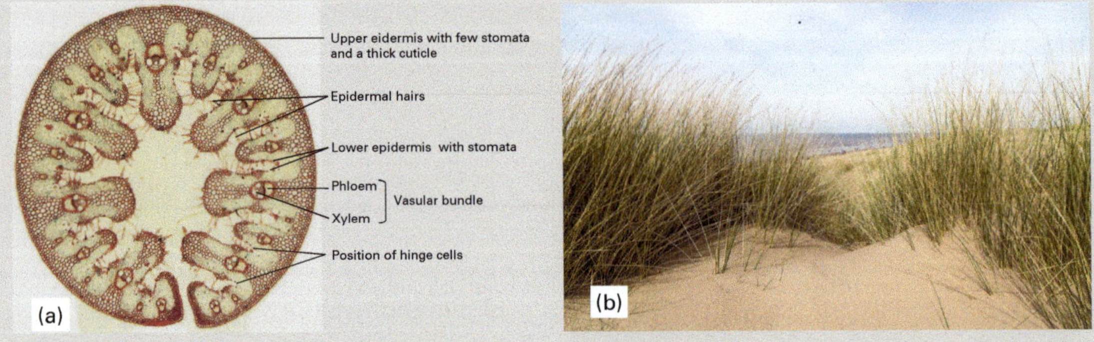

Figure 5.28 (a) Transverse section of marram grass (*Ammophila arenaria*) showing adaptations to prevent water loss; (b) marram grass on dunes (source: Shutterstock, Jason Wells)

Figure 5.29 The tropical shrub *Tibouchina urvilleana* has hairs covering its leaves and stems

by shading. In addition, some may also have **protective pigments** such as the yellow pigment **xanthophyll** and the purple pigment **anthocyanin** in their leaves (Figure 5.32c). These absorb harmful ultraviolet light and convert the intense ultraviolet light energy, which damages the photosynthetic apparatus, into heat. High ultraviolet light goes hand-in-hand with high altitudes, so this can also protect the plant against extreme cold. Some plants even produce high levels of vitamin C, which also protects against light damage.

Plants which thrive in **shade** have many structural adaptations to overcome the low light levels, for example on a woodland floor (Figure 5.33). Their leaves have a **narrow palisade mesophyll layer** (see p. 113) as the light does not penetrate so far into the leaf and the cells contain **more chloroplasts** and therefore more chlorophyll making them dark green. Chloroplasts can move up and down the cells according to the light level. They often have **large, thin leaves** which can trap more light by presenting a larger surface area and sometimes have deeply divided leaves or large holes in their upper leaves to allow light to pass to leaves below as the sun moves round. Some *Begonia* species even have specialized **epidermal chloroplasts** (iridoplasts) which enhance the capture of certain light wavelengths for photosynthesis. Many shade-loving tropical plants are also **evergreen** to make the most of the light all year round, whilst some temperate plants in deciduous woodland remain evergreen over winter to take advantage of higher light penetration (Figure 5.16c). Many plants can also have metabolic and behavioural adaptations to cope with shade (see p. 118).

▶ **Waterlogging and aquatic environments**
Plants which live in aquatic or marshy environments have to deal with anaerobic conditions (see p. 120) due to low oxygen levels (dissolved oxygen diffuses much more slowly than gaseous oxygen). Often nutrient levels are low too due to lack of microbial action which

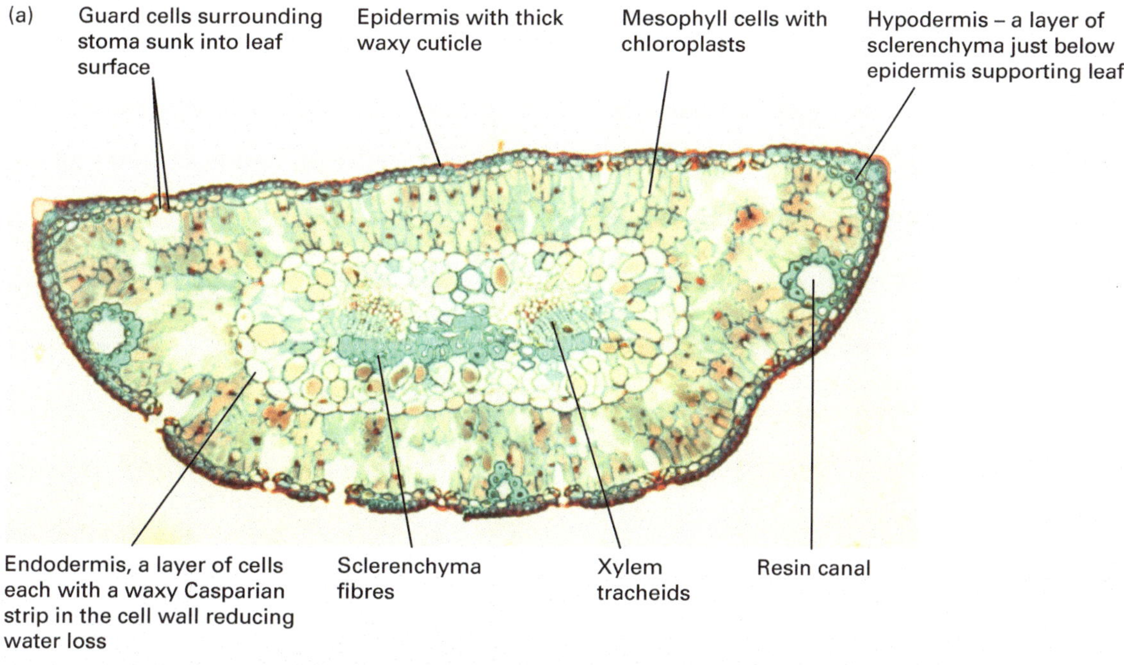

Figure 5.30 (a) Cross section of a Scot's pine (*Pinus sylvestris*) leaf, showing some adaptations to reduce water loss; (b) *Pinus* spp. needles showing reduced surface area

Figure 5.31 Plants adapted to drought: (a) leaves of *Plectranthus argentatus* with fine hairs which appear silver; (b) water storage leaves in the jade plant (*Crassula ovata*) which also has CAM metabolism

prevents recycling of nutrients in the soil (see p. 185), so insectivorous plants are often found in this habitat. Various strategies are employed, both physiological and structural. Many plants develop **aerenchyma** (Figure 5.34a), specialized parenchyma tissue (p. 51) with large air-filled spaces in submerged or waterlogged roots and floating leaves and stems, which may always be present or may develop in response to occasional flooding. *Nymphaea* spp. have most of their stomata on the upper surface of their floating leaves to obtain oxygen from the air and aerenchyma tissue in their stems and leaves to pass it to submerged tissues (Figure 5.34b). *Juncus effusus*, a marsh plant, has continuous air-filled channels in the stems and roots which enable diffusion of oxygen from the aerial parts of the plant to the roots. Many aquatic plants have **reduced or absent cuticles** so gases can diffuse more easily into and out of the leaves from the surrounding water. Water crowfoot

Alpine plants – life above the tree line

Alpine plants have a tough time! 'True' alpines are found between the tree line and the area of permanent snow in mountainous parts of the world such as the Alps, the Pyrenees, the Andes and the Himalayas, and also at sea level in high latitudes such as the Arctic. They are exposed to very low temperatures averaging 10°C and ground which freezes often at night and throughout the winter, which lasts from October to May. Strong winds, intense light without shade, dry air which holds little water, decreased air pressure which holds less carbon dioxide and a short growing season are their typical environment. Their water supply is frozen in the ground for most of the year and precipitation is in the form of snow rather than rain. The soils they grow in are easily eroded by the wind, and any organic matter breaks down very slowly. Consequently, soils are thin, rocky, free draining and nutrient poor and provide little opportunity for roots to anchor any plant.

The most obvious characteristic of alpine plants is their small size. They are low growing and their stunted nature traps humid air close to the ground where the temperature can be 15°C higher than the surrounding air. Often, they have a rosette habit, so that the growing point is protected close to the ground, or they form tussocks, mats or cushions which absorb heat and smooth out fluctuations in temperature and also protect the plant from the desiccating wind as it flows over them. In the winter they are covered by an insulating layer of snow. Their leaves are reduced in size and can be covered in a thick layer of hairs both above

Figure 5.32 (a) An alpine saxifrage (*Saxifraga frederici-augusti* subsp. *frederici-augusti*) showing rosette leaf arrangement and hairy leaves; (b) a hummock-forming alpine (*Dionysia* sp.); (c) purple leaf colouration, succulence and rosette habit in *Sempervivum* 'Rotund'

and underneath, protecting the stomata. They may have a thick waxy cuticle or be succulent with sunken stomata, to reduce water loss and protect the leaves from damaging and drying winds. They are often evergreen as it would take too much energy to shed leaves and produce new ones each year, and some have purple-coloured leaves, containing protective anthocyanin pigments, especially in winter. Many produce high levels of sugars in their tissues which act as 'antifreeze', preventing ice formation whilst others allow ice crystals to form but limit them to spaces between cells, preventing damage to the cells' membranes and their contents. Their flowering is limited to a few weeks when pollinating insects, such as midges, are active but is, nevertheless, spectacular!

The term 'alpine' is often used loosely, and many plants with similar growing requirements but described as alpines in garden centres can be termed 'rock plants'. The Alpine Garden Society definition as 'a plant which normally hibernates under snow' works well horticulturally! Alpine enthusiasts aim to provide an environment with good light, free-draining composts and ample ventilation to reduce temperature and humidity (alpine plants cannot tolerate wet conditions), often in a specialized alpine house.

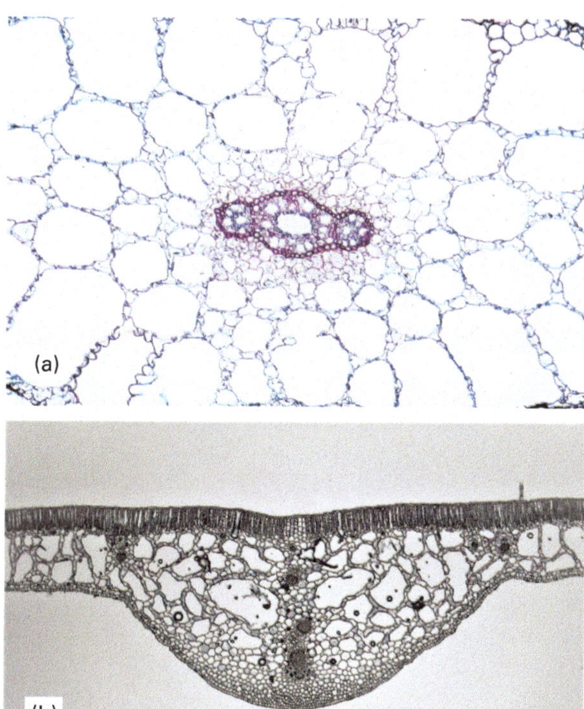

Figure 5.34 Aerenchyma tissue in (a) a cross section of an *Elodea canadensis*, a submerged freshwater plant (source: Shutterstock, D. Kucharski K. Kucharska) and (b) a cross section of a waterlily (*Nymphaea*) leaf (source: Berkshire Community College Bioscience Image Library – Angiosperm Leaf: The Hydrophytic Dicotyledonous Leaf of Nymphaea, CC0, https://commons.wikimedia.org/w/index.php?curid=70154965). The large air spaces aid buoyancy, store carbon dioxide for photosynthesis and provide oxygen for respiration to the underwater stems and leaves

5

Figure 5.33 Some evergreen plants adapted to shade: (a) Swiss cheese plant (*Monstera deliciosa*), a tropical climber with large, dissected leaves; (b) periwinkle (*Vinca major*), a temperate, shade adapted, ground cover plant; (c) large, thin leaves in a tropical climber

(*Ranunculus aquatilis*) grows elongated petioles in response to flooding which elevate the leaves above the floodwater, and it also has two different leaf types (**heterophylly**): large surface leaves to maximize photosynthesis and finely divided submerged leaves to prevent the plant being uprooted in fast-flowing water.

Further reading

Adams, C., Early, M., Brook, J. and Bamford, K. (2015) *Principles of Horticulture Level 3*. 7th ed. Routledge.

Allaby, M. (2006) *A Dictionary of Plant Sciences*. Oxford University Press.

Bidlack, J.E., Jansky, S. and Stern, K.R. (2020) *Stern's Introductory Plant Biology*. 15th ed. McGraw-Hill.

Capon, B. (2022) *Botany for Gardeners: An Introduction to the Science of Plants*. 4th ed. Timber Press.

Hickey, M. and King, C. (1997) *Common Families of Flowering Plants*. Cambridge University Press.

Hodge, G. (2013) *Practical Botany for Gardeners*. University of Chicago Press.

Royal Horticultural Society. (2013) *RHS Botany for Gardeners: The Art and Science of Gardening Explained & Explored*. Mitchell Beazley.

Zona, S. (2022) *A Gardener's Guide to Botany: The Biology Behind the Plants You Love, How They Grow, and What They Need*. Cool Springs Press.

The online material is accessible via the QR code and includes further information on many of the topics in the book, such as

- Vegetative propagation of plants
- Micropropagation
- Plant selection
- Planting & post-planting care
- Grafting

CHAPTER 6

Plant reproduction

Figure 6.1 Wind dispersed fruits (samaras) of *Acer palmatum* in autumn

This chapter includes the following topics:

- Flower structure
- Inflorescences
- Pollination
- Fertilization
- Fruits and seeds
- Asexual reproduction

DOI: 10.4324/9781003581260-6

The flowering plant represents the pinnacle of evolution in the plant world, and there is no doubt that in the ornamental garden it is the contribution of flowers which is often at the forefront of plant selection (Figure 6.2).

Flowers are the organs of **sexual reproduction** in flowering plants. Their structure forms the basis, in large part, of plant classification and reflects the different pollination mechanisms used and, in the case of animal-pollinated flowers, the need to advertise their wares! They are also where fertilization, the fusion of the male sex cell in pollen and the female sex cell in the ovule, takes place. Sexual reproduction leads to the

Figure 6.2 Range of flowers as organs of sexual reproduction having similar basic structure, but varying appearance, having adapted for successful pollination or by plant breeding: (a) *Iris chrysographes* 'Kew Black'; (b) *Eryngium giganteum* ('Miss Willmott's ghost'); (c) *Trollius chinensis* 'Golden Queen'; (d) *Rosa* 'L.D. Braithwaite'; (e) *Hemerocallis* 'Rajah'; (f) *Aquilegia fragrans*; (g) *Oenothera* 'Apricot Delight'; (h) *Helenium* 'Wyndley'; (i) *Helleborus × hybridus*; (j) *Nepeta nervosa*; (k) *Primula vialii*

development of seeds and fruits and is the means by which plants spread. It also brings about mixing of the genes contributed by each parent so that the offspring will be similar but not identical to the parent plants and to each other. This range of variation enables plants to withstand changes in environmental conditions as there will always be some individuals which are likely to survive. Flowers and their fruits and seeds also support wildlife in gardens, from invertebrates to birds and mammals, for which gardens are increasingly important havens.

In contrast, plants can also reproduce **asexually** (**vegetative reproduction**) through many natural means, such as runners, bulbs, and by layering or suckering, enabling plant dispersal (see Chapter 5). The resulting plants are genetically identical to the parent plant (clones). It is an alternative to sexual reproduction, useful if pollination is poor or seed production fails, but does not provide the genetic variation which is necessary for long-term survival and adaptation of the species. It is, however, very useful in vegetative propagation by growers where the aim is to maintain the desirable characteristics of the parent plant.

Sexual reproduction is the formation of new individuals through fusion of male and female sex cells (**gametes**). It results in variable offspring. **Asexual reproduction** is the formation of new individuals without fusion of gametes, resulting in genetically identical offspring.

Flower structure

Understanding flower structure is useful when identifying plants. However varied flowers appear, they all have the same basic structure with the flower parts arranged in four whorls (Figure 6.3). These are, from the outside in, the

- calyx (the sepals)
- corolla (the petals)
- androecium (the stamens).
- gynoecium (the carpels).

The **calyx** or ring of **sepals** initially encloses and protects the flower bud. The sepals are often green and can therefore photosynthesize. In some plants, for example *Fuchsia* spp. (Figure 6.14i) and hellebores (*Helleborus* spp.) (Figure 6.4), the sepals may be coloured to attract animal pollinators, whilst in wind-pollinated plants they are usually reduced in size.

The **corolla** or ring of **petals** may be small and insignificant in wind-pollinated flowers, for example in many tree species, or large and colourful in insect-pollinated species (Figure 6.14). The

6

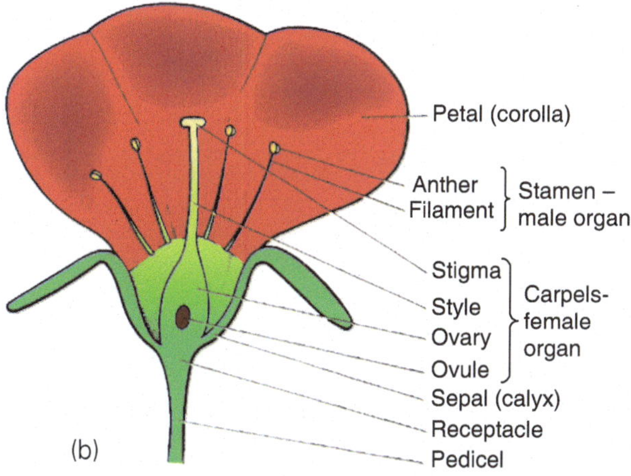

Figure 6.3 Flower structure: (a) flower of *Glaucium corniculatum*, a eudicot with four petals; (b) diagram of a typical eudicot flower to show structures involved in the process of sexual reproduction; (c) section through a pumpkin flower showing anthers (covered in yellow pollen) to the left, lobed stigma in the centre at the base and developing ovary with immature seeds on the right

petals may have **nectar guides** which are lines or differently coloured blotches on petals (often invisible to human eyes except in ultraviolet light), leading pollinators to the **nectaries** at the base of the petals (Figure 6.5). These have a secretory function, producing substances such as **nectar**, a sugary 'food' which attracts pollinating organisms.

The colour and size of petals can be improved in cultivated plants by breeding which may also lead to the multiplication of the petals or **petalody**, where male and/or female organs have been converted into petals, for example in many 'double-flowered' cultivars (Figure 6.6). This is often at the expense of pollen and nectar so such flowers are of no value to pollinators although useful for gardeners as they may continue to flower over a longer period because they cannot set seed.

Figure 6.4 Brightly coloured sepals of a hellebore attract a foraging honeybee in January when pollen and nectar supplies are scarce

- The **androecium**, the male organ, consists of **stamens**, which bear **anthers** that produce and discharge **pollen grains**, borne on a **filament**. The pollen contains the male sex cells or gametes.
- The **gynoecium**, the female organ, is positioned in the centre of the flower and consists of an **ovary** enclosing one or more **ovules** each of which contains an **ovum** (the female sex cell or gamete). The **style** leads from the ovary to a **stigma** at its top where pollen is captured. The basic unit of the gynoecium is the **carpel**, made up of a stigma, style and ovary. More evolutionarily primitive flowers such as buttercup (*Ranunculus* spp.) have many separate carpels but in most flowers the carpels are fused to form one large ovary, style and stigma. This is sometimes called the **pistil**.

The flower parts are attached to the **receptacle**, which is at the tip of the **pedicel** (flower stalk). Sometimes there are leaf-like structures called **bracts** at the base of the flower which can assume the function of insect attraction, for example, in poinsettia (*Euphorbia pulcherrima*), *Hydrangea* spp. and some *Cornus* species (see Figure 5.26).

In many monocots such as tulips and lilies, the outer two layers of the flower have a similar appearance, making the sepals and petals indistinguishable, in which case they are called **tepals** (Figure 6.7).

The symmetry of the flower is a useful identification aid. Flowers may be **regular** (**actinomorphic**), that is, symmetrical around a point, as in the lily flower in Figure 6.5a, or **irregular** (**zygomorphic**),

Figure 6.5 Flowers with nectar guides: (a) day lily (*Hemerocallis*); (b) *Digitalis stewartia*; (c) *Verbascum* 'Cotswold'

as in the foxglove in Figure 6.5b and antirrhinum in Figure 6.15 where the flowers are symmetrical along a line. This is often characteristic of a particular plant family such as in bean and pea flowers (Fabaceae) and the dead nettle family (Lamiaceae) which have irregular flowers.

A distinguishing feature of monocots and eudicots is the number of flower parts, particularly sepals, petals, carpels and stamens. In monocots these are usually in multiples of three (Figure 6.2a and e) whereas in eudicots, flower parts are in multiples of four (Figure 6.3a) or five (Figure 6.2i). Something to bear in mind when examining flowers for identification purposes is that many cultivated plants which have been produced through plant breeding are not necessarily 'true to type' compared with their original wild ancestors.

The flowers of most species have both male and female organs (**hermaphrodite**), but some have separate male and female flowers on the same plant (**monoecious** – meaning 'one house') (Figure 6.8) such as cucumbers (*Cucumis sativus*), courgettes (*Cucurbita pepo*) and many trees such as walnuts (*Juglans* spp.), alders (*Alnus* spp.) and birches (*Betula* spp.), whereas others produce male flowers on one plant and female flowers on another plant (**dioecious –** 'meaning two houses'), such as holly (*Ilex aquifolium*), willows (*Salix* spp.) and *Skimmia japonica*.

Figure 6.6 Petalody in a *Gardenia* flower

Most conifers are monoecious with male and female cones rather than flowers (Figures 2.7 and 6.9).

A plant possessing flowers with both male and female organs is **hermaphrodite.** Species with separate male and female flowers on the same plant are **monoecious.** Species which produce male flowers on one plant and female flowers on another are **dioecious.**

Inflorescences

The inflorescence is the arrangement of flowers on the flower stem. Inflorescences may be made up of a single flower (**simple**) or many flowers clustered together (**compound**). In compound inflorescences, each individual flower stalk (**pedicel**) is attached to a central stalk (**rachis**) and the stem supporting the inflorescence is a **peduncle**. Flowers are gathered together in a block, making a large impact to attract pollinators or providing a landing stage for them to walk over, for example, the capitulum of plants in the Asteraceae family (Figure 6.12). so that they expend less energy in moving from flower to flower. In wind-pollinated flowers, the inflorescence may contain many flowers grouped together in a catkin (a form of spike) to maximize release and capture of pollen, for example in trees such as hazel (*Corylus* spp.) and birch (*Betula* spp.) or by holding it above the leaves and stems as in grasses (Figure 6.10).

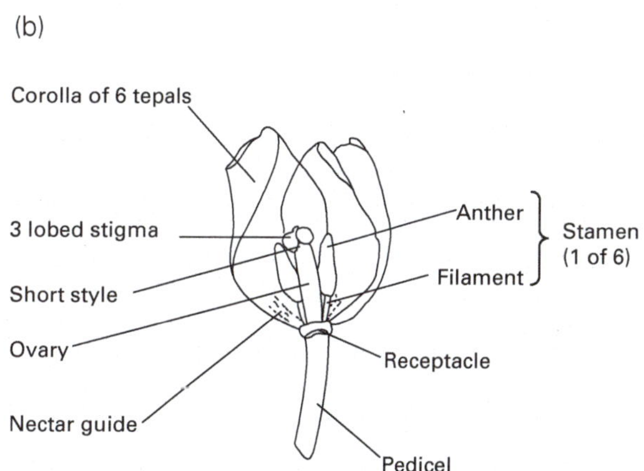

Figure 6.7 (a) Flower of *Tulipa*, a monocot with (b) diagram showing the typical structure of a monocot flower

6

Figure 6.8 Monoecy in courgette (*Cucurbita pepo*): (a) female flower showing an ovary (which will become the fruit) below the petals and a branched stigma within the flower. Stamens are absent; (b) male flower lacking an ovary (pedicel connects flower to stem) with stamens and pollen within flower but no stigma

Inflorescences are classified by the order in which flowers open and by the way in which they are arranged on the stem. They can be either **cymose** (cymes), where the stem terminates in a flower so that new flowers originate from the leaf axils further down the stem, or **racemose** (most inflorescences), where the youngest flowers are at the top of the stem. Inflorescence types are often characteristic of a particular plant family so a knowledge of these is useful for identification purposes. Some examples are described later and are illustrated diagrammatically in Table 6.1 with plant examples shown in Figure 6.11:

The **capitulum** (Figure 6.12), the characteristic daisy 'flower', is actually a very special inflorescence. It is made up of many individual flowers (**florets**) with fused petals, radiating out from the centre attached to a flattened receptacle. The outer flowers, called **ray florets**, often have their five petals fused together to resemble one long petal, whereas the inner or **disc florets** are small and tubular. The disc florets are often fertile and give rise to seeds, whereas the ray florets are usually infertile, think of a sunflower head (*Helianthus annuus*), and are colourful to attract pollinators. The tiny sepals are hidden at the base of each floret and may become the 'umbrella' (pappus) when the seed is dispersed on the wind as in dandelion (*Taraxacum*). The leafy bracts combine in an **involucre** supporting the whole inflorescence.

Figure 6.9 Male cones of blue cedar (*Cedrus atlantica* (Glauca Group))

Figure 6.10 Inflorescences which are designed to maximize pollen dispersal and capture by wind: (a) in birch (*Betula* spp.) male catkin on the left, female catkin on the right; (b) in grasses

Pollination

Pollination is the transfer of pollen (Figure 6.13), containing the male sex cell, from anther to stigma of either the same flower or different flowers and is the necessary first step in the process leading to fertilization. Fertilization is essential for seed and fruit production (although some commercial crops such as bananas, cucumbers and grapes have been bred to be seedless).

In **cross-pollination**, pollen transfer is between different plants with different genetic make-ups, while in **self-pollination** it is within or between flowers usually on the same plant, with the same genetic make-up. Cross-pollination, leading to cross-fertilization, ensures that variation is introduced into each new generation of offspring, making the population better able to survive adverse weather, pests or diseases. For the horticulturist, cross-pollination between plants with different characteristics can give rise to plants with new, sometimes desirable, features, and this is the basis of much plant breeding and hybridization. Cross-fertilization can also produce stronger plants since undesirable genes are removed through selection or masked through mixing with genetic material from other plants.

However, cross-pollinated plants are dependent on pollen being transferred between plants, for example by insects or wind, which may be risky, whereas self-pollination gives a more reliable outcome. Plants have mechanisms which favour either cross-pollination, for example in Brassicaceae (cabbage family), maize (*Zea mays*) and apple cultivars (see p. 95), or self-pollination, for example in some members of Fabaceae (bean family), tomato (*Solanum lycopersicum*) and sunflower (*Helianthus annuus*). In effect they 'choose' whether to play it safe or take a chance for greater rewards. Flower colour change is often a signal to pollinators that a

Table 6.1 Inflorescence types: in the diagrams the youngest flowers are indicated by the smallest circles

Type of inflorescence	Structure	Description	Plant examples
cyme		Flower forms on the end of the flower stalk so restricting the number of developing flowers to further down stem where they grow from the axils	*Geranium* spp., *Geum* spp., wood forget-me-not (*Myosotis sylvatica*), common chickweed (*Stellaria media*)
raceme		Stalked flowers with pedicels of the same length spaced out on a single rachis	Common in the pea family (Fabaceae) and orchid family (Orchidaceae); foxglove (*Digitalis purpurea*), wallflower (*Erysimumx cheiri*), sweet pea (*Lathyrus odoratus*), *Wisteria sinensis*
spike		Unstalked flowers (no pedicel) on a single rachis	*Verbascum olympicum*, *Gladiolus cardinalis*, *Acanthus spinosus*
umbel		Stalked flowers with the stalks (pedicels) attached at the same point on the main stem	Common in carrot family (Apiaceae) and onion family (Alliaceae); hogweed (*Heracleum sphondylium*), cow parsley (*Anthriscus sylvestris*), chives (*Allium sativum*), *Allium* 'Globemaster'
corymb		Flower stalks, spaced out along the rachis, are of different lengths so that the flowers are often all at the same level	Common in the cabbage family (Brassicaceae); elder (*Sambucus nigra*), candytuft (*Iberis sempervirens*)
panicle		A compound raceme with simple racemes arranged in sequence along the rachis	Common in the grass family (Poaceae); *Stipa gigantea*, lilac (*Syringa vulgaris*), *Rodgersia pinnata*
verticillaster		Unstalked flowers attached in a whorls at a node	Common in dead nettle family (Lamiaceae); *Phlomis fruticosus*, dead nettle (*Lamium maculatum*), mint (*Mentha spicata*)
capitulum		Disc-shaped inflorescence made up of many individual florets	Characteristic of the daisy family (Asteraceae); daisy (*Bellis perennis*), sunflower (*Helianthus annuus*), *Helenium*, dandelion (*Taraxacum*)

Figure 6.11 Inflorescence types: (a) **spike** – *Verbascum* spp.; (b–d) **raceme** – foxglove (*Digitalis* spp.), (c) veronica (*Veronica* spp.), (d) an orchid; (e) **panicle** – *Rodgersia* spp.; (f) **corymb** – elder (*Sambucus* spp.); (g) **umbel** – hogweed *Heracleum* spp.; (h) **capitulum** – *Inula* spp.; (i) **cyme** – *Geranium maderense*; (j) **verticillaster** – white dead nettle (*Lamium album*)

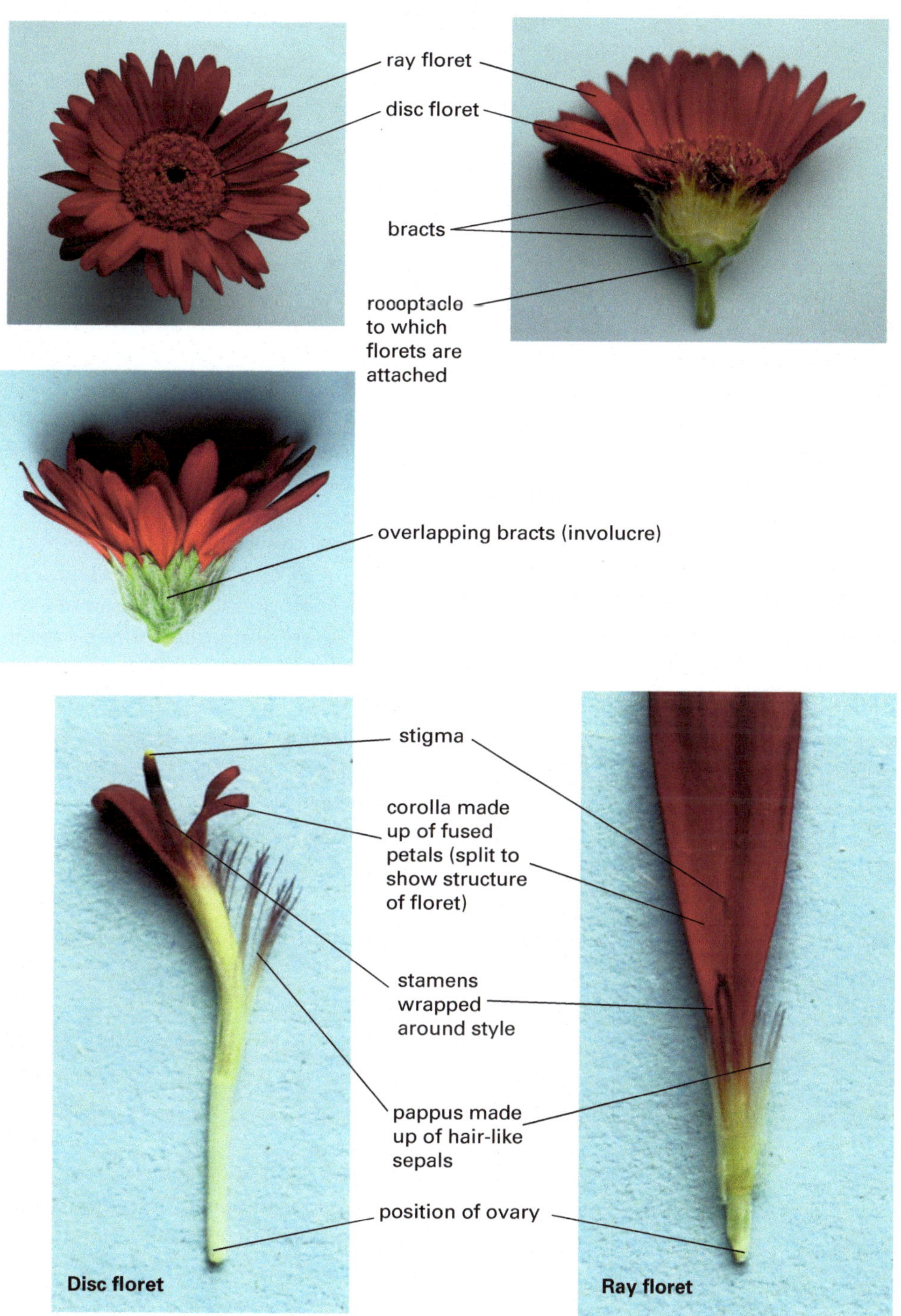

Figure 6.12 Structure of a capitulum in a *Gerbera* cultivar

flower is mature enough to be pollinated, for example in some *Geranium* spp., or that it has already been pollinated so not to visit. In horse chestnut (*Aesculus hippocastanum*) the nectar guides change from yellow to red within a day of pollination.

Some of the most important horticultural fruit crops are highly dependent on pollinators such as top fruit: apples, pears, apricots, plums, mangoes; and soft fruit: blueberries and raspberries. For some crops such as cocoa, avocado, kiwi fruit and pumpkins, they are essential. It has been estimated that 75% of the world's crop species are at least partially dependent on pollinators and 30% of global food production depends on pollinators.

Three examples where a knowledge of pollination characteristics is useful to horticulturists – apple, sweetcorn and F1 hybrid production – are described in the boxes in this chapter.

Animal pollination

Whilst some animal pollinated flowers such as orchids are highly specialized to utilize a particular

Pollen grains

Palynology is the study of pollen. The outer coat of a pollen grain (the exine) contains sporopollenin, one of the most indestructible structures in the natural world. It enables pollen grains to survive millions of years, and this is a useful tool in identifying types of plant cover in prehistory, in providing information about past climates and, also, in forensic investigations. Pollen grains come in many different shapes, and their surface is frequently beautifully sculpted in a pattern unique to a particular plant species. Somewhat surprisingly, even though pollen grains were first described over three hundred years ago, the reason why these myriad patterns exist are still largely unknown. One of the central roles of the pollen wall is believed to be protective. During the journey of pollen grains to stigmas, the wall shields pollen from harmful factors, such as ultraviolet radiation, pathogens and dehydration. Yet, wall patterns are unlikely to be involved in merely protection. Possible functions are helping to deliver pollen grains to stigmas, for example pollen transported by wind may have small bladders attached which act like sails to aid dispersal. Wall patterns may also mediate interactions between pollen and the surface of stigmas. The grains have indentations, furrows or apertures (three in eudicots and one in monocots) on their surface, where the exine is thinner which may allow the grain to expand or contract, helping it to dehydrate and rehydrate without losing viability. These are the points where the pollen tubes (see p. 103) emerge.

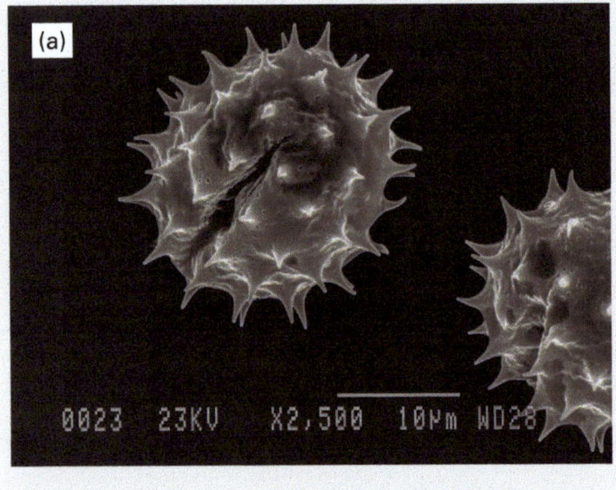

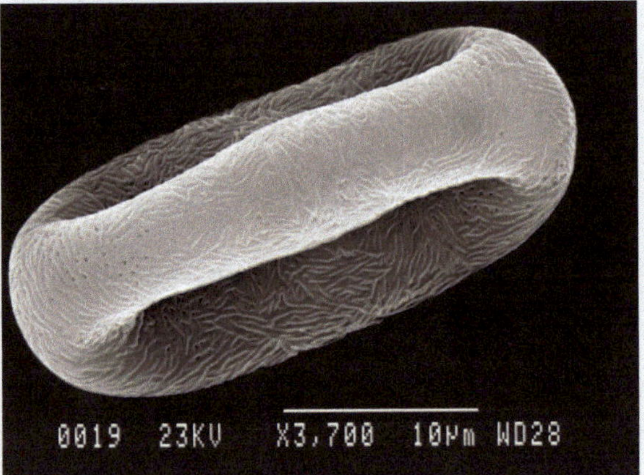

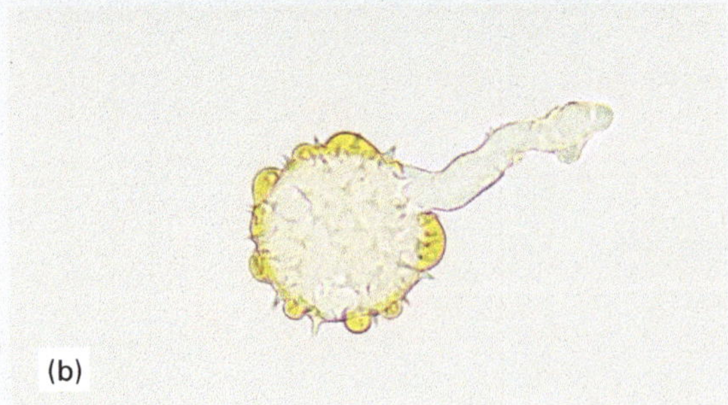

Figure 6.13 (a) Scanning electron microscope micrographs of pollen from daisy (*Bellis perennis*) (left) (source: Shutterstock, bearacreative) and cow parsley (*Anthriscus sylvestris*) (right) (source: Shutterstock, bearacreative); (b) a germinating pollen grain of *Sphagneticola trilobata* (source: Shutterstock, Ekky Ilham)

method of pollen dispersal, most are generalists. For example, cucumbers can be pollinated by bumblebees, honeybees, several species of solitary bee and flies, some wasps and butterflies. Over 60% of flowering plants are insect pollinated and 20% of insects are dependent on flowers for food. To attract animal pollinators, plants must offer a reward, advertise themselves in some way and transfer pollen to the pollinator effectively. To achieve this, many flowers tailor their structure and colour, their provision of scent and nectar, their flowering time and their position on the plant to favour particular pollinators.

The reduction in insect populations is a cause for concern. For example, bees, especially wild solitary bees and bumblebees, are in serious decline in the UK and globally. Here, we have already lost around 13 species and another 35 are currently at risk. Gardeners can encourage pollinators, especially insects, by planting a wide

Pollination and polyploidy in apples

Apple is an obligate cross-pollinating species, so two compatible cultivars, whose flowering times coincide, are essential to produce fruit. Apple cultivars are classified into 'pollination groups' based on when they flower, with the aim of planting two cultivars which belong to the same group, or adjacent groups, together, for successful pollination. A few apple cultivars are self-fertile and do not need a pollinator. An example is 'James Grieve', but even these fruit better if they are cross-pollinated.

In some cultivars such as 'Bramley's Seedling', a further complication is that cultivars may be triploid. Whereas most apples are diploid (they have two sets of chromosomes in their cell nucleus, one from each parent), triploids have three and triploidy gives rise to sterile pollen which is unable to pollinate another tree. The presence nearby of a suitable diploid pollinator cultivar such as 'Encore', which provides viable pollen at the same flowering time as 'Bramley's Seedling', would enable the 'Bramley's Seedling' to fruit, but a third diploid pollinator such as 'Hawthornden' would be necessary to yield fruit on all three trees. An alternative strategy for a private gardener is the inclusion of a pollinator onto the triploid tree, by means of a suitable graft (see p. 60), the result sometimes being called a 'family tree'.

Many plant species are **polyploids** and contain more than the usual two sets of chromosomes; up to octaploid (eight sets) can occur, as in cultivated strawberries. An increase in the size of cells to accommodate the extra chromosomes results in an increase in fruit and flower size, for example in chrysanthemums, fuchsias, strawberries, turnips and grasses. This is why 'Bramley's Seedling' is larger than many other apples. Polyploidy can occur spontaneously and has led to many variant types in wild plant populations or it can be artificially induced by the use of a chemical, such as colchicine.

range of plants which provide flowers over a long period through the year, from bulbs, for example snowdrops (*Galanthus nivalis*) and early daffodils (*Narcissus* cvs.) in late winter and spring through to autumn, with *Cyclamen hederifolium* and ivy (*Hedera helix*), always avoiding double flowers. Whilst native plants (see p. 151) are often seen as the ideal, insects will visit non-native plants too, and these are very useful additions to extend the season. Consideration should also be given to supporting insects throughout their life cycle, for example providing leaves for caterpillars to feed on and grass stems for overwintering eggs and larvae. With careful planning, suitable habitats for pollinating insects in gardens can be created, providing shelter and food (see Chapter 9).

Pollination is the transfer of pollen from stamen to stigma of a flower or flowers. **Fertilization** is the fusion of a male sex cell (the male gamete) from a pollen grain with a female sex cell (the female gamete) in the ovule to produce an embryo.

Examples of animal pollination (Figure 6.14) are:

- **Bees** favour sturdy flowers which are shallow, providing a landing platform, with short floral tubes whose length matches the length of the tongue of the bee species visiting it. They select brightly coloured white, yellow or blue but not pure red flowers which often have nectar guides (Figure 6.5). The nectar of bee-pollinated flowers is often 40%–70% sugar which provides a rich energy source for the bee. Flowers also emit heat, creating temperature arrays that mimic the colour patterns on petals. On average, heat spots are 4°C to 5°C warmer than the rest of the flower. Electrostatic charges on both bees and flowers may also aid pollination.

Bees also collect pollen, which is often sticky and scented, transferring it from their bodies to specially modified areas on their legs called 'pollen baskets'. Pollen provides protein in their diet. Bee-pollinated flowers often have a sweet mild scent. Sometimes the flowers can be highly adapted as in *Antirrhinum* with two projections on the lower lip which fit two hollows on the upper lip. Only a bee of just the right size and weight is able to trigger the flower to open (Figure 6.15). Bands of hair on the lower lip steer the bee's tongue to the nectar pouch at the base of the petals and the stigma and stamens are positioned to brush the bee's back, depositing and collecting pollen. In others such as foxglove (*Digitalis purpurea*) (Figure 6.11b), which are visited by bumblebees, the large spots indicate a landing stage and the bee crawls into the tube to locate the nectar at the bottom. A landing stage is also provided by bowl-shaped flowers such as in *Verbascum* (Figure 6.11a) and *Prunus* and enlarged lower petals or 'lips' as in *Lamium*.

(a)
(d)
(b)
(e)
(c)
(f)

Figure 6.14 Some flowers adapted to different animal pollinators: (a–c) **Bees**: foxglove (*Digitalis purpurea*) with spots indicating a landing platform and tubular flowers; small, blue tubular flowers of *Echinops* visited by honeybees; inflorescence provides a landing platform for a bumblebee dusted with pollen; (d) **Moths**: white flowers and scent in night-scented stock (*Matthiola*); (e–f) **Butterflies**: peacock butterfly feeding on pink tubular flowers of *Hebe*; large white butterfly feeding on florets in the capitulum of a *Zinnia* flower which provides a landing platform; (g) **Carrion flies**: flesh coloured petals and putrid scent of *Rafflesia*, a tropical parasitic plant; (h) **Hoverflies**: dull pale greenish white flowers of *Angelica*; (i–j) **Humming birds**: tubular corolla of *Fuchsia* which hangs down; bright red flowers of *Heliconia stricta*; (k) **Beetles**: leathery sepals of *Magnolia obovata*

In *Veronica* (Figure 6.11c), the bee grips two outstretched stamens on landing and clasps them to its body. Plant families commonly pollinated by bees (Figure 6.14a–c) include the Fabaceae (e.g. *Lathyrus*) and Lamiaceae (e.g. *Mentha*).

- **Moths** are usually night-flying insects so flowers tend to only open at dusk, during the night or at dawn. The flowers are white or pale green, and star shaped, so that they are easily located in low light and are often held horizontally or hang down with flat or swept-back petals for moths that hover, to feed. Stamens and stigmas may protrude outside the flower to contact the moth's head as it hovers. Their major attractant is a heavy, sweet scent typical in flowers such as tobacco plant

Figure 6.15 Antirrhinum flowers open when a bee of the right weight lands on the lower petal: a) external view; b) section through flower

(*Nicotiana tabacum*) and night-scented stock (*Matthiola longipetala*) (Figure 6.14d) which is often only produced at night or at dawn or dusk. Moths use the shape of the corolla to locate copious amounts of nectar at their base rather than nectar guides which would be less obvious at night, and sometimes the nectar is only produced when the moth arrives. Often the moth's tongue precisely matches the length of the corolla tube. Charles Darwin heard of an extraordinarily long, tubular flower from Madagascar (*Angraecum sesquipidale*) and predicted that a moth existed, then unknown, with a tongue of matching length. Forty years later the moth was discovered, and he was proved right. The moth was named *Xanthgopan morgani-praedicta*, meaning 'the one that was predicted'. Plants commonly pollinated by moths are found in the families Solanaceae (e.g. *Brugmansia*) and Caprifoliaceae (e.g. honeysuckle (*Lonicera*)).

- **Butterflies** use tightly packed, often flat-topped inflorescences or single flowers as a landing stage and feed on pollen and nectar as they walk across them, so flowers pollinated by butterflies tend to be massed together and held upright. Their long tongues, from half a millimetre to 30 cm long, can penetrate deep into corolla tubes or flower spurs whose length matches the butterfly's tongue, accessing nectar at the base of nectar guides. Research has shown that they are attracted to nectar which is rich in amino acids, the building blocks of proteins. Scents vary but bright colours are favoured including purple, red and pink; they have good colour vision and can detect a wider range of colour wavelengths than humans or bees. Examples include *Phlox* (Polemoniaceae), *Lantana* (Verbenaceae) and many members of the daisy family (Asteraceae) such as *Zinnia* (Figure 6.14e–f).
- **Flies** form two groups whose characteristics influence the types of flower they pollinate. Carrion and dung flies visit flowers which resemble the dead animal matter on which they feed. Consequently, these flowers have strong putrid scents resembling rotting flesh and their colour is a dull purple or brown. Nectar guides and nectar are absent with the flies solely feeding on pollen. Flowers may open by day or night, they have deep corolla tubes and often have appendages which trap the flies to ensure pollen is transferred to the fly's legs before the insect is released. The world's largest flower, *Rafflesia arnoldii*, has a hairy, brown-red flower and black spots which looks and smells like decayed flesh (Figure 6.14g). As the fly crawls across the flower it treads on 'translators', structures which wrap around the fly's legs together with lumps of pollen which are dragged out with the translators as the fly moves. In the next flower the pollen lumps fit exactly into the grooves in the flower's stigma.
  Hoverflies and bee flies (Figure 15.9a) are mainly day flying and have short tongues. Hoverflies mimic bees and wasps but have only one pair of wings and short antennae. They prefer shallow, open, regular flowers with little scent or nectar (Figure 6.14h). Flower colour is variable but they are often dull or light coloured such as ivy (*Hedera helix* in the family Araliaceae) and many flowers in the family Apiaceae (e.g. *Angelica sylvestris*). In *Arum maculatum*, the native cuckoo pint, moth flies (midges) are attracted by chemicals smelling of urine, which are produced at the tip of the flower's spadix at the top of the inflorescence. These are vaporized when chemical reactions in the flower tissue heat them up to 16°C above the surrounding temperature. The inflorescence is surrounded by a leaf-like spathe which has a small opening allowing the insects to enter. Inside the spathe the walls are slippery and the insect falls, tumbling past male flowers, where it is dusted by pollen, and falling onto the female flowers in the bottom. The walls have downward-pointing hairs which prevent the fly from climbing out, but during the course of the day, the hairs wither and the spadix becomes wrinkled, enabling the flies to escape.
- **Bird**-pollinated flowers have nectar but no nectar guides and contain little scent as birds lack a

sense of smell. They are often attracted to vivid colours, particularly red, and do not eat pollen. In flowers pollinated by hovering birds such as hummingbirds, the flower shape is generally regular with downward pointing corolla tubes and petals which may be folded back, out of the way. Often the stigma and stamens hang outside the flower. Hummingbirds have long curved beaks which penetrate deep inside the flower. Pollen is deposited on the head or beak of the bird as it probes the flower with its long and specially constructed tongue, which forms a tube to suck large amounts of nectar. The flowers often have furrows and tubes inside the corolla to guide beaks and tongues which may be specific to a particular bird species. Bird-pollinated flowers, which are often tropical or sub-tropical, open in the day when birds are active and produce copious amounts of dilute nectar; one hummingbird bird was shown to visit 1,311 larkspur (*Delphinium cardinale*) flowers in the space of six and a half hours. Another strategy is found in *Strelitzia* spp., where robust flowers permit birds to land, opening the anthers under the bird's weight. All bird-pollinated flowers are tough to withstand a bird's weight and probing beak, with thick petals and stiffened filaments and styles. Examples of bird-pollinated flowers include *Fuchsia* (Onagraceae) (Figure 6.14i), *Aquilegia* (Ranunculaceae), *Salvia* (Lamiaceae), *Strelitzia reginae* (Strelitziaceae) and *Hibiscus* (Malvaceae).

- **Beetles** are a large group representing about 40% of all known insects. They visit a wide range of flowering plants, especially those with tightly clustered small flowers such as *Achillea* ssp. and *Spiraea* spp. They are also one of the oldest pollinators having co-evolved with some of the earliest plants such as magnolia (*Magnolia* spp.) (Figure 6.14k), tulip trees (*Liriodendron tulipifera*), paw-paw (*Asimina* spp.) and water lilies (Figure 7.1). Beetles will eat through leaves and petals of flowers to reach the pollen, a food source, which sticks to their bodies. Flowers which are beetle pollinated therefore tend to have thick, leathery petals, tepals and leaves, to withstand the onslaught. They tend to be white, cream or pale green, and they have a strong spicy, sweet or musky smell which beetles rely on primarily rather than colour.

Bees in horticulture

The well-known social insect, the honeybee (*Apis mellifera*), is helpful to horticulturists. The female worker collects pollen and nectar as a supply of food for the hive and, in collecting it, the bee transfers pollen from plant to plant. Several crops, such as apple and pear, do not set fruit when self-pollinated. The bee therefore provides a useful function to the fruit grower. In large areas of fruit production, the number of resident hives may be insufficient to provide effective pollination, and in cool, damp or windy springs, the flying periods of the bees are reduced. It may therefore be advantageous for the grower to introduce beehives into the orchards during blossom time, as an insurance against bad weather. One hive is normally

Figure 6.16 (a) Bumblebee boxes provided in a glasshouse for the pollination of tomatoes; (b) buzz pollination of tomato by a bumblebee (source: Shutterstock, AJCespedes)

adequate to serve 0.25 ha of fruit. Blocks of four hives placed in the centre of a 1 ha area require foraging bees to travel a maximum distance of 70 m. In addition to honeybees, wild species, for example the potter flower bee (*Anthophora retusa*) and red-tailed bumblebee (*Bombus lapidarius*), increase fruit set, but their numbers are not high enough to dispense with the honeybee hives. Honeybees are less fussy in their choice of flower than bumblebees!

All species of bee are killed by broad-spectrum insecticides, and it is important that spraying of such chemicals be restricted to early morning or evening during the blossom period when hives have been introduced. Recently neonicotinoids, a group of pesticides which are used as seed dressings, have been implicated in the reduction in numbers of bees. In commercial glasshouses, the pollination of crops such as tomatoes and peppers is commonly achieved by in-house nest boxes of bumblebees (*Bombus terrestris*) (Figure 6.16a). In tomatoes (and also in kiwi fruit, aubergines and cranberries), **buzz pollination** by bumblebees ensures release of pollen through pores in the anthers. The bee grasps the anthers and vibrates her body to shake out the pollen (Figure 6.16b). This can be carried out manually by means of small brushes and 'electronic bees' where bumblebees are not available. Plant breeders may use also use blowflies in glasshouses to carry out pollination.

Wind pollination

In contrast to flowers pollinated by animals, wind-pollinated flowers lack scent, nectaries and nectar. They tend to be small and their corolla and calyx are coloured green or brown and reduced in size or even absent. Most have only one or two ovules per ovary, as the chances of intercepting more than one pollen grain are quite small. Their anthers and multibranched or sticky stigmas often hang outside the flowers to release and trap pollen efficiently. Unlike animal-pollinated plants, pollen is produced in large quantities and is small and light with a smooth surface to prevent it clumping together. It is released in dry weather when the breeze can carry it far, and in many trees, flowering occurs early in the spring when the leaves do not obstruct pollen distribution. Examples of wind-pollinated plants include grasses (Figure 6.17), plantains and rushes as well as many catkin-bearing tree species such as *Betula* and *Alnus* (Betulaceae), *Populus* (Salicaceae), *Corylus* (Corylaceae), *Fagus* and *Quercus* (Fagaceae) (Figure 6.10).

Conifers also use wind pollination to disperse copious amounts of pollen from the small male cones (Figure 6.9).

Water pollination

Water pollination is very rare in flowering plants as most freshwater aquatic plants have flowers which emerge above the water surface. Some examples are the freshwater species of *Vallisneria* and the marine plant *Zostera*. Pollen floats on the water's surface, drifting until it contacts flowers. An unusual example is in the tropical marine angiosperm *Thalassia testudinum*, where the male flowers release pollen in mucilage at night when invertebrate animals, which transfer it, are active.

Mechanisms to ensure cross-pollination and fertilization

Many plants have mechanisms which favour cross-pollination and cross-fertilization. However, sometimes cross-pollinating plants will switch to self-pollination towards the end of the flowering season as an insurance. Mechanisms which favour cross-pollination are:

- **Self-incompatibility**, which is a genetic mechanism that prevents fertilization taking place within a flower or between flowers on the same plant. Pollen landing on a genetically identical stigma fails to germinate or the pollen tube growth is very slow. This is common in the cabbage family (Brassicaceae).
- **Structural mechanisms** where flower structure encourages cross-pollination. An example is **heterostyly** in *Primula* where flowers have stigma and stamens of differing lengths: in thrum-eyed flowers, the anthers emerge further from the base of the flower than the stigma, so that insects rub against them when reaching into the flower tube; in pin-eyed flowers, the stigma protrudes from the flower and will catch the pollen from the same place on the insect body, so ensuring cross-pollination (Figure 6.18).
- **Staggered ripening** of the anther and stigma occurs in many plants. When the pollen is

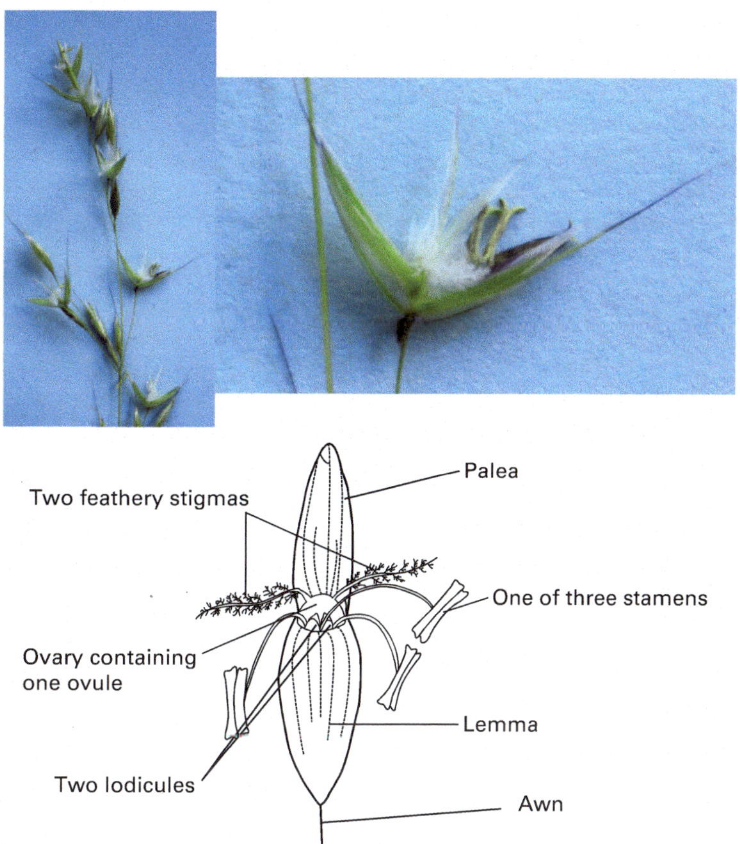

Figure 6.17 A grass flower (floret) highly specialized for wind pollination: small, green flowers without scent or nectar are enclosed in the palea and lemma, two leaf-like structures which protect it. The sepals and petals are reduced to two scale-like structures (lodicules). Two feathery stigmas on short styles and three long, flexible filaments with hinged anthers hang outside the flower. Grass pollen is produced in large quantities. Each pollen grain is small, smooth and has wing-like structures which promote dispersal by wind

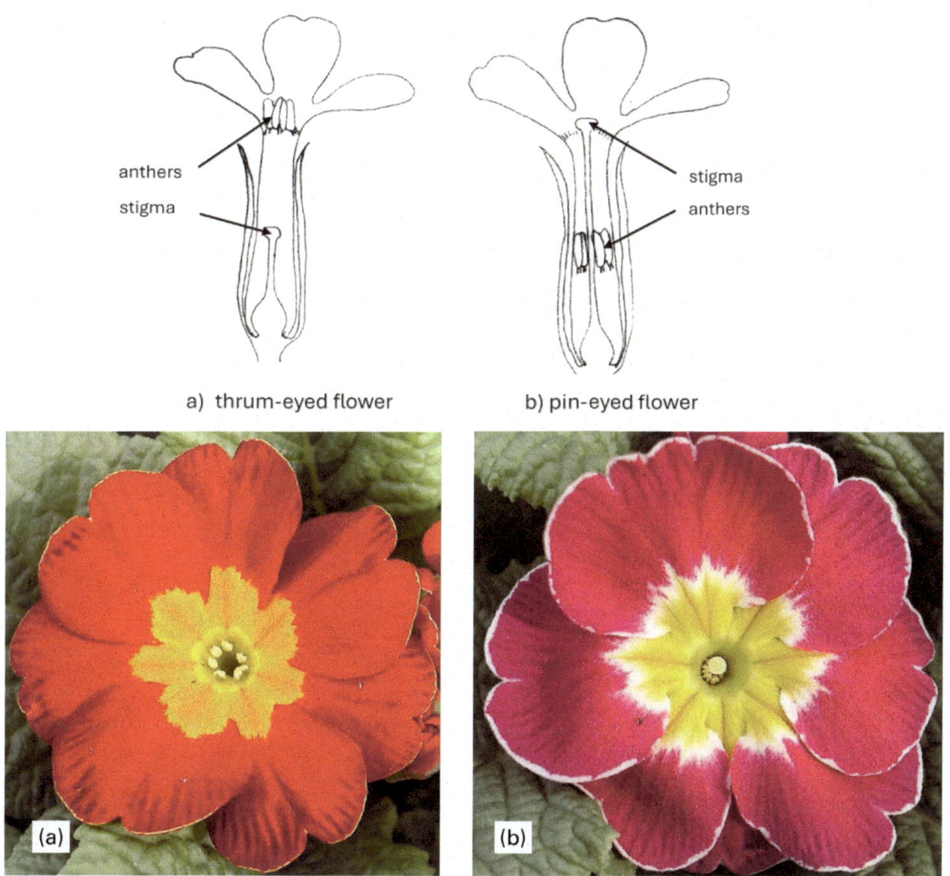

Figure 6.18 Heterostyly in *Primula*: (a) thrum-eyed flower; (b) pin-eyed flower

ready to be shed, the stigmas on the same plant are immature (**protandry**), for example in maize (*Zea mays*), or vice versa (**protogyny**) found in *Acacia* plants. Fertilization is only successful between flowers whose stigmas and stamens have matured at slightly different times and are therefore likely to be on different plants.

- **Monoecious** plants with separate male and female flowers on the same plant provide a greater chance of pollen being transferred between separate plants, especially in wind-pollinated plants or if the flowers mature at different times. In cucumber the male flowers often appear before the female flowers. However, **dioecious** plants, where male and female flowers are borne on separate plants, possess the ultimate mechanism. Cross-pollination is the only option.

Whilst many plants favour cross-pollination, in some plants self-pollination is the norm. In wheat, for example, pollination takes place before the flower opens.

Sweetcorn – an example of pollination in action

Maize (*Zea mays*) is an economically important grass with sweetcorn its most important form grown in horticulture. It is wind-pollinated and naturally cross-pollinating, being monoecious (Figure 6.19) with male flowers in a terminal inflorescence called a 'tassel' and a female inflorescence, which eventually forms the cob or 'ear'. Each flower in the ear produces long styles or 'silks' which may be up to 45 cm in length. A single tassel can produce between two and five million pollen grains a day, 20 to 30 thousand for each silk on the plant, and these can travel up to 500 m on the breeze. Maize is protandrous with pollen being shed before the silks are receptive, but there is some overlap, so some 5% of self-pollination can occur as a backup. The pollen grains are caught on the moist sticky stigmas and germinate to produce a pollen tube which travels down the styles (silks). They have an

Figure 6.19 Flowers in sweetcorn (*Zea mays*): (a) male flowers in the 'tassel' produce pollen; (b) female flowers, with leaves around cob removed, showing the 'silks' (styles) leading to the ovaries

amazingly rapid growth rate, up to 25 cm in a single day! The kernels in a cob are individual fertilized ovaries, and each becomes a fruit (although they are usually referred to as a 'seed' which is itself contained within the kernel). Maize is planted in blocks rather than rows to optimize the chance of intercepting wind-borne pollen and therefore fruit set in the individual cobs. Poor pollination results in empty or missing kernels on the cob.

Fertilization

When a pollen grain arrives at the stigma of a plant, it absorbs sugar and moisture from the stigma's surface and then germinates to produce a pollen tube. The pollen tube contains three nuclei: a pollen tube nucleus, which directs the growth of the pollen tube, and two male nuclei (the male gametes). These nuclei are carried in the pollen tube as it grows down inside the style to the ovary and enters the ovule through a small hole called the micropyle. After entering the ovule, one male nucleus fuses with the female gamete or ovum (fertilization) to form a **zygote**. The zygote undergoes repeated cell division of its young unspecialized cells before beginning to develop tissues through **differentiation** (see p. 53), forming the embryo within the mature ovule. The ovule becomes the seed whilst the ovary becomes the fruit in which it is contained.

The second male nucleus fuses with two extra nuclei (the polar nuclei) in the ovule, to form an **endosperm** which, in many seeds, acts as a storage tissue to feed the growing embryo and is only found in flowering plants. This **double fertilization** creating both the zygote and the endosperm is unique to flowering plants (angiosperms) (Figure 6.20).

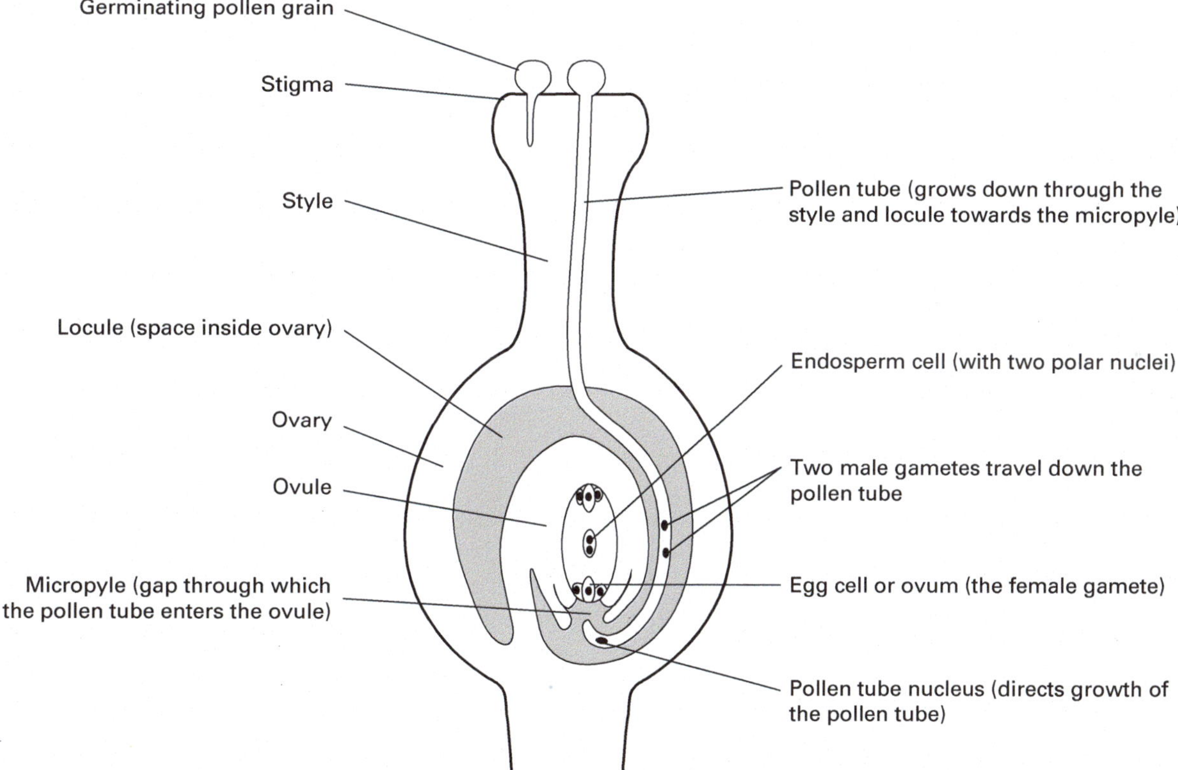

Figure 6.20 Double fertilization in angiosperms

F1 hybrids

When two plants are crossed (**hybridized**) their offspring are called the F1 (first filial) generation. In horticulture, though, the term F1 hybrid relates to a particular type of hybridization programme. F1 hybrid seed results from a series of controlled crosses (involving both cross- and self-pollination) by plant breeders. The parent plants are carefully bred to have a specific type of genetic make-up (pure breeding lines) before they are crossed to create the F1 hybrid. F1 hybrid seeds are important to the grower since, given a uniform environment, all offspring of the same cross will produce a

Figure 6.21 Uniformity of crop in F1 hybrid poinsettias

uniform crop because they are all **genetically identical** (Figure 6.21). Crops grown from F1 hybrid seed such as carrots, cabbage and Brussels sprouts can be harvested at one time ('once over harvesting') and they have similar yield characteristics, useful for commercial growers but not always for the gardener who may want to spread out their harvest. Similarly, F1 hybrid flower crops (e.g. *Petunia* cultivars) will have uniformity of colour, flowering time and flower size.

Another feature of F1 hybrids is **hybrid vigour.** F1 hybrid plants will display desirable characteristics to a greater extent than either parent, giving outstanding growth, especially in good growing conditions. For example, F1 hybrid carrots will be larger than the carrots produced by either of the parents. An analogy is 2+2=5! However, if the F1 plants themselves are crossed and produce seed, the plants that grow from them (the F2 generation) will have very diverse characteristics unlike the F1 generation. Some F2 seed is deliberately produced by breeders for flowering plants such as geraniums and fuchsias where a variety of colour and habit is required for bedding plant displays.

The impact of environmental factors on pollination and fertilization

Environmental factors such as temperature, humidity, rainfall, wind, light and nutrition can all have an important influence on pollination.

- **Impact on pollinators** – Insect foraging is reduced in high winds, rain and low temperatures, for example honeybees only fly when temperature is above 10°C. Wind and rain can remove flower scents and insect pheromones (signalling chemicals passed between animals) and can physically blow insects away. Rain can also damage insects, especially butterflies and moths; wetting can increase weight and the noise of rain can deter insects flying. Providing shelters such as hedges in gardens can help reduce the impact of wind. Artificial light sources are a particular threat for night-flying insects such as moths, attracting them away from plants and killing many. It also interferes with their response to, and release of, pheromones affecting mating and feeding behaviour and can increase susceptibility to disease. One study showed a 60% reduction in flying insects when artificial lights were introduced into a wildflower meadow. For wind-pollinated plants, if windspeed is too high or too low, pollination may also be less successful.
- **Impact on flowers and pollen** – Many wind-pollinated plants only release their pollen when the weather is sunny, warm and dry whilst flowers in the daisy family (Asteraceae) characteristically close when light levels are low and at night (hence 'daisy' deriving from 'day's eye'). Apart from damaging flowers, rain can degrade pollen and it may affect the electrostatic nature, temperature patterns and surface textures of petals. Low temperatures, especially frost, can damage flowers, and this is particularly a problem in early flowering crops such as pears and ornamentals such as camellias. Suitable siting of susceptible plants and provision of protection can help prevent this.
  High temperature and water stress can cause abnormal development of flowers, smaller flowers and fewer flower buds and can cause flower drop in many species. Pollinators are attracted to flowers which reward them with large volumes of nectar rich in sugar and pollen with a high protein content. Nectar volume and sugar content can be reduced when plants are water stressed and temperatures are high, whilst high humidity and rain can also reduce sugar content. Borage (*Borago officinalis*) plants that were well watered received twice as many visits by bumblebees.
  Pollen is particularly affected by humidity and temperature although the response varies

from plant to plant. In onion, low humidity decreases pollen viability. Temperature rise and water stress can also reduce the number of pollen grains, and can lead to abnormal pollen and anther development. In tomato (*Solanum lycopersicum*) a high day temperature of 32°C for 12 hours inhibits pollen production, and temperatures between 17°C and 24°C and low humidity prevented pollen attaching to the stigma. Pollen sterility due to environmental stress is a problem in many crops. Loss of pollen viability occurs in high temperatures and low humidity in maize (*Zea mays*), high humidity in olive (*Olea europaea*) and both high and low humidity in cucumber (*Cucumis sativa*). In maize, pollen release stops when the tassel is too wet or too dry and low humidity leads to drying out of silks, whilst in tomato and some night-pollinated and wind-pollinated species, high humidity prevents opening of the anther. In many plants, pollen dries when it is released to survive the period of transfer and rehydrates when it lands on a stigma, and if either is too rapid the pollen grain can be damaged. Nitrogen and phosphorus fertilizers can also influence the number and size of flowers and of pollen grains and the amount of nectar produced. A study showed that a high nitrogen, low phosphorus ratio given to cucumbers resulted in male flowers with more nectar and pollen leading to increased visits from bumblebees.

Climate change leading to increased temperatures and more extreme weather events is likely to affect pollination in the future. The distribution of plants and animals may change as the climate warms, so that plants and their pollinators may no longer be in the same geographical area. Changes in **phenology** (the timing of events in plant and animal life cycles) are particularly influenced by temperature, and milder winters have led to earlier flowering in native plants and in gardens. Some butterflies, which migrate, find that the flowers they feed on may have finished flowering by the time they arrive. For seed and fruit production it is important that plants and their pollinators synchronize their activities. Recent studies have shown that some plants and their pollinators have actually become more synchronous because plants historically lagged behind their pollinators. However, if global warming continues, they may become out of step in the future.

The seed

Plants must often survive through conditions that would be damaging to a growing seedling, so the seed is a means of protecting against extreme conditions of temperature and moisture and is therefore often the overwintering stage (see Chapter 3). The seed, together with the fruit, may also enable the embryo to be dispersed away from the parent plant and may have **dormancy mechanisms** (see p. 35) which prevent germination until conditions are favourable. A range of seeds is shown in Figure 6.22.

Figure 6.22 Seeds: a range of species. Top: runner bean; left to right: leek, artichoke, tomato, lettuce, Brussels sprout, cucumber, carrot, beetroot

In some plants, such as dandelion (*Taraxacum*), viable seeds can occur naturally without fertilization (**apomixis**), whilst fruits such as cucumber (*Cucumis sativa*) and pineapple (*Ananas comosus*) have been purposely bred to produce fruit without fertilization (**parthenocarpy**) and are therefore seedless.

> A **seed** is the structure that develops from the ovule after fertilization. A **fruit** is formed from the ovary wall usually following fertilization and encloses the seed.

Seed structure

The basic structure of a eudicot seed is shown in Figure 6.23.

The main features of the seed are:

- the embryo within the seed
- the testa on the outside of the seed.

The **embryo** is a small immature plant protected by a seed coat. It consists of a **radicle**, which will develop into the root of the seedling to take up water and nutrients, and a **plumule**, which develops into the shoot system, bearing leaves for photosynthesizing and eventually flowers for seed and fruit production. The region between the cotyledons and the radicle is the **hypocotyl**, while the short length of stem between the cotyledons and the shoot is termed the **epicotyl**. A single **cotyledon** will be found in monocotyledons, while two are present as part of the embryo of eudicots. The cotyledons may occupy a large part of the seed, such as in French bean (*Phaseolus vulgaris*), and act as the food store for the embryo (Figure 6.23). In other seeds, such as sweetcorn (*Zea mays*), the single cotyledon remains small, and the food store is provided by another tissue called the **endosperm** (see p. 103) which is not part of the embryo (Figure 6.24).

The **testa**, also known as the seed coat, is formed from the outer layers of the ovule after fertilization. It is waterproof and airtight and may contain germination inhibitors or be very tough, which enable seeds to stay dormant over winter. The **micropyle** is a weakness in the testa where water uptake occurs triggering germination and the radicle emerges through, whilst the **hilum** is a scar on the testa where the seed was attached to the fruit.

Seed germination and dormancy are discussed in Chapter 3.

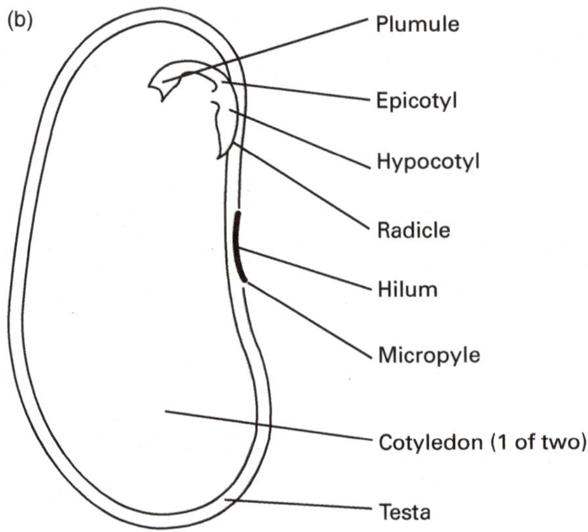

Figure 6.23 Eudicot seed structure: (a) germinating runner bean (*Phaseolus coccineus*) seed showing emerging radicle; (b) section through a French bean (*Phaseolus vulgaris*) seed

Food storage in seeds

In some species, such as grasses and the castor oil plant (*Ricinus communis*), the food of the seed is found in a different tissue from the cotyledons, which is called the endosperm. Plant food in either cotyledons or endosperm is often stored as the carbohydrate starch, formed from sugars as the seed matures – for example, in peas and beans. Other seeds, such as sunflowers, contain high proportions of fats and oils, and proteins are often present in varying proportions. These substances store energy in a very concentrated form, which is released through the process of respiration when the seed germinates, fuelling rapid growth (see Chapter 7). The seed is also a rich store of nutrients, such as phosphate, which it requires for seedling growth (see p. 197). This explains why seeds are so useful for humans, providing food and many other non-food products too.

The fruit

The development of a fruit involves the transformation of the ovary either into a juicy **succulent** structure, attractive to animals, or into a hard and **dry** structure. Fruits provide a means of protection and often a means of dispersal for the seeds they contain and may also contribute to delayed germination through dormancy (see p. 35). Some dry fruits split to release their seeds (described as **dehiscent**) while others rely on the fruit coat being broken down to release the seeds (described as **indehiscent**) (Figure 6.25). Fruit types are an important identification aid for plants and are often characteristic for a particular genus.

Some methods of dispersal include:

- **Explosive** or self-dispersed: the dry fruit splits open propelling the seeds into the air, for example brooms (*Cytisus* spp.), lupins (*Lupinus* spp.), sweet pea (*Lathyrus odoratus*), wallflower (*Erysimum*× *cheiri*), honesty (*Lunaria annua*), hairy bittercress (*Cardamine hirsuta*) and *Geranium* spp.

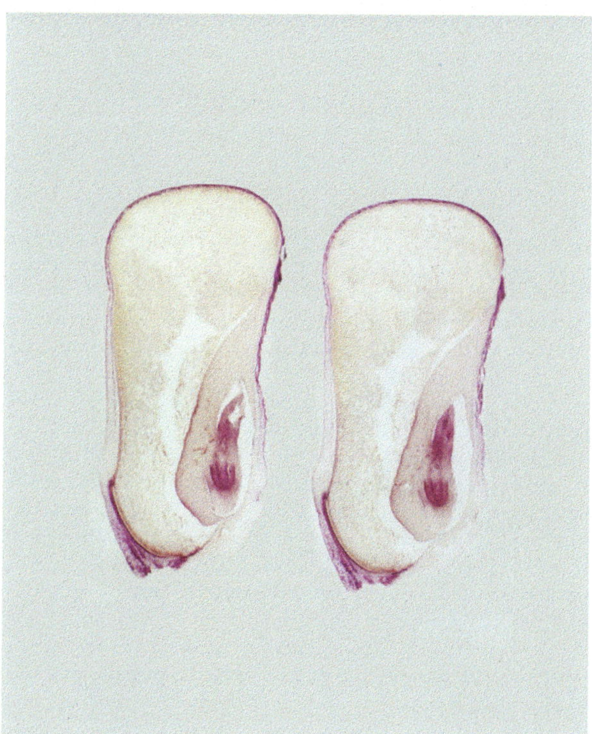

Figure 6.24 Structure of a monocotyledonous seed; sweetcorn (*Zea mays*). The bulk of the seed is the endosperm with a thin, single cotyledon separating it from the embryo to the right. The kernel is actually a fruit with the seed inside it. The fruit and the seed coat are fused together (source: Shutterstock, Jubal Harshaw)

- **Wind**: the seeds of poppy capsules (*Papaver* spp.) are shaken from small pores in the fruit as the plant sways in the wind like a church censer. Other fruits have tiny feathery parachutes attached, as in willow herb (*Epilobium* spp.), *Clematis* spp. and many members of the daisy family including groundsel (*Senecio vulgaris*), dandelion (*Taraxacum*) and thistles (*Cirsium* spp.), enabling them to drift on the wind. Many woody species such as lime (*Tilia* spp.), ash (*Fraxinus* spp.), sycamore and maples (*Acer* spp.) produce winged fruit which twists as it falls like a helicopter (Figure 6.1).
- **Animals**: mammals and birds can distribute fruits either externally or internally in three ways. Firstly, hooked fruits, for example, goosegrass or cleavers (*Galium aparine*) and burdock (*Arctium lappa*), become attached to an animal's fur and are transported away from the parent plant where they become detached. The sticky succulent fruits of mistletoe (*Viscum album*) attach to birds' beaks and are rubbed off onto trees where they germinate (**attachment**). Secondly, squirrels bury fruits such as nuts, for example oak (*Quercus robur*), beech (*Fagus sylvatica*) and sweet chestnut (*Castanea sativa*) in the ground far from where they were collected (**scatter hoarding**) forgetting to retrieve them later. Thirdly, succulent fruits, for example tomato (*Solanum lycopersicum*), blackberry (*Rubus fruticosus*), sloe (*Prunus spinosa*) and elderberry (*Sambucus nigra*), or those that are filled with protein, for example sorrel (*Rumex acetosa*), are eaten by birds and other animals, the seeds passing through their gut before being deposited elsewhere (**frugivory**).
- **Water**: many aquatic plants such as waterlilies (*Nymphaea* spp.) or those growing close to rivers and seashores use water to disperse their fruits. The fruits of coconut palms can travel thousands of kilometres in ocean currents. The introduced weed Himalayan balsam (*Impatiens glandulifera*) has an explosive mechanism to disperse its seeds, but they are also spread along waterways where they have become a serious threat to biodiversity in Britain and Ireland (see p. 7).

Some examples of fruit types and dispersal methods are illustrated in Figure 6.25.

Fruits

Succulent

Drupe, e.g. Sloe

Berry, e.g. *Viburnum*

Dry indehiscent

Samara, e.g. Sycamore

Lomentum, e.g. Trefoil

Cremocarb, e.g. Hogweed

Carcerulus, e.g. Hollyhock

Nut, e.g. Acorn

Dry dehiscent

Capsule, e.g. Poppy

Siliqua, e.g. Wallflower

Silicula, e.g. Honesty

Legume, e.g. Lupin

Follicle, e.g. Monkshood

Seed dispersal

Eaten by animals e.g. Blackberry

Hooked, e.g. Burdock

Winged, e.g. Ash

Parachute, e.g. Dandelion

Censer, e.g. Antirrhinum

Schizocarp, e.g. *Geranium*

Figure 6.25 Fruit types and methods of seed dispersal

Reproduction in non-seed producing plants

In higher plants, seeds are the means by which they colonize new areas and reduce competition from the parent plant and other seedlings. However, simpler plants such as mosses and ferns do not produce seeds, so other means of dispersal are needed. In these plants, during their life cycles, two stages of quite distinct types of growth occur. In **ferns**, the typical fern plant which we see growing is a vegetative (asexual) phase of the life cycle. Spores are released from the underside of the fronds and are dispersed by the wind (Figure 6.26). With suitable damp conditions, they germinate to produce a second sexual stage. Each spore grows into a tiny leafy structure in which male and female organs develop and release sex cells which fuse. The cell resulting from this fertilization gives rise to a new fern plant and the small leafy structure withers away (Figure 6.27). Ferns can be produced in cultivation by spores if provided with damp sterile conditions to allow the tiny spores to germinate and the new fern to grow without competition.

Figure 6.26 Sori, brown spore-producing structures on the underside of fern fronds: left: crested hart's tongue fern (*Asplenium scolopendrium* 'Cristata'); right: *Dryopteris erythrosora*

Figure 6.27 Germinating fern spores and plantlets

Further reading

Adams, C., Early, M., Brook, J. and Bamford, K. (2015) *Principles of Horticulture Level 3*. 7th ed. Routledge.

Allaby, M. (2006) *A Dictionary of Plant Sciences*. Oxford University Press.

Bidlack, J., Jansky, S. and Stern, K. (2013) *Stern's Introductory Plant Biology*. 13th ed. McGraw-Hill Education.

Chalker-Scott, L. (2015) *How Plants Work*. Timber Press.

Hickey, M. and King, C. (1997) *Common Families of Flowering Plants*. Cambridge University Press.

Hodge, G. (2013) *Practical Botany for Gardeners*. University of Chicago Press.

Holm, E. (1979) *The Biology of Flowers*. Penguin.

Royal Horticultural Society. (2013) *RHS Botany for Gardeners: The Art and Science of Gardening Explained & Explored*. Mitchell Beazley.

Zona, S. (2022) *A Gardener's Guide to Botany: The Biology Behind the Plants You Love, How They Grow, and What They Need*. Cool Springs Press.

The online material is accessible via the QR code and includes further information on many of the topics in the book, such as

- Propagating plants from seed
- Reproduction in conifers
- Reproduction in ferns
- Plant genetics and F1 hybrids
- Fruit development and classification
- Control of flowering

Plant growth

Figure 7.1 Giant lily pads of the giant waterlily (*Victoria amazonica*). A lily pad two metres wide can grow from a bud in just seven days

This chapter includes the following topics:

- What is meant by 'growth'
- Photosynthesis
- Leaf structure
- Aerobic and anaerobic respiration
- Factors affecting photosynthesis and respiration
- Variegation
- The balance between photosynthesis and respiration

DOI: 10.4324/9781003581260-7

What is meant by 'growth'? Growth is a difficult term to define because it really encompasses the totality of all the processes that take place during the life of an organism. However, it is useful to distinguish between the processes that result in an increase in size and weight, which we can call 'growth', and those processes that cause the changes in the plant during its life cycle, which can usefully be called 'development', as described in Chapter 3. Growth comes about by, firstly, cells dividing at the meristems (see p. 49), increasing the number of cells present. Secondly, these new cells expand due to turgor pressure (see p. 124) until they reach their final size which becomes fixed by the mature cell wall. To do this, all living organisms need food and energy and plants obtain these through **photosynthesis**, in which light energy from the sun is trapped, and, together with **water** and **minerals**, 'food' is created. **Respiration** is the process by which this food and energy is converted into a form which can be used by the plant. Because one process makes food and the other breaks it down, for plants to grow, the right balance between photosynthesis and respiration is essential.

The processes of photosynthesis and respiration are explored in this chapter, and the importance of water and of mineral nutrients and their transport is discussed in Chapters 8 and 13.

Photosynthesis

Photosynthesis is the process by which green plants manufacture food in the form of carbohydrates such as sugars and starch, using light as an energy source. Food is needed to build the plant's structure and, when broken down by respiration, provides energy to fuel its activities such as the manufacture of **proteins**, including enzymes (see p. 48) which speed up chemical reactions in cells; **cellulose**, used to build cell walls at meristems; and **oils and starches**, laid down in seeds to nourish the embryo when it germinates.

All the complex organic compounds, based on carbon, must be produced from the simple raw materials water and carbon dioxide. Green plants are able to do this through the process of photosynthesis (see producers p. 138). Many other organisms are unable to manufacture their own food and must therefore feed on already manufactured organic matter such as animals (see consumers p. 138). Since large animals predate smaller animals, which themselves feed on plants, all organisms depend directly or indirectly on photosynthesis as the basis of a **food web or chain** (see p. 138).

Photosynthesis involves the conversion of water and carbon dioxide into glucose (a simple sugar) and oxygen. Light energy from the sun is the fuel which drives the process and is captured by the

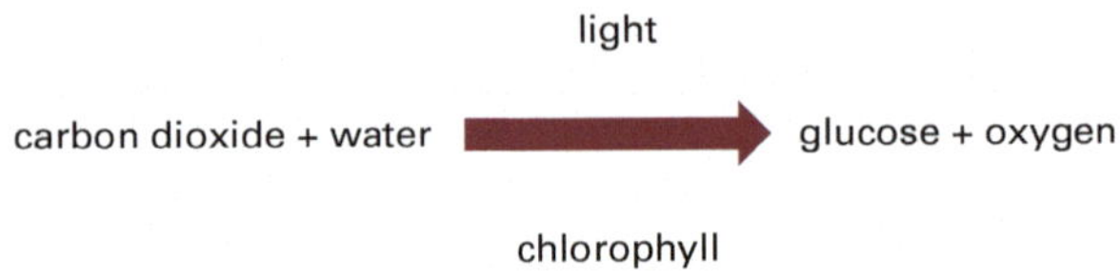

Figure 7.2 Equation for photosynthesis

green **chlorophyll** pigments and the yellow, orange and red **carotenoid** pigments in the **chloroplasts** (see p. 48), mainly in the leaf. Water is supplied by the roots and transported to the leaves while carbon dioxide is taken in by leaves from the air. The glucose produced is converted to another sugar, sucrose, which is exported from the leaves to other plant organs and may also be converted to starch (a large, insoluble carbohydrate) for storage, in roots, stems and other structures (see p. 72), until it is required. It can also be used as a starting point for many other chemicals in the plant. Oxygen is produced as a waste product and is released to the air. All the oxygen that we breathe on earth has been produced by plants through photosynthesis over millions of years. Although the overall equation for photosynthesis looks simple (Figure 7.2), it is actually a complex process involving two interdependent stages, the trapping of light energy and the conversion of carbon dioxide into other products (carbon dioxide 'fixation'). These take place in different areas within the chloroplast.

A most important enzyme!

The first step in the fixation of carbon dioxide is catalysed by an important enzyme called RuBisCo (Ribulose 1,5-bisphosphate carboxylase), which initially converts carbon dioxide molecules (with one carbon atom each) into molecules with three carbon atoms. This is called C3 metabolism. RuBisCo is an extremely important enzyme as it has been in existence as long as photosynthesis has, more than four billion years, and is the most abundant protein in leaves. Its genetic code, which has mutated over millions of years, is found in all green tissue, including green algae, and is used for DNA fingerprinting to establish the evolutionary relationships between plants (see p. 22). Strangely, oxygen has a detrimental effect on photosynthesis in C3 plants and actually reduces its efficiency through a process called **photorespiration**, leading to less growth and lower yields in many of the world's major crop plants. When photosynthesis first evolved there was no oxygen in the atmosphere but over time, through photosynthesis, it has increased to the 21% level we find today. This inhibition is especially significant when plants are grown in high temperatures and high light intensities. Some important tropical grasses such as maize and sugarcane have adapted their photosynthetic process and chloroplast structure (C4 metabolism) to overcome this.

The uptake of carbon dioxide in photosynthesis, and its conversion to glucose, plays a key role in locking up carbon dioxide (carbon sequestration). When fossil fuels are burned, they are in effect releasing the carbon dioxide which was trapped in them many millions of years ago. Growing long-lived plant species such as trees can lock up carbon dioxide in our present atmosphere in their woody tissue for many years to come and is an important tool in combating global warming.

Photosynthesis is the process in the chloroplasts by which green plants trap light energy from the sun and use it to produce glucose. The light energy from the sun is converted into chemical energy stored in the glucose. The raw materials are carbon dioxide and water, and oxygen is released as a waste product.

Leaf structure and photosynthesis

The leaf (Figure 7.3) is the main organ for photosynthesis in the plant, and its cells are organized in a way that provides maximum efficiency. The protective upper **epidermis** (see p. 49) is a thin transparent layer, without chloroplasts, permitting light to pass to the lower leaf tissues. The cylindrical **palisade mesophyll** cells are packed tightly together, vertically, under the upper epidermis. The many **chloroplasts** within these cells absorb light to carry out photosynthesis and can move to the top and bottom of the cells depending on light levels. The **spongy mesophyll**, below the palisade mesophyll, has a loose structure with many air spaces which allow for the two-way diffusion of gases. The carbon dioxide from the air is able to reach the palisade mesophyll and oxygen, the waste product from photosynthesis, can leave the leaf. The numerous **stomata** (see p. 128) on the lower leaf surface (positioned here to reduce water loss) are the openings to the outside, connected to spaces in the spongy mesophyll, through which this gas movement occurs. Each stomatal pore is surrounded by two guard cells which control the opening and closing of the pore. Many small vascular bundles (**veins**) within the leaf structure connect with the vascular

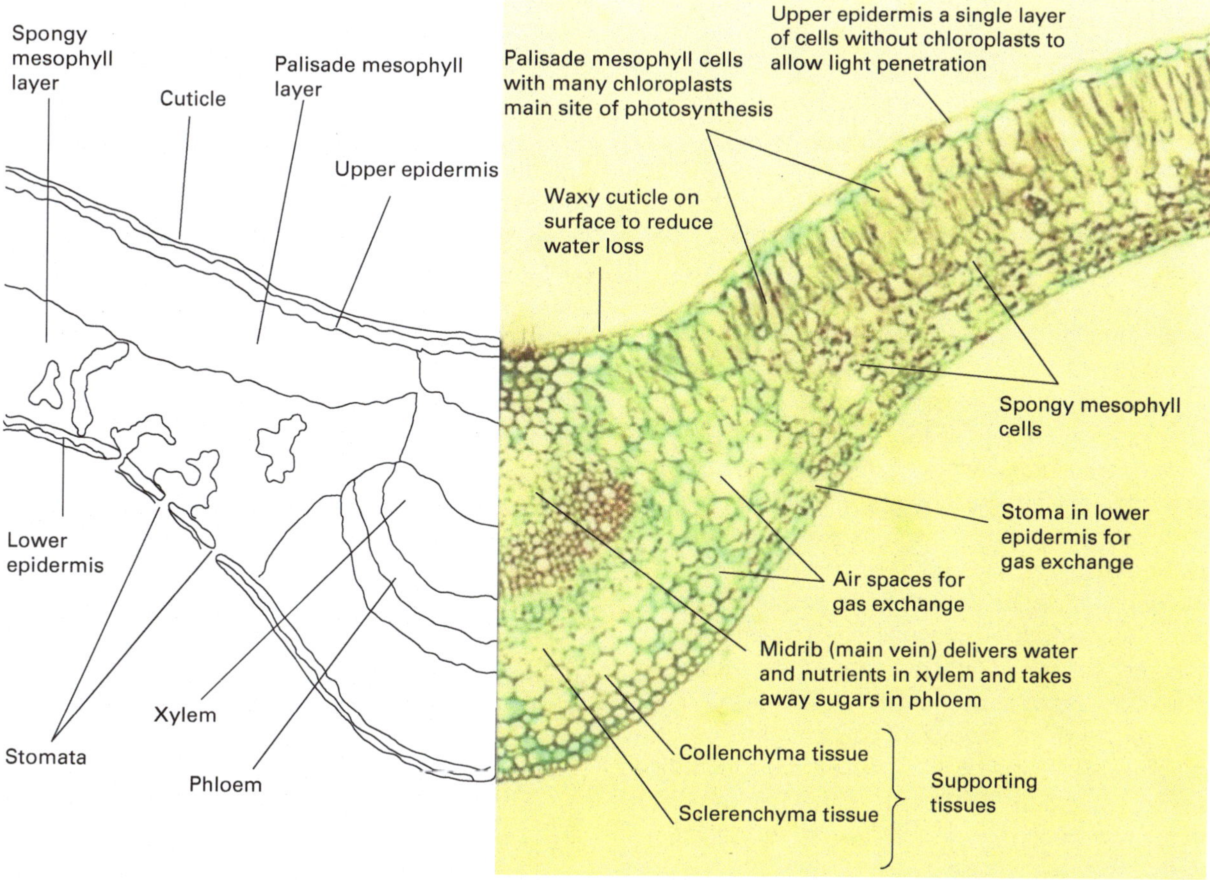

Figure 7.3 Cross section of a *Ligustrum* leaf showing how its structure is designed to optimize photosynthesis

tissue of the stem and root. They contain the xylem vessels that provide the water and minerals for the photosynthesis reaction and phloem sieve tubes for the removal of sucrose produced from photosynthesis to other plant parts.

The arrangement of leaves on the plant and the angle at which they are held maximizes light interception, as does the large surface area and thin structure of the leaf. A newly expanded leaf is most efficient in the absorption of light, but this ability reduces with age, so leaves may be constantly shed and replaced within a plant's life cycle, all at once in the case of deciduous plants or at the end of each leaf's useful life in evergreen plants.

Factors affecting photosynthesis

Photosynthesis depends on, or is influenced by, a number of key environmental factors:

- Carbon dioxide
- Light
- Adequate temperature
- Water (see Chapter 8)
- Mineral nutrients (see Chapters 8 and 13)

These factors can be manipulated by the grower to optimize photosynthesis and an understanding of their functions and how they interact can help support successful and healthy plant growth.

Carbon dioxide

In order to build up organic compounds such as sugars, plants must have a supply of readily available carbon. **Carbon dioxide** is currently present in the air in concentrations of around 420 ppm (parts per million) or 0.042%. It diffuses into the leaf through the **stomata** from the atmosphere. If no other factors are limiting, the rate of photosynthesis increases as the level of carbon dioxide in the air outside the leaf increases. The amount of carbon dioxide in the air immediately surrounding the plant can fall when planting is very dense, or when plants have been photosynthesizing rapidly, especially in an unventilated greenhouse. Ventilation can rectify this, replacing the carbon dioxide used up. Alternatively, in commercial growing, the atmosphere can be **enriched** in glasshouses by supplying carbon dioxide at levels above that in the atmosphere. In fact, the atmosphere within a greenhouse or polytunnel can be increased to levels well above ambient concentrations, typically three times greater, for example in lettuce production, up to 1,000 ppm (0.1%). This can result in the rate of photosynthesis increasing, leading to improvements in yield and quality of many glasshouse crops.

Light

In any series of chemical reactions where one substance combines with another to build a larger compound, energy is needed to fuel the reactions. In plants this energy is provided by light from the sun. As with carbon dioxide, the amount of light energy present is important in determining the rate of photosynthesis. Simply put, the more light there is, or the greater the **light intensity** supplied to the plant, the more photosynthesis can take place. Beyond a certain light intensity, however, the rate of photosynthesis levels off as the chloroplasts are fully engaged. This is called the **saturation point** and will vary from plant to plant. Shade lovers such as weeping fig (*Ficus benjamina*) have lower saturation points than those adapted to high light conditions. Light levels also affect stomatal opening – stomata close as light levels reduce, which restricts carbon dioxide uptake. Care must be taken in a greenhouse to maintain clean glass or polythene and to avoid condensation that restricts light transmission.

Light intensity is not normally a limiting factor (see p. 116) except where plants are shaded or in the winter and can be boosted by using artificial lighting (**supplementary lighting**) of a suitable wavelength, particularly in the winter when light becomes the rate-limiting factor.

The **duration** of lighting will naturally influence the length of time that photosynthesis can continue, longer in the spring and summer than during the winter months in temperate regions. Supplementary lighting can also be used to extend the duration of the daylight hours in winter.

Another important feature of plants is their **light compensation point**. This is the level of light at which the uptake of carbon dioxide used in photosynthesis is cancelled out by the release of carbon dioxide in respiration (see p. 119). As there is no net gain of carbon (in effect, the carbon taken up and converted to glucose is released again from glucose), there is no possibility of making new cells and increasing growth. In temperate regions, low winter light levels mean that plants are operating below the light compensation point, which is why growth rates are low and many plants 'shut down' for the winter, shedding their leaves and becoming dormant.

As well as light intensity and duration, **light quality** is important in optimizing photosynthesis. Photosynthesis only utilizes certain wavelengths of light, those in the red and blue parts of the visible spectrum. Pigments such as chlorophyll absorb light of these particular wavelengths, in this case the ones useful to photosynthesis, and reflect the rest such as yellow and green wavelengths, which is why chlorophyll appears green. Similarly, carotenoids absorb blue wavelengths and reflect yellow and red ones, appearing orange. If supplementary lighting is given in a greenhouse, the lamp chosen, as well as giving good light intensity, must produce the right wavelengths of light for photosynthesis (known as photosynthetically active radiation or PAR).

Lack of light can cause **chlorosis** (leaf yellowing) since light is needed in the final stage of chlorophyll synthesis and plants may show **etiolation** (see p. 118). Very high light intensities may destroy chlorophyll and retard photosynthesis, which can be seen as bleaching of the leaves, for example in plants that have been grown in shaded conditions and are suddenly transferred to full sun.

Light wavelengths and photosynthesis

Light, like other forms of energy, such as heat, X-rays and radio waves, travels in the form of waves, and the distance between one wave peak and the next is termed the wavelength. Light wavelengths are measured in nanometres (nm): 1 nm = one millionth of a millimetre. Visible light wavelengths vary from 800 nm (red light, in the long wavelength area) through the spectrum to 350 nm (blue light, in the short wavelength area). A combination of different wavelengths (colours) appears as white light. Photosynthetically active radiation or PAR contains wavelengths useful for photosynthesis, between 400 nm and 700 nm.

Other light-absorbing systems in the plant are responsible for developmental changes through the plant's life cycle. Blue light, with wavelengths around 400 nm, is important for vegetative growth, stimulating leafy growth and sturdy plants, and is involved in the directional growth responses to light (see phototropism p. 37). Red light, with wavelengths around 580 nm to 700 nm, controls flowering and other plant behaviours. For successful plant growth, therefore, artificial lighting contains a mix of red and blue wavelengths which aim to mimic sunlight, optimizing overall increase in plant material (through photosynthesis) and the correct development of the plant from early growth through to flowering and fruiting (through other light-absorbing systems).

Temperature

The complex chemical reactions that occur during the formation of carbohydrates such as glucose, from water and carbon dioxide, require the presence of special proteins called **enzymes** to accelerate the rate of reactions (see p. 48). Without these enzymes, little chemical activity would occur. Enzyme activity in living things increases with temperature from 0°C to 36°C and ceases at around 40°C when the enzymes break down irreversibly. This pattern is mirrored by the effect of air temperature on the rate of photosynthesis which increases with increasing temperature up to an optimum (this varies with plant species from 25°C to 36°C), above which it slows again. At high temperatures, stomata may close to reduce water loss (see p. 129) thus preventing carbon dioxide uptake, while at very high temperatures, enzyme are destroyed and leaves may be damaged, so photosynthesis ceases altogether. As with light, in temperate countries, low winter temperatures slow down photosynthesis and therefore slow the growth rate at this time of year.

To provide the optimum temperature conditions, greenhouses can be heated in winter using a range of methods such as thermostatically controlled electric heaters together with insulation such as bubble wrap. To reduce temperatures which are too high, shading using blinds, netting or washes applied to the glass together with ventilation or damping down (applying water to floor surfaces to evaporate) can be employed.

Water

Water is required in the photosynthesis reaction, but this represents only a very tiny fraction of the total water taken up by the plant. Nevertheless, lack of water can have very important **indirect effects** on photosynthesis. Firstly, water supply through the xylem is essential to maintain cell turgor (see p. 124) so that stomata remain fully opened for carbon dioxide movement into the leaf. In a situation where a leaf contains only 90% of its optimum water content, **stomata will close** to prevent further water loss, and this will reduce carbon dioxide entry to such an extent that there may be as much as a 50% reduction in photosynthesis. Secondly, loss of cell turgor will lead to **changes in leaf angle** in a wilting plant, which will reduce light interception. Wilting is most often associated with lack of water but can also be seen in waterlogged plants (see p. 125) where root damage prevents water uptake or when the ground is frozen (Figure 8.3). A visibly wilting plant will hardly be photosynthesizing at all, therefore it is essential that plants should be supplied with the correct amount of water if the rate of

Figure 7.4 Chlorosis due to magnesium deficiency in a lemon tree (*Citrus* × *limon*)

photosynthesis is to be optimized. Thirdly, plants responding to severe water stress will **reduce leaf area** and ultimately shed their leaves, with a reduction of photosynthesis overall (see p. 79).

Mineral nutrients

Minerals are needed by the leaf to produce the **chlorophyll** pigment that absorbs most of the light energy for photosynthesis and also for many of the steps involved in photosynthesis. Production of chlorophyll must be continuous since it loses its efficiency quickly. A plant deficient in iron, nitrogen or magnesium especially turns yellow (**chlorotic**) and loses much of its photosynthetic ability (Figure 7.4). Magnesium is a key component at the heart of the chlorophyll molecule, nitrogen is a structural component of chlorophyll and, importantly, also in proteins including enzymes. Iron and some trace elements such as zinc and manganese are essential for the enzymes involved in photosynthesis and other plant processes to work efficiently (see Chapter 13). Supplying the correct balance and amount of nutrients through application of fertilizers or other means is especially important in the vegetable plot and in lawns where plants are harvested or the grass mown and removed, since in these situations, nutrients are not returned to the soil. Similarly, plants in containers will eventually deplete the nutrients in the growing medium, so these will need to be replaced by feeding. See also Chapter 14.

Other environmental factors

Pests, diseases and disorders will affect the efficiency of photosynthesis in various ways, whether by simply reducing leaf area through damage or defoliation or more indirect actions such as disrupting water flow from the roots. These are explored in detail in Chapters 17 and 18. Trees and shrubs can be useful in combating **air pollution** in urban areas by trapping dust and smaller particulate matter on their leaves, especially if they have hairs or waxes on the leaf surface. However, this comes at a cost. Particulates in the air can negatively impact on photosynthetic efficiency in many ways such as reducing light absorption, blocking stomata or interfering with the photosynthetic process itself. Outdoors, **high winds** can damage leaves directly and plants also respond by closing their stomata to reduce water loss, so reducing photosynthesis. Providing shelter or windbreaks can help prevent this.

Variegation

Variegated leaves, where parts of the leaf are typically pale green, yellow or white (Figure 7.5a) are much prized in planting schemes. They can be a natural feature of the plant or the result of a genetic mutation occurring as the leaf is formed; in either case, the variegated areas lack chlorophyll. The leaf will have a lower rate of photosynthesis overall and the plant will have a slower growth rate. In the wild this would eventually lead to those individual plants dying out. The genetic make-up of tissues in the two areas of the leaf can differ, and this type of variegation is called a **chimaera**. In the aluminium plant (*Pilea cadierei*) and many other houseplants, a different type of variegation (blister variegation) is due to an air layer forming below the epidermis which reflects light (Figure 7.5b). In contrast, *Coleus scutellarioides* cultivars have bright red or purple variegation due to overproduction of these pigments at the expense of the green pigment chlorophyll. Variegation can also be due to virus infections as in some *Abutilon* cultivars. Variegation is often highly desirable horticulturally and is selected by growers. In many plants, to continue the variegation, plants must be propagated vegetatively (cloned) for example by cuttings or grafting. Some variegated plants are prone to 'reversion' where the variegation is lost and leaves are uniformly green. These stronger growing green leaves can rapidly take over the plant and should be pruned out as soon as they appear if the variegated form is to be retained (Figure 7.6).

The law of limiting factors

The **law of limiting factors** states that in a process influenced by many factors, the factor in least supply will limit the rate of a process overall. This can be applied to photosynthesis. If, for example, carbon dioxide levels fall in an enclosed system such as a greenhouse or conservatory, the rate of photosynthesis will decrease. In this situation, increasing light levels or temperature will not be advantageous if the carbon dioxide levels remain low. Similarly, if there is adequate carbon dioxide and a

Figure 7.5 Variegated leaves in (a) spotted laurel (*Aucuba japonica* 'Crotonifolia'), (b) a *Peperomia* cultivar with blister variegation

suitable temperature but light levels are reduced, as in the winter, there will be no advantage in increasing carbon dioxide levels or the temperature without giving additional light. In fact, if any one of these three factors (light, carbon dioxide or heat) is in short supply, then it will limit the rate of photosynthesis even though the other factors may be plentiful. It would be wasteful, therefore, to increase the carbon dioxide concentration, light or temperature artificially, if the other factors were not proportionally increased.

Figure 7.6 Reversion in (a) a large shrub *Pittosporum tenuifolium* 'Silver Queen'; (b) *Euonymus japonicus* 'Microphyllus Albovariegatus'. The attractive variegated leaves will soon be replaced by the green shoots if they are not removed

In winter most plants cease growing as temperature and light are limiting factors. Conversely, **in summer** when light intensity is above a plant's light compensation point and temperatures are higher, growth will resume, and atmospheric carbon dioxide becomes the rate-limiting factor.

The **law of limiting factors** states that the factor in least supply will limit the rate of a process, for example, in photosynthesis.

Plant size and growth rate

It is important for anyone planning a garden that the eventual size (in terms of both height *and* width) of trees, shrubs and perennials is considered. The impressive maidenhair tree (*Ginkgo biloba*) can grow to 30 m in height and is, therefore, not the plant to put in a small bed. Similarly, Leyland cypress (×*Hesperotropsis leylandii*), useful in rapidly creating a fine hedge, can also grow to 30 m and reach 5 m in width, to the consternation of even the most friendly neighbours, and its planting is now limited by planning rules.

The eventual size of a plant is recorded in plant encyclopaedias and on websites, and specialist nurseries will also give advice to potential buyers. It should be remembered that the final size of a tree or shrub may vary considerably in different parts of the country and may be affected within a garden by factors such as aspect, soil, shade and wind. Attention should also be given to the rate at which a plant grows. Yew (*Taxus baccata*) and star magnolia (*Magnolia stellata*) are two examples of slow-growing species.

Growing in low and high light environments

Since light is a key factor in their growth, plants have evolved to deal with both very low and very high light levels.

In shady habitats, such as next to a hedge or under trees, shading by taller plants reduces both the amount of light reaching plants below and changes the quality of the light (see p. 114). In a woodland, for example, when sunlight passes through or is reflected off leaves, the red light it contains is absorbed by leaves in the tree canopy. The light reaching plants beneath contains proportionately more far-red light. If the shaded plants are not shade tolerant, they may respond by showing **etiolated** growth – elongated, fast-growing, pale stems with few leaves which are positioned at the top to reach the light (see p. 35). Plants can measure the amount of red and far-red light in the sunlight they receive using a pigment called **phytochrome** and can actually tell if they are being shaded by another plant, in which case they can try to outgrow them, or simply by an inanimate object such as a rock or wall! Plants which are not adapted to shade must complete their main growth spurt when the canopy is open in late winter and early spring and then die down for the remainder of the year. They include many bulbous plants such as snowdrops (*Galanthus nivalis*), bluebell (*Hyacinthoides non-scripta*), winter aconite (*Eranthis hyemalis*) and *Crocus* spp. and cultivars (Figure 7.7). In the garden, these can be successfully planted under deciduous trees and some even in grass.

Plant species adapted to shade conditions, however, can survive in low light levels and do not need to respond with etiolated growth. Shade-tolerant plants include yew (*Taxus baccata*), winter box (*Sarcococcus* spp.), ivy (*Hedera* spp.), *Epimedium* spp., holly (*Ilex aquifolium*), *Hosta* spp. and cultivars, many ferns and leafy crops such as lettuce and spinach. Shade-loving plants adapt to reduced light levels by changing their metabolism and have **low light compensation points** and **low light**

Figure 7.7 Shade-intolerant plants flowering in January: (a) winter aconite (*Eranthis hyemalis*); (b) an early-flowering, clump-forming snowdrop (*Galanthus nivalis* 'Magnet')

saturation points enabling them to photosynthesize more efficiently in low light levels. They also have **low respiration rates**, so carbohydrates are broken down more slowly, often resulting in slower growth rates. Red colouration on the underside of leaves is thought to be a protection against damaging light wavelengths, for example in the *Peperomia* shown in Figure 7.5b. Shade adapted plants also have many structural modifications (see Chapter 5).

In contrast, plants growing in habitats with **high light intensities**, such as xerophytes (see p. 69) and alpine plants (see p. 81), also show a range of metabolic adaptations to protect them from damage. They have **high light compensation points** and **high light saturation points** to take full advantage of the higher light intensities. They may contain **protective pigments** such as the yellow pigment **xanthophyll** and the purple pigment **anthocyanin** which absorb damaging light wavelengths, especially at high altitudes where harmful levels of ultraviolet light occur (see p. 81). Their chloroplasts are able to **repair the damage** caused by high light levels. Structural modifications to reduce light interception are described in Chapter 5.

Some plants grown at high light intensities can perform well at low light intensities (and vice versa) if they are able to **acclimatize** gradually. For example, foliage plants used for indoor displays can be slowly introduced to lower light levels before they are positioned.

Respiration

In order that growth can occur, food manufactured by photosynthesis must be broken down in a controlled way to release the energy which was trapped from sunlight. This energy is used to build useful substances such as cellulose, the main constituent of plant cell walls, and proteins, for example, enzymes. It is also used to fuel cell division and the many chemical reactions that occur in the cell. **Respiration** is the process by which plant food in the form of carbohydrates (sugars and starch) is broken down to yield energy, releasing carbon dioxide and water as waste products. It takes place in the **mitochondria** of plant cells (see p. 48) and is often referred to as 'cellular respiration'. As with photosynthesis, respiration is a complex process with several stages which occur in different parts of the mitochondrion and the cell cytoplasm.

In order that the breakdown is complete and the maximum energy is released, **oxygen** is required in the process of **aerobic respiration** (Figure 7.8).

The energy released by aerobic respiration is stored in a chemical form in a substance called adenosine triphosphate (**ATP**), which can be transported to wherever energy is needed in the cell, in effect acting like a portable 'battery'. Some energy is also released as **heat**, similar to the cellular respiration which produces heat in our bodies to keep us warm. The energy requirement of cells within the plant varies. Reproductive organs can respire at twice the rate of the leaves for example, and in apical meristems, the processes of cell division and cell differentiation require high inputs of energy to create new cells.

It would appear at first sight that respiration is the reverse of photosynthesis. This is correct in that photosynthesis creates glucose as an energy-harvesting strategy, and respiration breaks down glucose as an energy-releasing mechanism. It is also correct in the sense that the simple equations representing the two processes are mirror images of each other. It should, however, be emphasized that the two processes have two notable differences. The first is that respiration in plants (as in animals) occurs in all living cells of all tissues, in leaves, stems, flowers, roots and fruits. Photosynthesis occurs predominantly in the palisade mesophyll tissue of leaves, and in some specialized stems, but not in other plant tissues such as the roots and flowers. Secondly, respiration takes place continuously whereas photosynthesis only operates when light is present; it cannot happen in the dark.

Aerobic respiration is the process by which plant food (glucose) is broken down to yield energy in the mitochondria of the cell. Oxygen and glucose are the starting materials and the waste products are carbon dioxide and water.
Anaerobic respiration takes place in the absence of oxygen. Glucose is partially broken down to form ethanol with a small release of energy.

Factors affecting respiration

Oxygen

Oxygen is essential for **aerobic** respiration, as it is needed to break down carbohydrates (glucose) fully to release the energy stored in them. It is analogous to the need for oxygen when a fire

glucose + oxygen → carbon dioxide + water + energy (ATP and heat)

Figure 7.8 Equation for aerobic respiration

7

burns, where the energy stored in the fuel (trapped from the sun by plants millions of years ago) is released as heat.

In the absence of oxygen, inefficient **anaerobic** respiration takes place in the cytoplasm of cells (see Figure 7.9). Incomplete breakdown of the carbohydrates produces alcohol (ethanol) as a waste product, with much energy still trapped in the alcohol molecule. If a plant or plant organ such as a root is supplied with low oxygen concentrations, such as in a waterlogged or compacted soil or in an overwatered pot plant, the consequent alcohol production within the cells may prove toxic enough to cause root death. Furthermore, since not all the energy trapped in the glucose molecule is released, it is insufficient for growth, repair or reproduction, only enabling the plant to 'tick over' until aerobic respiration can be restored. Plants that grow naturally in marshy or aquatic habitats (Figure 7.10) have developed strategies and adaptations to overcome the problem of anaerobic respiration (see Chapter 6).

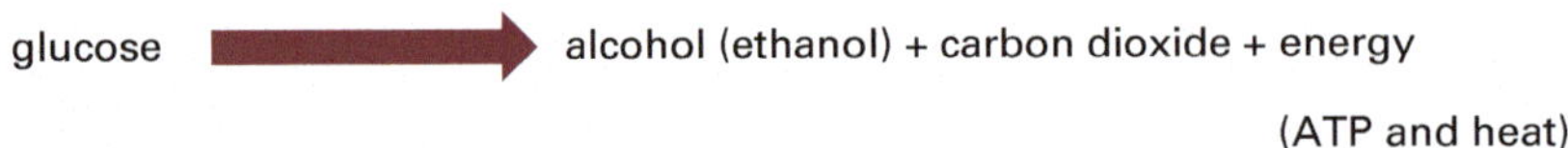

Figure 7.9 Equation for anaerobic respiration

Sometimes, anaerobic conditions can be advantageous. For the plant, it enables it to survive periodic inundations which would otherwise be fatal. For gardeners, the viability of stored seeds can be greatly increased if they are stored in the 'modified atmosphere' that is within sealed, airtight packets (see Figure 3.4). The oxygen is removed as the seeds respire and carbon dioxide levels rise which combine to reduce the respiration rate and inhibit germination, enabling longer storage.

In fruit and vegetable storage, inhibition of respiration is desirable to prevent produce going beyond the ripening stage and into senescence (see p. 42) with accompanying loss of quality. In addition, once harvested, photosynthesis ceases but respiration continues so carbohydrates will continue to be broken down and the dry weight of plants or produce will be reduced. Packaging may provide a 'modified' atmosphere similar to that of the seed packet with low oxygen and high carbon dioxide levels within. For some fruits such as apples, 'controlled atmosphere' storage in sophisticated large-scale airtight stores enables fine control of oxygen and carbon dioxide levels, which, along with temperature control, can extend the storage times from harvest in the autumn to well into the following year.

Temperature

As in photosynthesis, many of the reactions involved in respiration are speeded up by enzymes. The rate of respiration therefore shows a similar pattern of increasing rate with increasing temperature up to an optimum, beyond which the rate decreases. Plants adapted to high temperatures have a higher temperature optimum than those from temperate or cooler regions, but in general an optimum temperature is around 35°C. Low winter temperatures reduce growth rates as respiration slows, another reason for plants 'shutting down' for the winter, and of course, plants which grow in cold environments will always have slower growth rates all year round.

In some horticultural situations, a high rate of respiration is desirable – for example, in propagation of cuttings or seed germination. Here, new cell growth needs plenty of energy, so heat may be given to speed up respiration. Alternatively, where low respiration rates are required, such as in seed storage or to delay ripening and senescence (see p. 43) in stored produce such as fruit and vegetables or cut flowers, temperatures are reduced, sometimes with control of the gaseous atmosphere as well. Growers, distributors, retailers and consumers have developed a 'cool

Figure 7.10 Skunk cabbage (*Lysichiton americanus*) is a plant adapted to waterlogged soils and has many air filled spaces in its root and stem tissue (aerenchyma, see p. 83) which interconnect and enable oxygen to reach its roots (source: Shutterstock, Brookgardener)

chain' system to keep produce at consistently low temperatures, between 0°C and 10°C along the supply chain, from harvest to sale, minimizing waste and enabling longer shelf life through reducing respiration rates. As well as reducing respiration, cold storage can have other benefits. Cuttings stored at low temperature root more readily later, while strawberry runners kept in cold stores over winter maintain their quality and are also stimulated to flower when planted outside the following year. The cold treatment mimics winter temperatures in the field.

The balance between photosynthesis and respiration

The relationship between photosynthesis and respiration is crucial for plants and therefore gardeners and growers. Photosynthesis converts carbon dioxide into sugars whereas respiration does the reverse. If the rate of photosynthesis is too low, then all the carbon dioxide 'fixed' may be lost again in respiration, leaving no surplus sugars for growth. The point at which photosynthesis exactly matches respiration, that is, there is no net gain in carbon, is called the **compensation point** (see p. 114). Growers aim to provide light levels which keep photosynthesis rates above the light compensation point, otherwise their crops will not grow sufficiently and yields will be low.

Even if daytime light levels are above the light compensation point, at night respiration continues, so carbohydrates will also continue to be broken down.

The long-term effects of **climate change** on plant growth are complex and difficult to predict. Higher night-time temperatures could impact on plant growth and crop yields in the future as it would favour respiration. On the other hand, increasing carbon dioxide levels in the atmosphere due to climate change should theoretically increase the rate of global photosynthesis, and increased carbohydrate levels in leaves have indeed been shown in field experiments. However, it was also found that stomata reduced in size in response to increasing carbon dioxide levels in order to reduce transpirational water loss (see p. 129) which would potentially slow the uptake of mineral nutrients from the soil. Reduced levels of mineral nutrients were found in the crops studied, possibly because of this effect (see p. 130). Lower protein levels and greater insect damage were also seen. Another possible effect is that climate change, which has led to less cloud cover in Britain and Ireland and therefore an increase in the amount of light received, could benefit photosynthesis. At a plant community level, rising carbon dioxide levels may lead to changes in the composition of plant populations (see p. 134), as some species are more successful at using the increased carbon dioxide for growth than others.

Further reading

Adams, C., Early, M., Brook, J. and Bamford, K. (2015) *Principles of Horticulture Level 3*. 7th ed. Routledge.

Allaby, M. (2006) *A Dictionary of Plant Sciences*. Oxford University Press.

Bidlack, J.E., Jansky, S. and Stern, K.R. (2020) *Stern's Introductory Plant Biology*. 15th ed. McGraw-Hill.

Capon, B. (2022) *Botany for Gardeners: An Introduction to the Science of Plants*. 4th ed. Timber Press.

Chalker-Scott, L. (2015) *How Plants Work*. Timber Press.

Hodge, G. (2013) *Practical Botany for Gardeners*. University of Chicago Press.

Ingram, D.S., Vince-Prue, D. and Gregory, P.J. (2008) *Science and the Garden*. Blackwell Science.

Royal Horticultural Society. (2013) *RHS Botany for Gardeners: The Art and Science of Gardening Explained & Explored*. Mitchell Beazley.

Zona, S. (2022) *A Gardener's Guide to Botany: The Biology Behind the Plants You Love, How They Grow, and What They Need*. Cool Springs Press.

The online material is accessible via the QR code and includes further information on many of the topics in the book, such as

- Basic chemistry
- Carbon chemistry
- Photosynthesis and respiration (advanced)

CHAPTER 8

Transport in plants

Figure 8.1 Apple (Cox on M1 rootstock) excavated at 16 years to reveal distribution of roots. Note the vigorous main root system near the surface, with some penetrating deeply (source: Dr E.G. Coker)

This chapter includes the following topics:

- Diffusion and osmosis
- Uptake of water
- Water movement in the plant
- Transpiration and root pressure
- Mineral and nutrient uptake
- Sugar movement in the plant
- The relationship between photosynthesis, respiration and transpiration

DOI: 10.4324/9781003581260-8

Plants, like people, are dependent on the transport of substances around their structure. Sugars made in the leaves by photosynthesis must be transported to roots, stems, shoots and flowers where they can be used for immediate growth or stored until they are required for future growth. Water is essential for many processes in the plant and, since it is absorbed through the roots, it has to be moved to cells in other plant organs where it is essential for support, as a medium for chemical reactions and as a raw material for processes such as photosynthesis. The two 'superhighways' along which sugars and water flow, the phloem and xylem, respectively, are also responsible for moving other vital substances such as essential mineral nutrients and plant growth regulators (hormones) around the plant body. In this chapter we examine how water is moved both through the whole plant and from cell to cell and how it is lost from the leaves. The transport of sugars and mineral nutrients is also described. The structure of xylem and phloem is discussed in Chapter 4.

As **water** is the major constituent of any living organism, the maintenance of optimum water content is a very important part of plant growth and development. There is a tendency sometimes to overwater, but probably more plants die from lack of water than from any other cause.

Movement of substances in the plant

Diffusion and osmosis

Two ways in which substances move in the plant are described as follows:

- **Diffusion** is a process whereby molecules of a liquid or a gas move from an area of high concentration to an area of lower concentration of the diffusing substance (Figure 8.2a). For example, sugar in a cup of tea will diffuse through the tea without being stirred – eventually! Examples of diffusion in the plant include the movement of gases such as water vapour (see transpiration p. 129), carbon dioxide and oxygen (see photosynthesis and respiration, Chapter 7) into and out of the leaf. Osmosis is a special kind of diffusion where water is the diffusing substance.
- **Osmosis** is defined as the movement of **water** from an area of high water (low solute) concentration to an area of lower water concentration (higher solute concentration), through a **selectively permeable membrane**, such as the cell membrane (Figure 8.2b). Osmosis is in effect the diffusion of water across the cell membrane and is the method by which water enters cells. A selectively permeable membrane allows passage of some dissolved substances but not others, and the term 'solute' refers to the substances dissolved in the water.

Diffusion is the movement of a substance from a high concentration to a lower concentration. Osmosis is the movement of water from a high water (low solute) concentration to a low water (high solute concentration) across a selectively permeable membrane.

When water moves into a cell, the cell swells like a balloon. This inner pressure, caused by the water pushing outwards, is called **turgor pressure** and is very important in providing support to young plants and non-woody herbaceous plants. The plant and its individual cells stay upright like a stack of inflated balloons. Turgor pressure is also the way in which new cells enlarge, contributing to growth by causing the cell to swell until the cell wall prevents further expansion. Without the cell wall the cell would explode!

The pathway of water movement through the plant falls into three distinct stages:

- water uptake from the soil by the **roots**
- movement up the **stem** in the xylem
- movement across the **leaves** and loss to the air by transpiration.

Root and stem anatomy are described in Chapter 4 and leaf anatomy in Chapter 7.

Water uptake from the soil

It is the function of the root system to take up water and dissolved mineral nutrients from the growing medium, and this system can be extensive (Figure 8.1). Most of the water uptake takes place in the **root hair zone** where the root surface area is greatly enlarged by the presence of thousands of root hairs (see p. 55).

Soil water enters the root in two ways. It can pass directly into the root hair cells of the epidermis across their **cell membranes** by **osmosis** or alternatively, it can be absorbed by the cell walls and move between the root hair cells through the adjacent walls.

Whereas the cell wall is permeable to both soil water and its dissolved inorganic minerals, the cell membrane freely allows water through but is selective about passage of other dissolved molecules, somewhat like a sieve. Sucrose, for example, is too large a molecule to cross the membrane. A greater concentration of dissolved substances such as minerals and sugars (solutes) is usually maintained inside the cell compared with the soil water outside

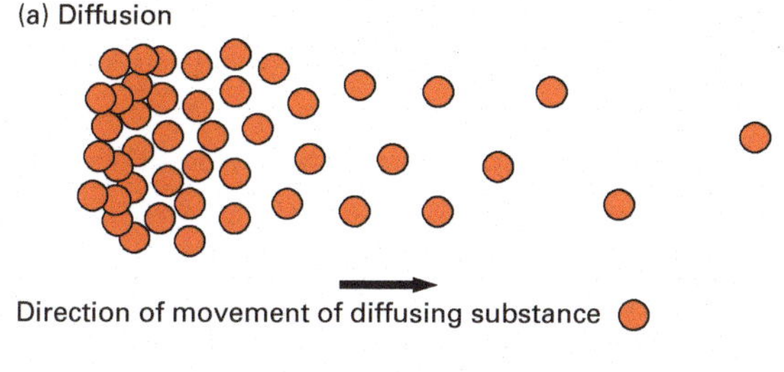

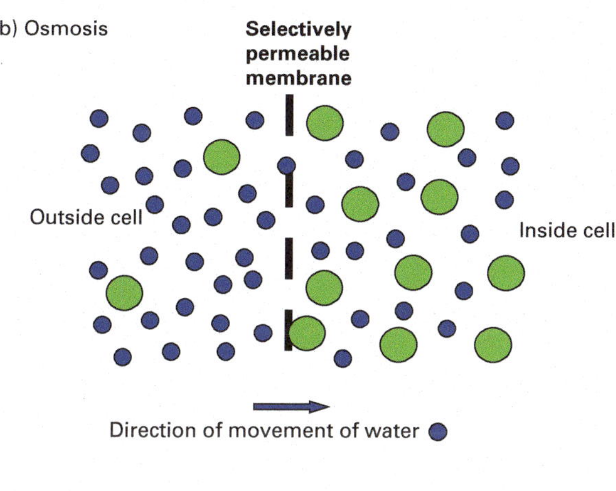

Figure 8.2 Diagrammatic representations of (a) diffusion and (b) osmosis

the cell. The water concentration outside the cell will therefore be greater than inside the cell so water will move in by osmosis (Figure 8.2). The greater the difference in concentrations of water, the faster water moves into the root cells.

If water is not available at the roots (see permanent wilting point, p. 178) or if the plant is losing water from the leaves faster than it can be replaced, water will cease to enter the cell and will eventually start to move out of the cells. Turgor pressure is reduced, and the plant will **wilt** (Figure 8.3). This will not be a problem if the water supply is only reduced temporarily; the plant will be able to recover once water loss is reduced (for example at night when stomata are closed) or water supply is restored. However, if water loss continues, the cells may then become **plasmolysed**, a situation where the cell contents shrink away from the cell walls and linking plasmodemata (see p. 47) between cells are broken, leading to irreversible damage and cell death. Plasmolysis can also occur if there is a build-up of salts in the soil, for example, where too much fertilizer is added causing **root scorch**. Water moves out of the root cells, osmosis is reversed, because the solute concentration is now greater outside the cell than inside. Such situations can be avoided by applying the correct dosage of fertilizer to soils and leaves, where patches of plasmolysed cells appear as leaf scorch if foliar feeds are too concentrated (see p. 201).

Figure 8.3 Reversible wilting: (a) the *Primula* is wilting because the water in the pot is frozen (physiological drought) – the roots are unable to replace the water lost by the leaves through transpiration so turgor pressure is lost; (b) a few hours later the water in the pot has thawed, turgor pressure is restored and the plant has recovered

Functions of water

The plant consists of about 95% water, which is the main constituent of **protoplasm** or the living matter of cells. When the plant cell is full of water, or turgid, the pressure of water enclosed within a membrane or vacuole acts as a means of **support** for the cell and therefore the whole plant, so when a plant loses more water than it is taking up, it may wilt. Aquatic plants are supported largely by external water and have very little specialized support tissue. In order to survive, any organism must carry out complex chemical reactions, such as photosynthesis and respiration, described in

Chapter 7. Raw materials for these chemical reactions must be transported and brought into contact with each other by a suitable medium; water is an excellent **solvent**, that is, many substances are able to dissolve in it. One of the most important processes in the plant is **photosynthesis**, and a small amount of water is used up as a raw material in this process. Water may also be used for **seed and fruit dispersal** in aquatic plants, and in more primitive plants such as mosses and liverworts, water is needed for **reproduction**.

Movement of water in the roots

The pathway of water movement through the root is shown in Figure 8.4.

Initially water crosses the epidermis, as described previously. It then meets the **cortex** layer, which is often quite extensive, and moves across it to reach the transporting tissue that is in the centre of the root (see p. 52). Water movement is relatively unimpeded as it moves through the intercellular spaces and the latticework of cell walls (**the apoplast**), although some will also pass into the root cells by osmosis and then from cell to cell through the plasmodesmata (**the symplast**) (see p. 47).

The central region, the **stele**, is separated from the cortex by a single layer of cells, the **endodermis** (see p. 53), which has the function of controlling the passage of water and minerals into the stele. A waterproof strip (the **Casparian strip**) containing the wax, suberin, forms part of the cell wall of the endodermal cells and prevents water from moving between the cells. All the water now has to pass across the endodermal cell membranes and into the cells of the endodermis by osmosis. The cell membranes also act as a control point for mineral uptake as only certain minerals, dissolved in the soil

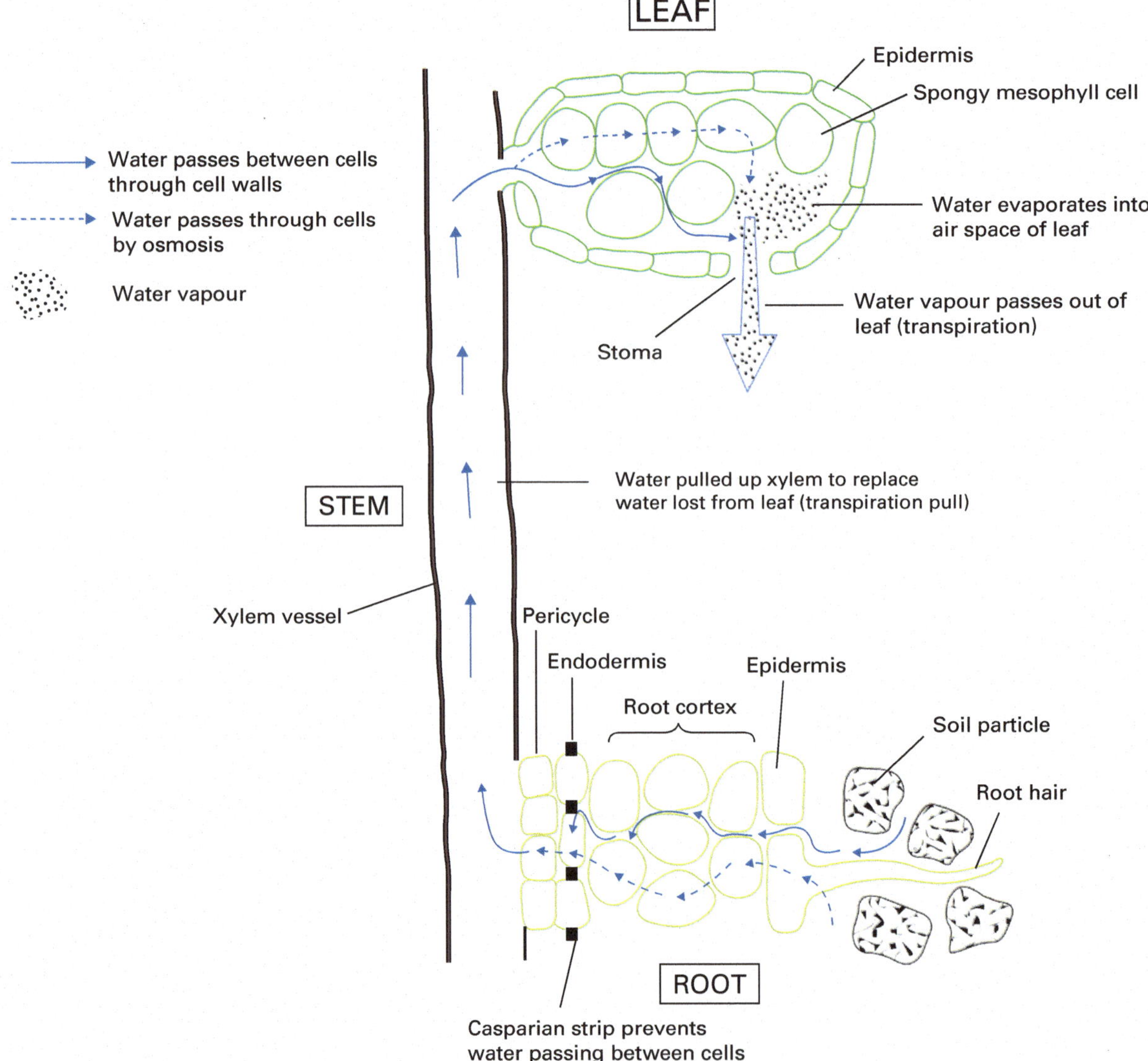

Figure 8.4 Pathway of water movement from the soil through the plant to the atmosphere

water, can cross them. Water passes through the endodermis and pericycle to the **xylem** tissue (see p. 53), which transports the water and dissolved mineral nutrients up the stem to the leaves.

Movement of water up the stem

Water always moves in one direction, up the stem from the roots to the leaves and flowers.

It is 'sucked' up through the **xylem vessels** and **tracheids** (see p. 50) of the stem (Figure 8.4) by a process called **transpiration pull**, that is, as water is lost from the leaves it is replaced by water which is drawn up the stem carrying dissolved minerals with it (see later). The evaporation of water from the cells of the leaf means that, in order for the leaf to remain turgid, which is important in keeping stomata open for efficient photosynthesis, the water lost must be replaced by water in the xylem. Pressure is created in the xylem and water moves up through the stem and leaf petiole by suction, provided the water forms a continuous column. Water moves freely through the xylem vessels and tracheids aided by **pits** in the cell walls and because no cytoplasm is present in the dead cells to slow it down (see p. 50). The water molecules 'stick' to the xylem walls (**adhesion**) and each other (**cohesion**), and water's high **tensile strength** together with strengthening lignin in the cell walls helps maintain the water column under great pressures. If the water column is broken, water can no longer move up. The high pressure in the water column means that dissolved air can sometimes be 'sucked' out of the water forming bubbles (**cavitation**) in the xylem vessels and tracheids, which can break the continuous column of water and prevent water uptake, like bubbles in a straw. This can be demonstrated when the stem of a flower is cut. Air moves into the xylem and may restrict the further movement of water when the cut flower is placed in a vase of water. However, by cutting the stem under water, the column is maintained and water continues to enter and pass up the plant.

Cavitation is especially a problem on hot sunny days when transpiration rates are high, although some water can bypass the bubbles through pits, or tyloses can block them (see p. 51). At night, transpiration ceases and the air in the bubbles can redissolve in the water columns. By the end of the season, many xylem vessels and tracheids are non-functioning because of accumulation of air bubbles and so woody plants need to continue to make new xylem tissue each year to replace this. The tremendous pressures involved can be demonstrated experimentally in trees by measuring the daily fluctuations in stem girth which it causes.

Water can also move up stems a short way by the process of **root pressure**. This takes place in stems particularly at certain times of year. Watery sap can often be seen to exude from the cut stumps of woody shrubs and trees, but in this situation the leaves have been removed, so transpiration is minimal and transpiration pull cannot operate. Another mechanism must be causing the movement of water up the stem xylem tissue, which is root pressure (Figure 8.5).

Even though transpiration is not taking place, water continues to pass into root cells by osmosis pushing water up the stem. Water can only be forced up the stem a short distance (and out of a cut stem) by this process and certainly not to the top of trees. It is probably important as a **'priming' mechanism** in the spring when continuous columns of water need to be restored to kick-start transpiration as the leaves emerge. Root pressure can be enough to force some air out of the columns of water in the xylem vessels and tracheids and repair the damage done by air bubbles. Some trees such as *Betula* spp. and *Acer* spp. show particularly strong root pressure and should not be pruned in late winter or early spring as they will lose a large amount of sap (Figure 8.5). However, some *Acer* species (sugar maples *Acer saccharinum*) are tapped commercially for their sap at this time of year in a controlled way, to produce maple syrup.

Under conditions of high humidity and low transpiration, **guttation** (Figure 8.6) may occur. This can be seen in small herbaceous plants close to the moist soil, for example in the early

Figure 8.5 Sap exuding from the xylem tissue of a freshly cut birch (*Betula pendula*) stem in late winter due to root pressure

8

Figure 8.6 Guttation in the leaf of (a) cucumber and (b) tomato

morning when light levels are low, and stomata are almost closed or in a well-watered but unventilated greenhouse. Water is forced by root pressure up through the plant and out of the leaves and appears as water droplets around the leaf margins which are sometimes mistaken for dew. Some plants have special groups of cells called **hydathodes** around their margins to facilitate this. Sometimes in these conditions, more water is forced into the leaves than is being lost to the air, and the more delicate cell walls in the leaf may burst. This condition is known as **oedema** (see p. 309). It can be seen in *Pelargonium* as corky patches on the underside of leaves, and can occur in other glasshouse crops and outdoor plants such as *Eucalyptus* spp. and *Camellia* spp.

Transpiration is the evaporation of water vapour from leaves and other plant surfaces. **Root pressure** is the osmotic pressure that builds up in the root systems of plants which forces water upwards in the xylem tissue. **Capillarity** is the movement of water against gravity in narrow tubes such as xylem vessels.

Water can move up very narrow xylem vessels by **capillary action** due to **surface tension, cohesion** and **adhesion** as is seen in a straw placed in a beaker of water. The distances moved are very small so this contribution to water movement is probably insignificant.

Xylem tissue transports water and dissolved mineral nutrients from the roots up the stem to the leaves and other plant organs.

Movement of water in the leaf

On reaching the leaf, water is distributed in the fine network of veins and passes out of the xylem (Figure 8.4). It flows between the leaf cells through their cell walls (the apoplast) and also passes from cell to cell by osmosis and through plasmodesmata (the symplast) as in the root. Eventually it **evaporates** from the cell surfaces into the air spaces of the leaf mesophyll, especially the spongy mesophyll. From here, the water vapour diffuses out of the leaf into the surrounding air (see Figure 8.7) through pores in the leaf called **stomata**. It always moves in this direction because there is a lower relative humidity, that is, a lower concentration of water in the surrounding air compared with inside the leaf. The loss of water vapour from the leaf is called **transpiration** and the flow of water up the xylem is the **transpiration stream**.

Stomata

Both monocot and eudicot plants have stomata through which water vapour, carbon dioxide and oxygen pass. Most are found in the epidermis of leaves, but they are also present in young stems and in flower petals. In a eudicot leaf, most stomata are generally distributed across the lower leaf surface where it is cooler and more humid, whereas in many monocots with upright lance-shaped leaves the stomata are distributed equally above and below. In aquatic floating leaves (Figure 7.1), stomata are on the upper surface.

The stomatal complex is made up of a pair of **guard cells** surrounding a central pore or **stoma** (Figures 4.3 and 8.7a). Behind the stoma is the **substomatal cavity** surrounded by the air spaces and cells of the **spongy mesophyll**. **Guard cells** of eudicots are typically kidney-shaped, whereas those of monocots such as grasses are dumbbell-shaped. Guard cells have a number of peculiar characteristics. The cell wall surrounding the pore is thicker than the outer wall so that when water enters the guard cell by osmosis, it swells unevenly and the stomatal pore opens. When water passes out of the guard cells, turgor pressure drops and the pore closes. Guard cells have few plasmodesmata connecting them to the surrounding epidermal cells, making them very sensitive to slight turgor changes. They also often contain chloroplasts (unlike the surrounding epidermal cells) which provide energy to regulate stomatal opening. The plant controls the water flow into and out of the guard cells by changing the levels of solute (in this case potassium) within them (Figure 8.7b).

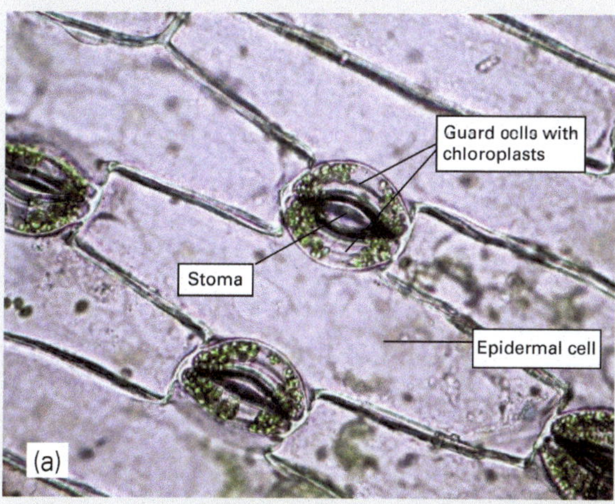

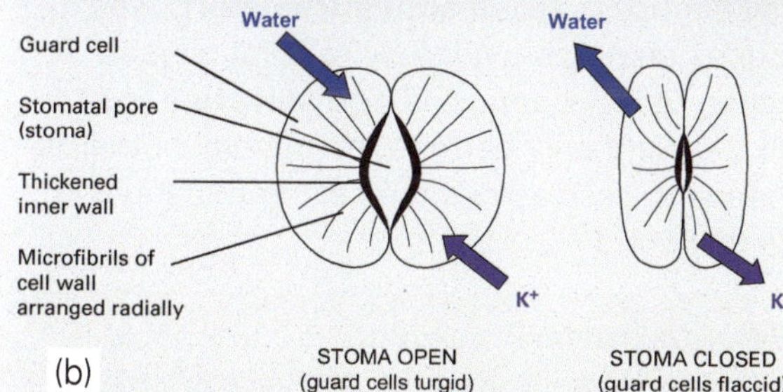

Figure 8.7 (a) Stomata on the surface of a leaf; (b) features of guard cells and the mechanism of stomatal opening; K^+ refers to potassium ions

Transpiration is the evaporation of water vapour from the leaves and other plant surfaces.

Transpiration

Any plant takes up a lot of water through its roots. For example, a tree can transport about 1,000 litres a day. Approximately 98% of the water taken up moves through the plant and is lost by transpiration; only about 2% is retained as part of the plant's structure, and a yet smaller amount is used up in photosynthesis. The seemingly extravagant loss through leaves is due to stomata having to be open for carbon dioxide uptake for photosynthesis. Whilst transpiration mainly takes place through the stomata of the leaf, it can also occur through stomata on stems of young or herbaceous plants, on petals and through lenticels on fruit and woody stems.

A remarkable aspect of transpiration is that water can be pulled ('sucked') such a long way to the tops of tall trees. Engineers have long known that columns of water break when they are more than about 10 m long, and yet tall trees such as the giant redwoods (*Sequoiadendron giganteum*) pull water up 100 metres from ground level! This apparent ability to flout the laws of nature is probably due to the small size of the xylem vessels, which greatly reduce the possibility of the water columns collapsing.

A number of environmental factors affect the rate of transpiration:

- humidity and windspeed
- temperature
- light
- carbon dioxide
- soil factors.

Humidity and windspeed. If the air surrounding the leaf is still, humidity builds up as water vapour is released by plant leaves through the stomata. The rate of diffusion of water vapour will be much reduced and the rate of transpiration will decrease. In contrast, **moving air** around the leaf will reduce the surrounding humidity so will increase the rate of transpiration unless stomata close in high windspeeds, when transpiration will cease.

Temperature. As temperature rises, evaporation of water and diffusion of water vapour out of the stomata is speeded up, so the rate of transpiration increases. In addition, the size of the stomatal pore increases with temperature up to around 30°C. At high temperatures, though, the stomata will close and transpiration stops.

Light. Stomata open and close in response to light. In the daytime, stomata open to allow uptake of carbon dioxide for photosynthesis as light is available. At night, stomata close because photosynthesis is not taking place and transpiration is reduced. An exception is in CAM plants which are adapted to arid environments (see p. 81) and reverse their stomatal opening to conserve water during the day.

Carbon dioxide. If the plant has enough carbon dioxide for optimal photosynthesis, at levels above this the stomata will reduce in size to limit transpiration water loss. Theoretically, this is most likely to happen in a carbon dioxide enriched greenhouse, although small changes may take place as atmospheric carbon dioxide levels rise (see p. 121).

Soil factors. If the uptake of water from soil or compost is limited, then transpiration will slow. This could be due to a low soil moisture content, too much fertilizer or salt in the root zone, a poorly developed root system or root pathogens such as *Pythium* spp. or *Rhizoctonia* spp. (see Chapter 18). Low moisture content in soils can be detected by root tips which manufacture a plant growth regulator (hormone) called **abscisic acid** which is transported to the leaves in the transpiration stream and triggers rapid stomatal closure.

The **function of transpiration** has been much debated. Because plants are obliged to keep their stomata open to take in carbon dioxide for photosynthesis, water loss through transpiration is inevitable – it is often described as a 'necessary evil'. However, plants always aim to balance their water uptake and loss and will try to reduce transpiration water loss providing this does not negatively impact on photosynthesis and growth. Some level of transpiration is necessary to deliver minerals in the transpiration stream to plant leaves and shoots. In certain circumstances such as very high temperatures and low humidity, it is thought that transpiration has a cooling effect on the leaf, much as sweating does in humans.

The gardener can use a knowledge of how transpiration is affected by environmental factors to prevent water stress, optimize growth and maintain good plant health. In addition, recognizing the many structural adaptations that plants use to reduce transpiration and withstand low water supplies (see Chapter 5) can help gardeners to choose the most suitable plants for the existing and future conditions in their gardens. This will become increasingly important with predicted increases in temperature due to **climate change**.

Root:shoot ratio

Plants aim to keep a balance between their leaf and root areas to control their water content. If their leaf area is too large, the root cannot take up enough water to replace the water lost through transpiration and they will wilt. If their root area is too large, then the plant is expending too much energy on root growth at the expense of photosynthesis and growth of the whole plant. Plants therefore naturally adjust their growth to keep this balance, for example, by producing smaller leaves (due to reduced cell expansion) or by shedding leaves if transpiration rates are too high or water supply is short (see p. 79). Gardeners also interfere with this ratio when carrying out some horticultural operations such as pruning branches or roots or by transplanting plants when roots are inevitably damaged. Often the leaf area of cuttings or transplanted shrubs needs to be reduced to redress the balance until the plant has established a new root:shoot ratio.

Mineral nutrient uptake and movement in the plant

Essential mineral nutrients are inorganic substances necessary for the plant to grow and develop (see Chapter 13). They are dissolved in the soil water and are taken up when this is absorbed by the root. At the endodermis, nutrients must cross the cell membrane (see earlier) if they haven't already done so and enter the cells. Since the concentration of nutrients inside cells is almost always greater than the concentration in the soil water, uptake is **against a concentration gradient** (like rolling a ball uphill). Mineral nutrients therefore cannot enter the cell by simple diffusion and have to be taken in by a process called **active transport**, which requires energy provided by respiration (see Chapter 7). This uptake is also **selective** (see p. 48); the plant only permits the mineral nutrients it needs to pass across the cell membrane and rejects others.

Essential mineral nutrients are inorganic substances necessary for for unrestricted growth and development.

Active transport is the movement of a substance into a cell across the cell membrane against a concentration gradient. It requires energy.

Mineral nutrients are taken up predominantly by the extensive network of fine roots that grow in the top layers of the soil (Figure 8.1) where, together with water and oxygen, they are most abundant (see p. 169). Damage to the roots near the soil surface by cultivation should be avoided because it can significantly reduce the plant's ability to extract nutrients and water. Care should be taken to ensure that trees and shrubs are planted so their roots are not buried too deeply, and many advocate that the horizontally growing roots should be set virtually at the surface to give the best conditions for establishment. Over-enthusiastic hoeing of weeds can damage crop roots near the soil surface and also cause increased loss of water from the soil by bringing more moisture to the surface. In contrast, compaction of the soil surface or waterlogging can lower oxygen levels, reducing aerobic respiration and the energy needed to extract nutrients (see anaerobic respiration p. 120).

Having crossed the roots, mineral nutrients are transported up the xylem to the leaves in the transpiration stream and are also redistributed in the phloem to other plant organs such as flowers and fruits.

Movement of sugars in the plant

Phloem tissue transports sugars made in photosynthesis from the leaves to other plant organs where it is used in respiration to release energy or stored as starch for later use.

Phloem tissue (see Chapter 4) is responsible for transporting sugar (sucrose) manufactured from the glucose, produced by photosynthesis in the leaves, to other plant organs (as well as minerals and plant growth regulators). Sucrose is a food supply for the release of energy through respiration in plant cells. Unlike xylem, where the direction of flow is always from the root to the leaves, sugars can flow in the phloem sieve tube cells both up and down the plant, moving to the plant organs where it is needed, such as growing points, shoots, roots, flowers, fruits and storage organs. Loading the sugars in the phloem, like the uptake of minerals in the roots, is an **active process** requiring energy, as the concentration of sugars in the phloem is much greater than in the leaf cells. The **companion cells** which accompany each **sieve tube cell** (see p. 50) are thought to control this process, some of which are specialised **transfer cells** responsible for phloem loading with many mitochondria to provide energy. The flow can be interrupted by the presence of disease organisms such as in clubroot (see p. 293). Aphids insert their mouthpieces directly into phloem sieve tubes to feed on the highly concentrated sugary sap (Figure 8.8). The pressure in the sieve tube cells can be so great that the sap is actually forced through the body of the aphid and exuded as droplets, the 'honeydew' making leaves sticky. Whilst honeydew is a foodstuff for some bees, it can also cause a secondary fungal infection called sooty mould, which blackens the surface of affected leaves.

Figure 8.8 Pea aphid (*Acyrthosiphon pisum*) feeding on a stem (source: Shipher Wu (photograph) and Gee-way Lin (aphid provision), National Taiwan University – PLoS Biology, February 2010 (CC BY 2.5 licence))

The balance between photosynthesis, respiration and transpiration in horticulture

Gardeners strive to establish a balance between these three plant processes to maximize yield and quality in their crops and ensure healthy plants in their borders. Through photosynthesis, plants capture carbon dioxide and sunlight energy (see p. 112) which are used for growth, reproduction and many other plant processes. Growing conditions should always seek to optimize this process. Whilst respiration is necessary to release energy (see p. 119), the carbon 'fixed' in photosynthesis must always exceed that lost by respiration, therefore providing a surplus of carbon, if the plant is to grow (see light compensation point p. 114). If respiration exceeds photosynthesis, then more carbon will be lost (as carbon dioxide) than gained. Similarly, water loss due to transpiration must be controlled to keep cells turgid, enabling them to expand fully (see growth p. 112), to maintain open stomata for carbon dioxide uptake and to prevent wilting, reduction in leaf size and leaf shedding. By manipulating the factors which affect photosynthesis, respiration and transpiration, namely light, temperature, the gaseous environment, windspeed, humidity and soil, growers can optimize the relationship between these processes.

In **open ground**, plants should be grown in areas suited to their light requirements. Shade-tolerant ornamentals and leafy crops like spinach and raspberries can grow in a degree of shade which, although reducing light levels and therefore reducing photosynthesis, will have the advantage of reducing transpiration. Most crops, however, require good light with accompanying higher transpiration rates. Other means of reducing transpiration such as lowering windspeed with shelterbelts and windbreaks (which also can help reduce water stress through reducing soil water loss) in conjunction with good soil structure and soil water management can help limit the detrimental effects of transpiration. Pruning fruit trees to maintain an open centre can help increase air circulation within the tree and prevent build-up of humidity, which can encourage fungal diseases. Spacing of plants in the border or in the greenhouse influences humidity too. If planting density is high, evaporation from the shaded soil or compost surfaces will be reduced which may reduce water stress, although if humidity builds up, fungal disease can be a problem (see p. 289). Too close spacing will also prevent light reaching lower leaves, reducing photosynthesis. On the other hand, if planting density is low, light will be wasted, and soils or composts may dry out. This balance can be carefully maintained as plants grow, particularly in the vegetable garden, by correct spacing of seeds and thinning of seedlings.

In **greenhouses, polytunnels and conservatories** there is greater opportunity to control these processes. In summer, high temperatures will raise leaf temperatures, so increasing the rate of transpiration. This can cause water stress especially as water uptake tends to lag behind transpiration, making plants vulnerable during the hottest part of the day. **Shading** reduces temperature but also reduces photosynthesis. Increasing humidity by **damping down** or **misting** (spraying water) will reduce transpiration, by raising humidity and lowering temperature through evaporative cooling. Exchanging cool air outside with warm air inside by **ventilation** lowers leaf temperature both by reducing the temperature next to the leaf and by the air movement carrying warm air away from it. Ventilation also replenishes carbon dioxide supplies for photosynthesis as these are quickly used up in a closed greenhouse in summer when light levels are high and plants are photosynthesizing rapidly.

In winter, light levels need to be maintained above the compensation point for net growth, sometimes with the use of supplementary lighting (see p. 114), and this can also be used for houseplants as indoor lighting will not have the right range of wavelengths useful for photosynthesis. However, to optimize photosynthesis by increasing light in this way, temperature, carbon dioxide and water must not be rate limiting (see law of limiting factors p. 116). Increasing temperature increases respiration as well as photosynthesis, and respiration can outstrip photosynthesis leading to less growth than expected.

In **vegetative propagation**, for example semi-ripe, softwood and leaf cuttings, photosynthesis, respiration and transpiration must also be controlled while roots develop. In this situation it is important firstly to stimulate root initiation and growth as quickly as possible and secondly to reduce transpiration water loss until the cutting has developed an adequate root system. Cutting back the leaf area (see root and shoot ratio p. 130), increasing the humidity around the aerial parts and keeping the air temperature cool will reduce water loss. However, respiration needs to be optimized to supply energy and carbon for rapid growth of new root cells, so heat can be applied specifically to the rooting area. Light will be needed by the aerial parts for photosynthesis to maintain growth of the cutting but should not be at a level that gives rise to high transpiration rates. A mist unit is the ideal environment providing bottom heat through soil or bench cables together with humidity and some shade above.

Further reading

Adams, C., Early, M., Brook, J. and Bamford, K. (2015) *Principles of Horticulture Level 3*. 7th ed. Routledge.

Bidlack, J.E., Jansky, S. and Stern, K.R. (2020) *Stern's Introductory Plant Biology*. 15th ed. McGraw-Hill.

Capon, B. (2022) *Botany for Gardeners: An Introduction to the Science of Plants*. 4th ed. Timber Press.

Chalker-Scott, L. (2015) *How Plants Work*. Timber Press.

Hodge, G. (2013) *Practical Botany for Gardeners*. University of Chicago Press.

Ingram, D.S., Vince-Prue, D. and Gregory, P.J. (2008) *Science and the Garden*. Blackwell Science.

Lack, A.J. and Evans, D.E. (2005) *Instant Notes in Plant Biology*. Taylor & Francis.

Royal Horticultural Society. (2013) *RHS Botany for Gardeners: The Art and Science of Gardening Explained & Explored*. Mitchell Beazley.

Zona, S. (2022) *A Gardener's Guide to Botany: The Biology Behind the Plants You Love, How They Grow, and What They Need*. Cool Springs Press.

The online material is accessible via the QR code and includes further information on many of the topics in the book, such as

- Water is unusual
- Water potential
- Phloem transport (advanced)

Chapter 9

Ecology and garden wildlife

Figure 9.1 Veteran black walnut tree (*Juglans nigra*) at Dyrham Park in Somerset, encrusted with mosses and lichens

This chapter includes the following topics:

- Studies of garden wildlife
- Ecological principles and gardens
- Interactions between organisms, competition and cooperation
- Biodiversity
- Gardening for wildlife

DOI: 10.4324/9781003581260-9

Wildlife in the countryside is under pressure as never before from many causes such as changing agricultural practices, use of pesticides, urban spread and development, and introduced species. In 2023 the **UK State of Nature Report** was published and showed that nearly 19% of all species have declined since 1970 and one in six species are threatened with extinction. The value of gardens for wildlife has begun to be appreciated in recent years, and they are seen more and more as an important contribution to conservation of our natural heritage. An understanding of the ways in which wildlife can be encouraged and supported, while retaining all the desirable qualities of a garden, is useful to know.

To know what makes a good garden for wildlife, it is helpful to understand some ecological principles.

Studies of garden wildlife

Although gardening for wildlife has become popular in recent years, one of the first published studies took place some time ago. From 1972, Jennifer Owen recorded wildlife in her garden in Leicester. She provided a range of heights and growth forms in her planting, grew flowers for insects and plants with fruits for birds. Apart from a few concessions (minimal pruning and clearing, dead heading to prolong flowering and never using pesticides), she managed her garden conventionally and described it as 'a typical suburban garden . . . neat, attractive and productive' with a lawn, some herbaceous borders, a rockery and fruit and vegetables. In the first 15 years, a huge diversity of organisms was found, some 422 species of plants, 1,602 insects, 121 other invertebrates and 59 vertebrates representing a large proportion of the fauna of Britain and Ireland. After 30 years she recorded 2,024 insect species with 20 new to Britain and 6 new to science. She commented, 'gardens would seem to be of considerable significance in conservation'.

Since then, studies carried out by the University of Sheffield across UK (Biodiversity in Urban Gardens in Sheffield or BUGS project) from 1999 to 2007 have evaluated urban gardens for wildlife in several cities. They too found an astonishing range of plants and animals – for example, 42% of native plants were found in a combined garden area of two football pitches. They concluded surprisingly that small gardens were just as good as large gardens, city centre gardens were as good as suburban gardens, and all gardens were good for wildlife whether specifically managed for this or not. Trees were found to be the most important factor and piles of logs were beneficial, while artificial methods of attracting wildlife such as bird boxes and insect hotels varied in their success. They concluded that any reduction in urban gardens would impact on 'biodiversity, conservation, ecosystem services, and the well-being of the human population'. The BUGS project was the first large-scale study showing the importance of domestic gardens for urban biodiversity. Since then, there have been numerous studies into the value of domestic gardens for wildlife and people.

An understanding of the basics of **ecology** can help us design and manage gardens in a way that supports all the organisms that live there. In effect, it is the recipe for a healthy, wildlife-friendly garden. Broadly speaking, ecology takes over where the study of individual organisms ends. It investigates the relationships between the organisms themselves and the environment they live in. Ecologists therefore study groups of plants and animals (**populations**) living together in a **community** which, together with their non-living environment, form **ecosystems**. On a global scale, major regional communities of organisms form well-recognized **biomes**.

The term '**ecology**' is formed from the Greek words *oikos*, meaning a house or place, and *logos* meaning knowledge or understanding.

Ecology and gardens

Communities

A **community** is a group of populations in a given area or **habitat.** A **population** is defined as a group of individuals of one species which interbreed, for example, all the groundsel (*Senecio vulgaris*) plants in a garden, or all the individual hedgehogs within a certain area.

A community is a group of plant and animal species (populations) living within a particular area (**habitat**). 'Wild' communities are defined either by the habitat in which they occur, for example a 'lake community', or by a particular plant species which is dominant, for example a 'grassland community'. A **microhabitat** is on a much smaller scale such as under a log or against a wall and its community may differ considerably from that of the larger habitat it occupies. Gardens are themselves a type of habitat and often have many microhabitats which increase the variety (see **biodiversity**, p. 145) of organisms that live there.

Populations may be restricted to a very specific habitat – for example, marsh willow herb (*Epilobium palustre*) is only found in slightly acidic ponds. However, some populations can occupy a wide range of habitats such as blackberry (*Rubus fruticosus*) which occurs in heathland, woodland, hedgerows and open fields (Figure 9.2).

In natural communities, a particular environment will often have a specific group of plants growing there which are not found together anywhere else; the plants are said to form **plant associations**. For example, in a chalk habitat, bee orchid (*Ophrys apifera*) is often found together with greater knapweed (*Centaurea scabiosa*) and salad burnet (*Sanguisorba minor*). In the very high rainfall acid bogs of northern Britain and Ireland, *Sphagnum* mosses and common cotton grass (*Eriophorum angustifolium*) are found together with sundew (*Drosera anglica*) (see p. 49) and bog myrtle (*Myrica gale*) (Figure 9.3).

Figure 9.2 (a) Broad-leaved (common) arrowhead (*Sagittaria sagittifolia*) grows only in waterlogged conditions (source: Shutterstock, olko1975); (b) blackberry (*Rubus fruticosus*) grows in many different conditions

Figure 9.3 (a) Bee orchid, found in chalk grassland (source: Shutterstock, Andi111); (b) cotton grass, found in acid bogs; (c) *Sphagnum* moss found in acid bogs

9

Figure 9.3 (Continued)

Although gardens are not natural communities, the same applies in that groups of plants often do well together. A hot, dry border, for example, will successfully accommodate a range of plant species if it reflects their original habitat. Understanding the habitat to which a particular plant or group of plants is adapted enables us to give it the right conditions in a garden setting, increasing the chance of success. Since many garden plants are exotic species (see Table 9.1), that is, introductions collected from many very different habitats worldwide, their origins should be taken into account.

Communities can sometimes be '**closed**' in that they receive only minimal contact with outside organisms and materials. A small, isolated island community in the middle of a lake would be an example, or a collection of plants in a conservatory. However, most communities, are '**open**', that is, organisms and materials can move in and out freely. For example, wind-borne plant seeds can travel between gardens, and birds and other animals migrate widely across their respective ranges and 'belong' to no single garden (Figure 9.4).

A **biome** is a large geographical area or **global community** of distinctive plants and animals who have adapted to a **particular climate**. Major biomes include desert, grassland, tropical and temperate forests, arctic and alpine tundra. Sometimes gardeners can seek to recreate a particular biome, for example in an alpine house or a tropical conservatory, by providing the appropriate climatic conditions for their plants to flourish (see p. 81).

Figure 9.4 Marine shoreline – an example of an 'open' community

Niches

The position or role of each species within its habitat is called its **niche**. An organism's niche includes both its physical and its biological environment, as well as how its needs and behaviour vary with time. For a plant, the **physical environment** in which it grows will reflect its requirements such as:

- light or shade
- a particular soil pH
- resources such as a good water supply or a high level of a nutrient such as iron.

Its **biological environment** could include any pests, diseases or predators which feed on it. For example, rosette plants such as daisies originally adapted to survive being grazed by having their growing point at ground level. In lawns the grazing animal has been replaced by a mower, but the daisies still occupy a particular niche in the grassland we call a lawn. An organism's niche may also be described **in time**. In the case of an annual plant, it flowers in spring and dies at the end of the season, so occupies its niche only part of the year.

A fundamental concept of the niche is that no two species can occupy an identical niche. If they do, sooner or later one will dominate and the other will die out because of competition for resources. However superficially it may appear that organisms occupy the same niche, in real life they rarely do, they will find ways of sharing. For example, consider a group of weeds all growing together in the vegetable plot. Although they all grow in the same place, they all have different requirements and behaviours. Some, such as ephemerals, start their main growth period very early in the year and may produce several generations in a single season, surviving as seed in between and throughout the rest of the year. Annual weeds have a single life cycle in a season and produce seed in the autumn, while perennial weeds overwinter, often through use of underground rhizomes and other perennating organs (see p. 71). These different weed populations therefore vary the time of year when they have greatest need of resources and have their own food supplies as seeds or roots when competition is fiercest.

Other plants, which on the face of it grow together, may mine nutrients at different depths in the soil enabling them to survive alongside each other. For example, lettuces with shallow roots can be successfully intercropped with carrots which are deeper rooting. Yet others may need full light for growth whereas their neighbours can tolerate shading – sweetcorn which requires good light intercropped with shade-tolerant spinach would be an example. As in a jigsaw puzzle where each piece has a unique shape and only fits in one place, each species has its own unique niche in its environment, and although many plants tolerate a range of conditions, the nearer the gardener gets to providing the ideal conditions for a plant's particular niche, the more likely they are to establish a healthy plant.

Ecosystems

An ecosystem is a community of living organisms which, together with their non-living environment, operate as a unit.

An **ecosystem** is composed of all the living organisms (the biotic component) and the non-living environment (the abiotic component) they inhabit, functioning as a unit. Implicit in the term is the idea that all these components react together to form an integrated, balanced and self-sustaining system. Non-living factors include the type, structure and pH of soils; climatic conditions such as rainfall, light, wind and temperature; and topography including altitude, slope, aspect and degree of exposure. Thus, a garden ecosystem in a coastal area in the south-west of mainland Britain and Ireland will be quite different compared with a garden ecosystem in the uplands of the Pennines. Understanding the living and non-living components of each ecosystem will help in the successful choice of plants for a particular garden.

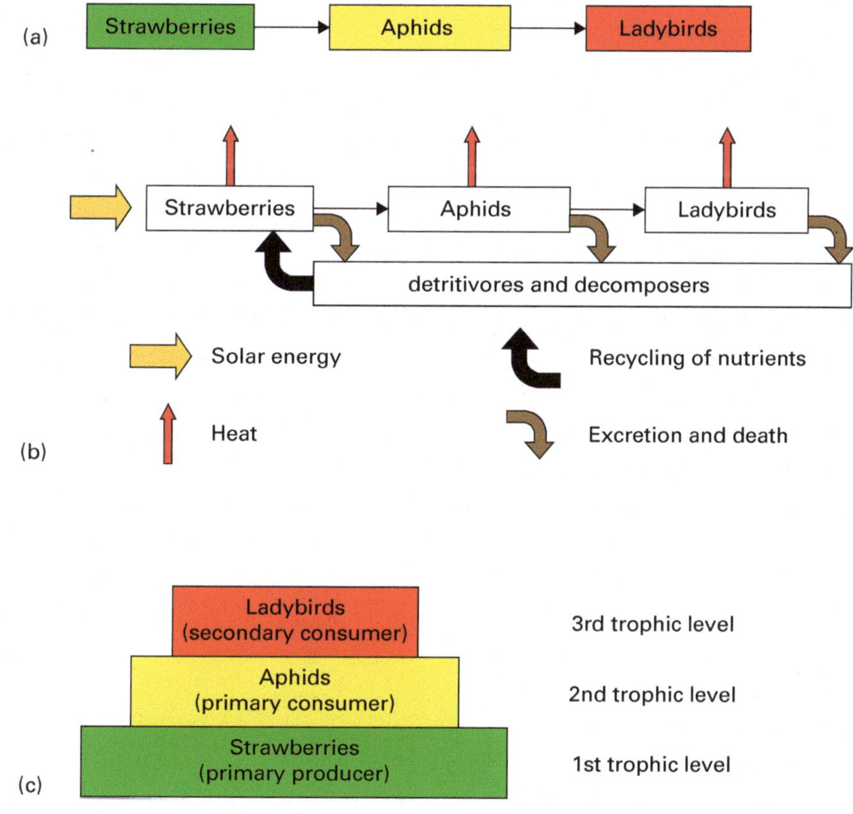

Figure 9.5 Food chains: (a) a simple food chain found in a garden – the strawberry is the primary producer, the aphids the primary consumer and the ladybirds the secondary consumer; (b) flow of biomass and energy through the food chain illustrated in (a) – biomass and energy (heat and chemical energy) are lost from the food chain and nutrients are recycled by detritivores and decomposers; (c) an ecological pyramid representing the loss of biomass and energy at each trophic level

Food chains and webs

Charles Darwin is said to have told a story about a village that produced higher yields of hay than the nearby villages because it had more old ladies. Darwin reasoned that the old ladies kept more cats than other people and these cats caught more field mice. Field mice are important predators of wild bees and since these bees were essential for pollination of red clover (and clover improved the yield of hay), the increased number of bees increased the hay yield. This is an example of a **'food chain'** which highlights the fact that plants and animals in an ecosystem have important feeding relationships which affect the overall success of the system (Figure 9.5a).

Plants always form the first link in the food chain. They are the **primary producers**, manufacturing their own food through the process of photosynthesis (see Chapter 7). Organisms which feed on plants are termed **primary consumers**. In turn, organisms which feed on the primary consumers are called **secondary consumers** (Figure 9.5c). A habitat may include a third (tertiary) level and even a fourth (quaternary) level of consumers but rarely more. A garden example could be a rose bush (the primary producer) on which aphids feed (the primary consumer) which in turn are eaten by birds (the secondary consumer). In practice, the rose bush may support several primary consumers including, for example, a fungus causing black spot or caterpillar larvae or slugs, some of which themselves may be consumed by a range of secondary consumers, such as hedgehogs. In this way food chains are interconnected, forming complex **food webs** (Figure 9.6). Because of this interdependency, removal of any one of the organisms in a food web can have a profound effect on the whole ecosystem. So, in a garden, for example, if we remove the aphids from the rose bush completely, we could also be reducing the numbers of birds.

A primary **producer** manufactures its own food from simple molecules (e.g. green plants photosynthesizing). A **consumer** feeds on living organisms.

Biomass and energy flow

The weight or volume of living plant and animal material in an ecosystem is called its **biomass**. The flow of biomass in a food chain as organisms feed on each other can be represented as a **pyramid** (Figure 9.5c). At the base is the primary producer (strawberry), resting on this is the primary consumer (aphids), and above this is the secondary consumer (birds). Each level in the pyramid is termed a **trophic level**. The pyramid shape reflects the loss of matter from the food chain because whenever biological material is consumed at each level, some matter is lost as waste, through death of organisms and through release of gases and water in respiration (see Chapter 7). In our strawberry example, this means that the total biomass of strawberry material must always be more than the biomass of aphids which feed on it which, in turn, is always more than the biomass of the birds which feed on the aphids. In this way, if we reduce the strawberry biomass available in the garden (say by planting fewer strawberries), this will have an even greater impact on the number of birds because the birds' biomass is comparatively smaller.

In practice, much of the biomass lost as waste and dead organic matter is recycled by a whole group of organisms called **decomposers** and **detritivores** which feed on dead organic matter (see Chapter 12). As well as the flow of biomass, there is also a

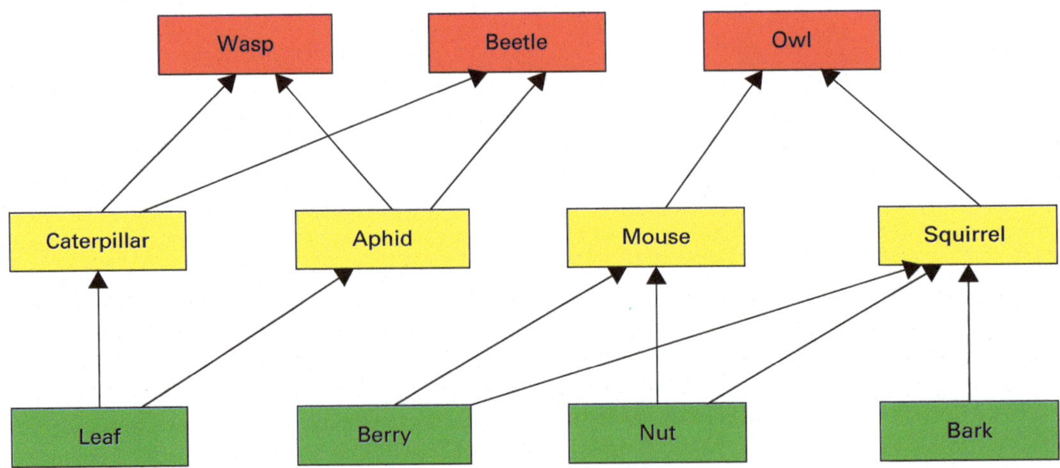

Figure 9.6 A simple woodland food web showing how species interconnect. The bottom level is made up of plants (primary producers). Both primary consumers such as mice (the second level) and secondary consumers such as predatory beetles (the third level) may feed on several different food sources. Removal of any species in the food web can have an effect on one or more of the other species.

loss of **energy** through the system (Figure 9.5b). The process of photosynthesis (see Chapter 7) enables the plant to convert sunlight energy into chemical energy, which is stored in the biomass of the plant. As the plant is eaten by primary consumers, approximately 90% of the energy trapped in the leaf is lost either by respiration, or by heat in the consumer's body, or by waste which is excreted by the consumer and by death of organisms. Since matter and energy are lost from the food chain at each trophic level, this explains why only a limited number of levels are possible. Eventually a point is reached where there is insufficient biomass and energy to sustain a further level of organisms (which is why large carnivores are comparatively rare).

Gardens are highly productive and can produce a large amount of plant biomass, and the greater the biomass, the more consumers it will support. A densely planted border, for example, is estimated to produce 0.5–2.0 kg of biomass per square metre per year. Therefore, in a garden the aim is to provide as rich a variety of organisms as possible, with a high biodiversity (see p. 145) at every level together with as much plant biomass as can be achieved to maintain a healthy and functioning ecosystem.

Succession

Figure 9.7 Examples of primary succession: (a) vegetation colonizing a mountain landslide; (b) marram grass (see p. 79) on sand dunes; and secondary succession (c) colonization of an open glade by birch trees (*Betula pendula*). All are species

9

Communities of plants and animals change with time. Within the same habitat, the species composition will change, as will the number of individuals within each species, because they will alter the environment in which they live. The process of change is known as 'succession'. **Primary succession** (Figure 9.7) starts from uncolonized rock, where plants move in to grow on it. **Secondary succession** results from disturbance of an existing habitat, for example, tree and shrub removal or burning off vegetation. In this case, existing plants present as seeds in the soil can then recolonize the area. This kind of succession is common in Britain and Ireland where most land surfaces have been covered by vegetation at one time or another.

Succession is a sequence of changes in the composition of plants and animals in an area over time. The **pioneer community** is the first in the sequence and the **climax** community is found at the end.

Succession involves a characteristic sequence of plant types which each change the environment in some way enabling the next group of plants to become more successful. The first species to establish make up the '**pioneer**' community (Figure 9.7). Often these early colonizing plants characteristically spread rapidly, mature quickly, produce large quantities of seed and extend the period over which they germinate. In fact, they have typical weed characteristics. Such species are often referred to as **opportunistic**.

For example:

- on exposed rocks pioneer species will be lichens and mosses since they do not require depth of soil to grow and can cling to the rock face. As they die, their organic matter accumulates, enabling larger plants to move in (see soil formation, p. 162). This is an example of primary succession.
- in a woodland where a space opens up, for example through clearance or due to a fallen tree, plants that require light to grow such as birch (*Betula pendula*) are the pioneer species. These need light for seed germination and are fast growing and short lived. As they grow, they act as 'nurse' trees providing some shade for other tree species enabling them to establish. The birches are eventually shaded out by those trees as the canopy closes over (Figure 9.7). This is an example of secondary succession.
- in the vegetable garden or in a planted border, wherever light reaches the soil the pioneer species would be ephemeral and annual weeds whose seeds lie on the surface or are exposed when the soil is disturbed. These would die out as surrounding plants grow, or vegetables are planted, and light is excluded.

A typical succession from a recently cleared area might be pioneer species (small herbaceous plants) growing from the soil seedbank, removing water from and consolidating the soil. They die and contribute to soil formation, changing its depth and structure and sometimes its pH (see p. 203). This means that larger and deeper rooting herbaceous plants, for example bracken, willow herb, foxgloves and tall grasses, can establish and outcompete the pioneer species for light, water and nutrients, eventually taking over. In their turn, they change the soil composition and their environment enabling the next successional

Figure 9.8 A mature woodland in the New Forest, Hampshire, in spring, representing the natural climax community in Britain and Ireland. Note the limited range of vegetation beneath the canopy (source: Shutterstock, DRPL)

stage to take over. This involves larger, more deeply rooting plants, shrubs and climbers such as bramble and honeysuckle outcompeting plants in the previous stage. Eventually, these stages are kept in check or shaded out by trees. The final stage is described as the '**climax**' community (Figure 9.8) which is stable and no further succession takes place. The climax vegetation in Britain and Ireland is deciduous woodland. If no interventions are made by humans or large herbivores, all habitats, including ponds, meadows and scrub, would eventually reach this point.

Through the stages of succession there is usually an increase in the number of species found, both plant and animal, meaning an increase in biodiversity. These form increasingly more complex food webs, although often in the climax community eventually just a few species dominate.

In gardens we constantly interfere with natural succession. If succession is halted at the beginning of the sequence, biodiversity will be low. For example, this could be through constant cultivation or repeated mowing of a lawn. However, if garden interventions are made so stages in succession can establish or be maintained, this will support a greater degree of biodiversity than if succession was left to continue. For example, by allowing some weeds to cover bare soil, or by planting densely or by cutting parts of the lawn higher or less frequently, biodiversity will be increased.

Interactions between organisms

Mutualism

Within a garden there is constant interaction between all the organisms that live there. When very close they are '**symbiotic**'. Sometimes the relationships are beneficial to both partners (**mutualism**). An example is the nitrogen-fixing bacteria which live in the nodules of some plant roots such as peas and beans (legumes) or alders (Figure 9.9). The plant roots provide a home for the bacteria and sugars for them to utilize, while the bacteria trap nitrogen gas in the atmosphere and convert it to nitrates, a form of nitrogen which the plant can use (see p. 241). Other examples of mutualisms are mycorrhiza (an association between plant roots and fungi) which take the place of root hairs in many plants, and lichens (an association between fungi and cyanobacteria) (see Figure 2.20).

In other relationships, the pairing is beneficial to one partner but has no effect on the other – for example, climbing plants and **epiphytic** ferns which live on trees, giving them greater access to light for photosynthesis (Figure 9.10). In Britain and Ireland, these are common in the temperate rainforests of the west.

Predation

Predation (Figure 9.11) is a relationship which is harmful to one of the partners and beneficial to the other, often where one partner (the predator) consumes all or part of the other individual (the prey or host). Predation includes:

- **herbivores** which feed on plants, for example, aphids, vine weevils, slugs, rabbits and deer, many of which are plant pests
- **carnivores** which feed on animal tissue, for example ladybirds feeding on aphids, nematodes eating bacteria, hedgehogs eating slugs
- **parasites** in which the predator has a very close but detrimental relationship with the host, often living inside the host's tissue, for example viruses, fungal diseases, some parasitic plants such as *Rafflesia arnoldii* (see p. 97) and some biological controls.

(see Chapters 17 and 18).

Some plants such as mistletoe (*Viscum album*) can manufacture their own food through photosynthesis but extract water and mineral nutrients from the host so are only partly parasitic (**hemiparasite**) (Figure 9.12).

Figure 9.9 An example of a mutualistic relationship – root nodules in soybean (*Glycine max* (source: Shutterstock, Tomasz Klejdysz))

9

Figure 9.10 Epiphytic ferns growing on tree branches: (a) common polypody (*Polypodium vulgare*) on oak (*Quercus robur*); (b) *Lepisorus thunbergianus* on *Prunus mume* in Japan

Mutualism is a relationship between two organisms which is beneficial to both. **Predation** is a relationship which is harmful to one partner (the host) and beneficial to the other (the predator). **Parasitism** is an extreme form of predation where the predator often lives within the host.

Predators and their prey follow closely linked population cycles (**predator–prey cycle**). When there is abundant prey, there is plenty of food for predators and their numbers increase. This leads to a reduction in the number of prey which then leads to a reduction in the number of predators as their food supply shrinks. In a balanced ecosystem, the prey population is never completely removed as some prey individuals are able to hide in pockets in the garden and survive to reproduce. Once the predator population decreases, the prey population can expand causing the cycle to start again. In a garden, for example, beneficial insects such as ladybirds (predators) control the numbers of harmful insects such as aphids (the prey) (Figure 9.11). Understanding predator–prey cycles is important in the use of biological control of plant pests (see Chapters 15 and 17). In addition, if broad-spectrum pesticides are used which kill all insects, they can interfere with these natural cycles causing disastrous results. For example, ladybird numbers might be reduced to such a low level they cannot recover as quickly as the aphids, so the latter grow to even greater numbers than before. In this way, the use of chemical pesticides frequently causes minor pests to become serious problems by disturbing the natural controls that keep them in check.

Competition and cooperation

Interspecific competition is competition for resources between individuals of different species. **Intraspecific competition** is between individuals of the same species.

Organisms compete for shared resources such as light, moisture, nutrients and space, reducing their growth and their potential for reproduction. Competition can be by **exploitation** of the shared resource, for example, when two plants compete for water in a dry environment, the more successful organism is the one which is able to survive on less. It can also be by **interference** where one organism directly affects the other, for example the introduced shrub *Rhododendron ponticum* (see p. 7) produces chemicals which prevent seeds germinating beneath their canopies (see Figure 1.10b). Organisms overcome the effects of competition by moving to new sites through seed and fruit dispersal in the case of plants, or by seeking new territories in the case of some animals.

Figure 9.11 Examples of predators: (a) muntjac deer – a large herbivore; (b) larvae of the seven-spot ladybird which feeds on aphids; (c) rose aphids feeding on a rose bud (source: Shutterstock, AlessandroZocc); (d) a broomrape, (*Orobanche* spp.) a parasitic plant – note the absence of chlorophyll (source: Shutterstock, zdenek_macat)

The degree of competition between plants as they grow will depend very much on the spacings between them, whether this is in the garden border, in the vegetable plot or in the greenhouse. Competition can be **intraspecific**, that is, between individuals of the same species, and this will be the case in plants grown in a **monoculture**, where only one species is present, as in many horticultural situations. For example, if carrots are sown too close together, their roots will be small. Sometimes this can be an advantage, as in carrots grown for canning, where closer spacings are

Figure 9.12 (a) Mistletoe (*Viscum album*) – a hemiparasitic plant; (b) mistletoe in a cider orchard in Herefordshire

used deliberately. Small, even-sized carrots may be specified by supermarkets so spacings can be designed to achieve this. Usually though, close

Figure 9.13 Crown shyness in stone pines (*Pinus pinea*); note the interlocking branches and the distance between the crowns of individual trees in the canopy (source: Shutterstock, Nadezhda Kharitonova)

planting will be detrimental. Overcrowded plants may be more susceptible to fungal diseases due to poor air circulation, they may grow leggy with weak stems or seed production may be reduced if they are shaded and competing for light. Growth may be poor if roots are competing for soil, water and nutrients. Three ways by which gardeners overcome these problems as plants grow is by transplanting seedlings from trays into pots, increasing the spacing of pot plants in greenhouses and hoeing out or 'thinning' a proportion of young vegetable seedlings from a densely sown row. Crops grown in deep bed systems, in which a one metre depth of well-structured and fertile soil enables deep root penetration, reduce competition by rooting at different depths, allowing plants to be grown closer together.

Competition can also be **interspecific**, between two different species. When designing a mixed border, it is important to allow for the future size of the plants (see p. 118) and place them so they do not compete with each other for resources (see 'niche' earlier).

Plants can sense when they are being crowded by other plants and grow accordingly (see p. 118). This is most obvious in the 'crown shyness' phenomenon seen in some tree species where the trees grow together without touching each other, to occupy the space most efficiently (Figure 9.13).

Where competition is removed completely, growth can take place unchecked. A single plant growing in isolation with no competition is as unusual in horticulture as it is in nature. However, specimen plants such as leeks, marrows and potatoes, lovingly reared by enthusiasts looking for prizes in local shows, grow to enormous sizes when freed from competition. In landscaping, specimen plants are placed away from the influence of others so that they not only stand out and act as a focal point, but also can attain perfection of form.

Recently, studies of old growth forests in Germany and British Columbia have shown that trees share resources such as water, sugar and nutrients through their underground networks of mycorrhiza, both between the same species and between different species They also send chemical, hormonal and electrical signals above and below ground. Older trees can support the growth of younger seedlings on the shady forest floor in this way. These controversial findings suggest that far from competing with each other, they are actually **cooperating** to ensure increased health, more photosynthesis and greater resilience in the face of disturbance.

Gardening for wildlife

Biodiversity

When we think of gardening for wildlife, it is very easy to just imagine a garden full of flowers and the three 'B's – birds, butterflies and bees – and certainly we want to encourage them. But a healthy garden which is good for wildlife contains much more. It supports a wide range of interdependent organisms, as we have seen previously, most of which we never see but on which many more visible organisms depend. Invertebrates (organisms which have no internal skeleton) include not just bees and butterflies but other insects such as flies and lacewings, beetles, spiders, snails and slugs, earthworms and microscopic worms, which all have an important role in food chains. We have some 2,400 native moth species, an important food source for bats, but only around 60 butterflies which are regularly seen in Britain and Ireland. By providing suitable environments within the garden and managing it appropriately, we can create the right conditions for complex food webs to become established, thereby increasing biodiversity. Fundamentally, it is the primary producers, the plants, which hold the key. If there is a good range of these and plenty of them, the ecosystem's food chains and webs will be well supported.

In addition, the presence of decomposers and detritivores in a healthy soil will cycle organic matter through the ecosystem (see carbon cycle p. 185). To support the widest biodiversity a garden must be designed to provide three factors for all the organisms that live there:

- food
- shelter
- breeding sites.

Some of the ways by which we can increase and maintain biodiversity in a garden are outlined next.

What do we mean by biodiversity?

Biodiversity is generally thought of in terms of numbers of different species in a particular habitat, but the definition also has a broader scope. Biodiversity can refer to habitats themselves, a wide range of habitats in a particular location such as a garden will result in greater biodiversity. At the other end of the scale, biodiversity can apply to the range of genetic variability within a species too, for example, the number of cultivars listed for a particular horticultural species such as apple.
When we speak about conservation of biodiversity, we therefore mean conserving **the genetic variation within a species, the total number of species present** and **all the habitats** they live in.

Traditional orchards and biodiversity

'Traditional' orchards are probably the most biodiverse agricultural habitat we have. They can broadly be distinguished from modern commercial orchards by being widely spaced, planted in permanent grassland, often grafted on vigorous rootstocks and trained as standards or half standards and managed with low intensity and chemical input (Figure 9.14). They contain a mixture of fruit and nut cultivars and their trees are allowed to reach a veteran stage. Most will have been commercial orchards originally. When combined with fallen and dead wood, meadows, scrub, ponds and hedgerows they provide a rich mosaic of habitats and microhabitats for wildlife.

Fruit trees have a short lifespan, rarely more than 120 years old, compared with many woodland trees. They acquire their veteran characteristics in a relatively short time of 60 to 70 years. As they age, they provide hollow stems and branches, cracked bark and cavities which collect moisture and support a range of invertebrates, including the rare noble chafer beetle (*Gnorimus nobilis*), small mammals, nesting bats and birds. Fungi, mosses and lichens soon set up home on fruit tree branches and fallen fruit is a food source for mammals including threatened species such as hedgehogs, and migratory birds such as fieldfares. Grassland in old orchards will have been undisturbed for years with no fertilizer inputs or cultivation so will have reduced fertility which favours wildflowers and a good soil structure which will support the mycorrhizal fungi networks, which have been shown to benefit nearly 80% of plants. Traditional orchards will not have had high pesticide inputs, if at all, which will benefit insects, especially pollinators.

As well as supporting wildlife, traditional orchards are a genebank for the future. The range of apple cultivars grown in commercial orchards is very limited but heritage trees, though mostly dating from the middle of the 19th century onwards, can provide useful genetic material for future breeding programmes against pests and diseases or to introduce new desirable characteristics in apples. The East of England Apples and Orchards Project propagates heritage varieties and is an example of *in situ* conservation (see p. 10).

A recent research project, Orchards East, carried out by the University of East Anglia between 2017 and 2020, documented historic and existing orchards in six eastern counties in England. They found that the area of orchards had fallen dramatically by around 85% since the 1950s, and although traditional orchards were given priority status in the UK Biodiversity Action Plan (2008), many have little statutory protection (in some circumstances they can be given Tree Preservation Orders). Fortunately, the growing popularity of community orchards, if properly managed, can go some way towards redressing the balance.

Figure 9.14 A 'traditional' orchard at Shenley Park, a former mental hospital. The trees were planted in the 1950s. Note the wide spacing, wildflower meadow and colonization of the bark by mosses and lichens

Garden structure

A good wildlife garden tries to provide the greatest range of habitats within the space available. In nature, woodlands have the highest biodiversity of any terrestrial habitat. Vertical structure is important; a woodland has typically at least three layers in its structure (Figure 9.15a) with an upper canopy of tall trees, a shrub layer beneath, a herbaceous layer at ground level and ground-hugging vegetation which provide habitats for a wide range of plant and animal species. Some of them, such as ivy (*Hedera* spp.) and grey squirrel, will access all levels whilst others are limited to specific layers such as herbaceous plants, beetles and voles on the woodland floor or tawny owl, purple emperor butterfly and oak processionary moth in the canopy. The plants in a woodland have to cope with a severe reduction in light, most of it being absorbed by the canopy when it is in leaf and have developed a range of strategies to succeed (see pp. 82 and 118).

The most biodiverse part of any wood is the **woodland edge,** which provides a wide range of habitats over a short distance. Here, the light ranges from very shaded under the tree canopy to open and well-lit on the very edge where naturally it would join a path, glade, clearing or meadow area. Furthermore, because of the light penetration and vegetation, the soil moisture content will vary across the edge too. Gardens can successfully imitate the tiered structure and the woodland edge in a deep border with trees at the back, shrubs of varying height in the middle and herbaceous planting and ground cover next to a lawn, path or paving.

Some garden features are especially useful to increase biodiversity:

- **Trees**. A tree can be thought of as a habitat in its own right, with several microhabitats within it. There is a variety of light and shade through the canopy, the leaves, flowers and fruit provide a food source for organisms, and it provides shelter and areas for roosting and nesting birds and overwintering sites for insects. The bark, especially in an old tree, also provides its own microhabitat, with breeding sites and shelter for many insects in its furrows. Old trees also provide support for lichens and epiphytic plants and even small pools of water for many organisms to utilize or live in (Figure 9.1 and Figure 9.16). Rotting wood and leaves are a food source for soil organisms below. Log and brash piles in a garden can provide a similar microhabitat to fallen trees in a woodland so should be left on the ground. Most conifers support a lower biodiversity than do broadleaved trees. Nevertheless, a single tree, however small, has been shown in studies to provide one of the most beneficial habitats

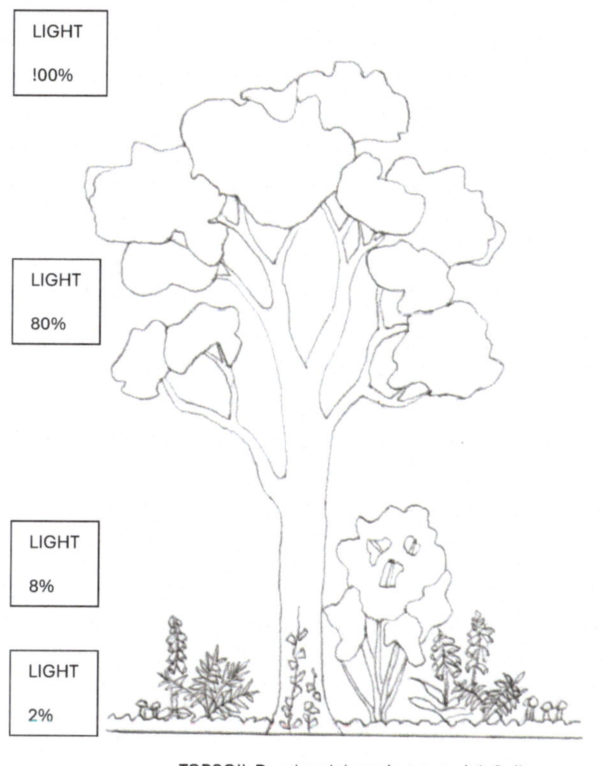

Figure 9.15 (a) Diagram of vegetation layers in a typical deciduous woodland; note the reduction in light through the layers (source: Adapted from 'Biological Science' by Green. N.P.O, Stout, G.W. and Taylor D.J. and 'Woodland Habitats' Read, H.J. and Frater, M.); (b) a multilayered garden structure with trees, shrubs, perennials, ground cover and areas of light and shade

in a garden. For example, ornamental birches (*Betula* spp.) potentially host more than 150 species of butterfly and moth.

- **Ponds** provide an additional important habitat (Figure 9.17). As with woods, the pond edge is the most biodiverse part of the pond because it is shallow and warm and light can penetrate easily. A small pond therefore has greater wildlife potential than a large lake where much of the area is deep water and less biodiverse. In suburban gardens, many small ponds are better than a large pond covering the same total area, and this also means that the loss of any one pond through development has less impact

Figure 9.16 A veteran oak tree; ancient trees provide a wealth of microhabitats for many organisms

Figure 9.17 A garden pond with a good range of marginal plants (source: Shutterstock, Goglio Michele)

overall. The area between the highest and the lowest water levels (the **draw-down zone**) is the most biodiverse, so sloping pond edges in ponds which are not frequently topped up are the best arrangement. A good range of marginal plants will provide cover for organisms to hide, feed and breed at the water's edge and the addition of some stones or rocks increases the range of microhabitats (Figure 9.18). Ideally, ponds should not contain fish which would feed on pond-dwelling organisms.

Figure 9.18 Dragonfly nymphs live underwater then climb up the waterside plants to hatch into adults

- **Boundaries**. As with borders, a hedge is basically a woodland edge in miniature and provides a good range of habitats for many organisms. Hedges can be important for wildlife in three ways:
  - Firstly, as a **habitat** they provide a food source, shelter, breeding and territorial sites for a whole range of organisms from invertebrates to birds and mammals. A mixed hedge containing a range of deciduous and evergreen species and bearing flowers and fruits is ideal, although a single species hedge will be better than a fence or wall. If hedges are not an option, a sunny wall or fence will support a range of invertebrates, and if climbers are grown up them then the habitat is enriched.
  - Secondly, hedges act as **wildlife corridors** connecting habitats together and enable organisms such as hedgehogs and other small mammals to travel further in search of food, shelter or breeding sites and also provide a route for prey to flee from predators. They are commuting highways for bats, butterflies, other insects and small mammals between feeding and roosting sites. Where habitats become fragmented, for example due to removal of hedgerows or urban development, the populations in them become threatened as they are unable to escape predators (Figure 9.19). In addition, when populations are isolated from each other, they can become inbred, so genetic problems such as increased susceptibility to disease or reduced fertility can occur. A collection of suburban gardens with hedges can therefore be seen as one large habitat because they are interconnected. They can often form wildlife corridors themselves linking larger green areas such as parks or fields to each other.

- Thirdly, the hedge itself creates **shelter** in the garden for other plants and **protects soils from erosion and moisture loss** enabling a wider range of plants to be grown, along with the organisms they support, and a more biodiverse soil community which is essential for a healthy soil (see p. 182).
- **Meadows**. While lawns are an attractive feature in gardens, they incorporate only a few plant species, supporting a limited range of organisms. They produce no flowers and are poor cover for invertebrates. Studies show that even a small area of long grass is beneficial for invertebrates for shelter and breeding. Incorporation of flowers as plug plants among the lawn grasses to imitate a meadow will improve wildlife value or existing lawns can be mown less frequently to allow common lawn 'weeds' such as daisies (*Bellis perennis*), selfheal (*Prunella vulgaris*) and dandelions (*Taraxacum*) (see p. 000) to establish. These can all tolerate close mowing and provide pollen and nectar for invertebrates.

A meadow can mean different things to different people but is always a mix of grass and flower species. Meadows can be created in gardens on a small scale, in borders or less formal areas. Annual meadow mixes contain annual flower species such as cornflower (*Centaurea cyanus*), corn marigold (*Glebionis segetum*) and field poppy (*Papaver rhoeas*). They rely on the plants self-seeding for continuation. Perennial meadow mixes contain perennial plants such as meadow buttercup (*Ranunculus acris*), common knapweed (*Centaurea nigra*) and red campion (*Silene dioica*) which will set seed but will also overwinter and continue growth in the following spring. Often a mix of the two will be sown initially to give a burst of colour in the first season. It is important to reduce competition from vigorous grasses, and the site should be thoroughly cultivated before sowing to prevent existing grasses taking over. Wildflowers thrive best in full sunlight and on infertile soils so light sandy soils are more suitable than heavy clay soils but there are meadow mixes available for various soil types and situations. When the meadow is cut, usually once a year in late summer or autumn, the clippings should be removed to prevent them returning fertility back to the soil. If space allows, different areas of the garden can be mowed at different times to give a mosaic of meadow habitat.

Figure 9.19 (a) Disrupted 'green corridors' in farmland where hedgerows have been removed; (b) farming has also left fragments of disconnected woodland

Wildflower seed should be sourced within Britain and Ireland from reputable suppliers and with known **provenance**, especially when growing native plants. Provenance is important because it keeps the local genetic strains pure when cross-pollination occurs. It also means that the seed is adapted to local conditions and therefore has a greater chance of survival compared with seed grown in a different locality which may have different climatic conditions.

What is 'provenance'?

Provenance refers to the actual geographical location a particular plant was collected from, often in the form of seed. It is not the same as the **origin** of the plant species. It has implications for genetic purity, as local plants will be more closely related, and also for its potential survival. For example, bay trees (*Laurus nobilis*) come from and are native to the Mediterranean, but plants grown in Britain and Ireland for several generations (those with local provenance) will have evolved to adapt to the different climatic conditions here. Therefore, plants grown from seeds or cuttings taken from local plants will have a better chance of survival, particularly regarding hardiness, compared with imported plants. Similarly, native wildflower seed grown and harvested in Britain and Ireland will show greater genetic similarity with local wild plants and be better adapted to conditions here, compared with the same seed species grown or collected from abroad.

Yellow rattle

Yellow rattle (*Rhinanthus minor*) is named after its large seeds which rattle in their pods when dry (Figure 9.20). It was also called 'catch all' by farmers as it was a weed of hay meadows, which reduced yields and was spread in haymaking. The plant is an annual hemiparasite that feeds on the roots of coarse grasses and leguminous plants but can also photosynthesize. It can reduce grass biomass by up to 80% and allows less competitive wildflower species to establish and compete more effectively. Yellow rattle seed is often incorporated into wildflower seed mixes.

Figure 9.20 (a) Yellow rattle (source: Shutterstock, M. Schuppich); (b) note the species-diverse area in foreground containing yellow rattle compared with the area of coarse grasses behind

- **Bare soil** is enjoyed by some invertebrates so leave a few patches if possible here and there.
- **Plant selection**. As primary producers, plants are the foundation of the living component of a garden. They provide food in the form of nectar, pollen, seeds, fruits and green leafy material so should be chosen to supply these useful 'products' over the longest period of time, in particular by including winter-flowering plants such as *Viburnum tinus*, *Mahonia*, snowdrops (*Galanthus nivalis*) and winter aconite (*Eranthis hyemalis*) (Figure 7.7). Ivy (*Hedera helix*) (Figure 9.21) is one of the best plants for wildlife in the garden. It provides evergreen foliage for nesting and shelter for small birds and insects and its leaves are a food source for some butterfly species. It flowers late in the season when there are few other nectar and pollen sources available and its berries ripen in early spring providing a food source for many birds.

Cultivated plants with double flowers should be avoided as the male and female flower parts are converted into extra petals unlike the wild forms (see petalody p. 88), and so they may be sterile and lacking pollen and nectar (Figure 9.22). Grasses, bamboos and ferns are poor food sources for invertebrates, although they do provide good cover.

Whether planting should be native or non-native is a subject of much debate. Native plants are often considered the 'best'. This is because they have evolved alongside the organisms that use them over a long period of time so support the greatest number, of which many will be dependent on a particular plant and adapted to it. For example, the oak tree (*Quercus robur*) is often cited as supporting over 400 species of invertebrates. However, these will not all be found on a single tree at any one time. Furthermore, most invertebrates are not limited to one tree species; they are generalists rather than specialists. Even within Britain and Ireland, some species are only truly native in some localities. Scots pine (*Pinus*

Figure 9.21 A mature ivy with flowers and developing fruits

Figure 9.22 (a) A single paeony (source: Shutterstock, Tony Baggett) and (b) a double paeony, flowered forms in a paeony; note the stamens present in the single form (source: Shutterstock, Kristine Rad)

Table 9.1 Some definitions relating to plant origins

Native	Plants which were present at the end of the last Ice Age when mainland Britain and Ireland separated from the rest of Europe.	oak (*Quercus robur*), bluebell (*Hyacinthoides non-scripta*), lady fern (*Arthyrium filix-femina*)
Near-native	Originating outside Britain and Ireland but from the northern hemisphere.	onion (*Allium* spp.), coneflower (*Echinacea purpurea*), thyme (*Thymus* spp.), lamb's ear (*Stachys byzantina*)
Naturalized	Plants that have been introduced by humans and have spread and now reproduce in the wild. Often invasive.	butterfly bush (*Buddleja davidii*), Himalayan balsam (*Impatiens taprobanica*).
Exotic	Plants that have been introduced by humans more recently from outside Britain and Ireland and are dependent on human management to survive.	lavender (*Lavandula* spp.), *Berberis thunbergii*, tulip tree (*Liriodendron tulipifera*), *Rhododendron ponticum*
Cultivated	Plants that do not exist in the wild anywhere. Usually cultivars.	*Narcissus* 'Tete-a-Tete', *Rosa* 'William Lobb', *Papaver* (Oriental Group) 'Patty's Plum'

sylvestris), which is a 'native' tree, is only native to Scotland, that is, records show it was growing there at the end of the last Ice Age. In southern England it has been planted. Similarly, there is evidence that sycamore (*Acer platanoides*) may have been native in north-west England originally, but elsewhere it is an introduced species and is frequently removed. Even native trees such as oak (*Quercus robur*) may have been planted from European sources sometime in the past, so would be genetically different from their predecessors. In effect, any tree is useful whether native or not and numerous naturalized and exotic plants such as lavender (*Lavandula angustifolia*), buddleja (*Buddleja davidii*), borage (*Borago officinalis*), *Hylotelephium spectabile*, *Cotoneaster* spp. and *Echium* spp. are known to be excellent for wildlife.

The majority of organisms are probably not selective about the plants they use, and it is what they can provide that counts rather than what they are. The RHS 'Plants for Bugs Study' showed that for flying pollinating insects the best habitat has both native and near-native species, with some exotics to extend the season (Table 9.1). For other invertebrates living on the plants such as caterpillars, native and near-natives were preferred and the more ground cover available, the greater the benefit. For ground-dwelling invertebrates such as ground beetles, dense year-round vegetation cover combined with plantings containing native and near-native plants and small patches of bare ground were beneficial.

Lists of plants suitable for attracting wildlife to the garden are easily obtainable. Often they focus on butterflies and bees, but it should be remembered that plants and planting should encourage *all* wildlife to provide the greatest variety of organisms, supplying a food source for higher trophic levels in the food chain (Figure 9.23).

Garden management

Managing a garden to encourage wildlife is not difficult nor should it result in a garden which is overgrown or unpleasant to look at. Many of these management approaches follow that of **organic gardening**: reduced pesticide use, minimal soil disturbance and a focus on good soil health, mixed planting rather than monocultures and a general aim of growing with nature rather than against it. Organic gardening relies on a good range of natural

Figure 9.23 Some native plants suitable for different habitats in a garden: **pond edge** (a) yellow flag iris (*Iris pseudacorus*); **damp meadow** (b) snake's head fritillary (*Fritillaria meleagris*), (c) ragged robin (*Lychnis flos-cuculi*), (d) white butterbur (*Petasites albus*); **woodland with dappled shade or glade** (e) foxglove (*Digitalis purpurea*), (f) daffodil (*Narcissus pseudo-narcissus*); **open ground** (g) teasel (*Dipsacus fullonum*)

predators for pest control and soil organisms for fertility. Some general principles are:

- **Prune to maximize flower and fruit** production and deadhead herbaceous plants to prolong flowering, so extending the supply of pollen and nectar.
- **Do not use pesticides** because they interfere with the balance of organisms in the garden and disrupt food chains and webs (see earlier). Pesticides may remove 'beneficial' insects, such as bees or those which feed on 'harmful' species. Even if the pesticide is targeted at a specific pest species, that may be a food source for other organisms. Pests may also become resistant to a pesticide and will no longer be controlled. It is worth tolerating some plant damage in return for improving biodiversity in the garden.
- **Minimize interference** with plants, shrubs and trees and carry out operations at a time of year which is least harmful. For example, hedges are an overwintering site for many invertebrates and a nesting site for birds in spring, so pruning times should take this into account. In fact, there is evidence that pruning hedges in summer when nesting is over prevents excessive growth compared with pruning at other times of year. Pruning hedges in the winter may disturb overwintering organisms such as small mammals and invertebrates.
- **Resist the urge to tidy** in the autumn and winter. It is important to leave seed heads and dead material for food and shelter as long as possible rather than simply tidy up at the end of summer, especially if this leaves bare soil. Winter flower and seed heads can be attractive in their own right (Figure 9.24).
- **Encourage a good population of soil organisms** by allowing organic material such as composted matter, fallen leaves grass clippings and herbaceous plants to return to the soil, recycling nutrients, providing warmth and shelter and protecting the soil from erosion.
- **Do not overcultivate** as this can damage soil structure and organisms. Use green manures in the vegetable plot and mulches to protect the soil over winter.
- In a small garden where natural vegetation is limited, **insect hotels** can be provided in a sunny location (Figure 9.25). By using a range of materials, solitary bees, butterflies, predatory insects and other invertebrates can be provided with shelter and refuges especially over winter.

Figure 9.24 (a) Flowerheads left on hydrangeas over winter; (b) teasels against purple *Muehlenbeckia astonii* in January provide structural interest and food for birds

Figure 9.25 An insect 'hotel' (source: Shutterstock, Pack-Shot)

The concept of **rewilding** (or **wilding**) is the subject of considerable discussion currently. It involves habitat restoration, reinstating natural processes and sometimes reintroducing species which have been extinct for hundreds of years, in particular top predators and large herbivores, over a large area. The success of projects, such as those at Knepp in Sussex and the Wilder Blean project in Kent, show the benefit to nature of recreating ecosystems largely untouched by humans. Whether this is applicable on a smaller scale in gardens, and what is meant by these terms in a garden setting, is interesting and a lively matter for debate.

Garden birds

One of the most visible forms of wildlife in the garden are birds, and for many people this is one of the key benefits of a wildlife-friendly garden. Over half the adults in the UK feed birds in their gardens according to the Royal Society for the Protection of Birds (RSPB). Measures taken to encourage birds with feeders (Figure 9.26) and nest boxes (Figure 15.6) undoubtedly attract birds to the garden and help to support their numbers. Winter feeding can often lead to earlier laying and increased breeding success and for some species, such as song thrushes, breeding populations in gardens are now better than those of farmlands. Adult birds of other species will feed on supplementary sources of food such as peanuts, but their fledglings require a high protein diet of invertebrates such as spiders and worms. Blue tit fledglings, for example, feed largely on caterpillars. Unless the garden can supply these foods (or the gardener is prepared to provide them artificially), birds may breed earlier in the year but their broods may not survive. It has also been suggested that feeding birds may also increase the spread of disease and may affect natural selection through influencing reproduction and behaviour, enabling birds which are weaker genetically to survive. Therefore, while encouraging birds into the garden undoubtedly has many benefits for both birds and humans, ecological principles remind us that this is no substitute for also providing natural food sources, shelter, roosting, cover and nesting places too.

Figure 9.26 Blue tit at a bird feeder

Further reading

Adams, C., Early, M., Brook, J. and Bamford, K. (2015) *Principles of Horticulture Level 3*. 7th ed. Routledge.

Baines, C. (2023) *RHS Companion to Wildlife Gardening*. Frances Lincoln.

Buczacki, S. (1986) *Ground Rules for Gardeners*. Collins.

Buczacki, S. (2010) *Garden Natural History*. Collins New Naturalist Library. Collins.

Doberski, J. (2024) *The Science of Garden Biodiversity*. Pimpernel Press Ltd.

Frater, M. and Read, H.J. (1999) *Woodland Habitats*. Routledge.

Mackenzie, A., Ball, A.S. and Virdee, S.R. (1991) *Instant Notes in Ecology*. Taylor & Francis.

Owen, J. (1991) *The Ecology of a Garden: The First Fifteen Years*. Cambridge University Press.

Owen, J. (2010) *Wildlife of a Garden. A Thirty-Year Study*. Royal Horticultural Society.

RHS. (2019) *Royal Horticultural Society Plants for Pollinators*. www.rhs.org.uk/science/pdf/conservation-and-biodiversity/wildlife/plants-for-pollinators-plants-of-the-world.pdf
Ricklefs, R.E. (2000) *The Economy of Nature*. W.H. Freeman.
Simard, S. (2022) *Finding the Mother Tree: Uncovering the Wisdom and Intelligence of the Forest*. Penguin.
State of Nature Report. (2023) https://stateofnature.org.uk/wp-content/uploads/2023/09/TP25999-State-of-Nature-main-report_2023_FULL-DOC-v12.pdf
Stewart, D. (2023) *A Gardener's Guide to Sustainable Gardening*. The Crowood Press Ltd.
Taylor, D.J., Green, N.P.O., Stout, G.W. and Soper, R. (1997) *Biological Science 1 & 2*. CUP.
Thompson, K. (2006) *No Nettles Required*. Eden Project Books.
Townsend, C.R., Begon, M. and Harper, J.L. (2008) *Essentials of Ecology*. Blackwell.
Webster, E., Cameron, R. and Culham, A. (2017) *Gardening in a Changing Climate*. RHS.
Williamson, T. and Barnes, G. (2021) *The Orchards of Eastern England*. University of East Anglia.
Wohlleben, P. (2016) *The Hidden Life of Trees: What They Feel, How They Communicate*. Greystone Books.

9

The online material is accessible via the QR code and includes further information on many of the topics in the book, such as

- UK biodiversity policy
- Garden invertebrate studies
- Threats to biodiversity
- Citizen science studies
- The benefits of green spaces

CHAPTER 10

The root environment

Figure 10.1 Sedimentary rocks showing the characteristic layers originally laid down horizontally but buckled during subsequent upheaval in the Earth's crust

This chapter includes the following topics:

- The plant's requirements
- Origin of soils
- Soil formation and development
- Composition of soils
- Soil texture
- Soil structure
- Topsoil and subsoil
- Cultivations
- Creating a seedbed

DOI: 10.4324/9781003581260-10

The plant's requirements

Most of us are familiar with plants doing well in one place but not in another. Indeed, in some situations, it's our specially selected plant that has performed badly. Invariably, it is not 'the right plant in the right place' (see p. 233). It can be because of nutrient problems which are dealt with in Chapter 13 or because of the effect of weeds (Chapter 16), pests (Chapter 17) or pathogens (Chapter 18). Often it can be attributable to a poor rooting environment, that is, the physical properties of the soil.

Although out of sight, roots play a vital role in supplying water and nutrients to the plant. They must access a large enough volume of soil to supply the plant's needs and reach a depth that helps to maintain a water supply as the surface layers dry out. Within a full season, a plant growing well in open ground may develop some $500-1000\,\text{km}$ of root, with most plants penetrating at least half a metre below the surface. This vast root system is usually more than is required to supply the plant in times of plenty, but the extent of the network is indicative of what is needed in unfavourable conditions. It also serves to remind us of what we undertake to provide when we restrict root growth by growing plants in containers (see Chapter 14).

Plants do not grow until the growing medium is warm enough; usually above 5°C for most temperate plants and 10°C for those from tropical areas. Optimum temperatures for a range of plants is given in Table 3.1. Growing media also need to be free of harmful substances which can include some nutrients applied in excess quantities (see osmosis, p. 124).

The growing tip of the root wriggles through the growing medium following the line of least resistance. Roots enter cracks that are about 0.2 mm in diameter, which is about the thickness of a pencil line. Once into these narrow channels the root can overcome great resistance to increase its diameter. Nonetheless, compacted soils severely restrict root exploration which in turn limits plant growth. When this happens action should be taken to remove the obstruction to root growth (see soil structure, p. 167).

The root ball normally provides the **anchorage** needed to secure the plant in the soil. Plants, notably trees with a full leaf canopy, become vulnerable to the effect of the wind if their roots are shallow, for example roots over rock strata close to the surface, in loose material or in soil softened by high water content. Transplants are very susceptible to wind rocking until their roots have penetrated the surrounding soil; the plant may be left less upright and with damaged roots unless secured. Furthermore, water uptake remains limited after transplanting (see p. 62) until the delicate root hairs damaged in the process are replaced.

To take up water and nutrients and to grow the root must have an energy supply. Efficient energy production is only possible if **oxygen** is available (see respiration, p. 119). Consequently, the soil around the root must contain air as well as water. To ensure the supply of oxygen is constantly replenished, and for the carbon dioxide to be taken away, there needs to be good gaseous exchange between the atmosphere around the root and the soil surface (see soil structure, p. 167). A lack of oxygen or a build-up of carbon dioxide will reduce the root activity and anaerobic bacteria (see p. 119) will proliferate and produce toxins. In warm summer conditions roots can be killed after just a few days in waterlogged soils. There are plants that can grow successfully in waterlogged soil, or in water, because they have adaptions that make that possible, for example aerenchyma (see p. 83). Plants can also grow very successfully in water if the water around the roots is kept oxygenated (see hydroponics, p. 213).

A naturally fertile soil or well-managed one under cultivation provides the plant with:

- water
- air (oxygen)
- nutrients
- anchorage.

Origin of soils

Soils form in the layers of rock fragments on the Earth's surface. The parent rocks that provide the mineral material for soil formation are weathered by physical, chemical and biological forces.

Parent rock is the rock from which a soil is made.

Weathering is the breakdown of rocks.

Erosion is the movement of rock fragments and soil.

Most soils in Britain and Ireland are derived from the weathering of rocks. The remainder are peats which are made solely from organic matter, that is, dead plants (see p. 185).

Rocks

There are three main types of rock – igneous, sedimentary and metamorphic (see also Support Material).

Igneous rocks are those formed from the Earth's molten material. All other rock types, as well as soil, are ultimately derived from them. When examined closely, most igneous rocks can be seen to be a mixture of crystals. **Granite** is one of the most common and contains crystals of quartz, which are white and shiny; feldspars, grey or pink; and micas, shiny black (Figure 10.2). Many of these crystalline materials have a limited use in landscaping; they are more commonly used in monuments and building facades where the larger grained igneous rocks are sought after. The weathering by rain (carbonic acid) gives rise to 'rotted' granite; the feldspar and mica is broken down to clay particles, potassium and other soluble minerals that are washed away and the inert quartz is released from which much sand is derived.

Figure 10.2 **Granites** are made up of quartz, mica and feldspar crystals. There are several types of feldspar, and it is the alkali feldspar that gives the pink colouring to some types (as shown)

Figure 10.3 Many **limestones** are made up of the shells of sea creatures; in some cases these fossils can readily be seen with the naked eye

Sedimentary rock (see Figure 10.1) is derived from accumulated fragments of rock such as the huge deposits of sand in deserts. Many have been formed in the sea or lakes to which agents of erosion such as rivers and the wind carry the weathered rock particles. Layers of sediment build up and, under pressure and slow chemical change, eventually become rock strata such as

- **Sandstones.** These vary enormously in colour – browns, yellow, red and pink as well as black, grey and white – and in texture and hardness. The variation depends on the nature of the source material (mainly quartz and feldspars), the iron oxides present and the conditions in which they were laid down (e.g. hot dry desert or in water). There is usually strong indication of having been deposited in layers (strata), making them attractive to use in rock gardens.
- **Limestone** (see Figure 10.3). Organisms which die in the sea and accumulate on the seabed may become one of many types of limestone which contain at least half its content as calcium carbonate (the rest is often clay). The shells of the organisms are often quite evident, but some are derived from algae. Strata (layers) are less obvious in limestones. Some are almost pure calcium carbonate like chalk whilst others have significant amounts of other material present to provide a range of colours. There are also limestones that are made up of weathered limestone ('recycled' calcium carbonate) where shell is not evident, whilst others are derived from calcium carbonate in solution.

Metamorphic rock. These are formed from igneous or sedimentary rock subjected to high heat and/or pressures, for example slate formed from shales and marble from limestones.

Figure 10.4 Weathering and erosion by water. Note that the faster moving water on the outside of the bend removes soil and rocks (weathering) and carries them until the water slows down (erosion). The larger stones that were carried downstream when the stream was in spate are left on the inside of the bend where the water is moving more slowly. Note the meandering stream has cut into the alluvial soil deposited across the valley floor in earlier times

Physical weathering

- **Water**. Moving water, whether in streams, rivers or the sea, can carry particles in it, and the faster it moves the more it can carry. This leads to the rocks it flows over being abraded. As the speed of the water increases it carries disproportionally more particles, both in total quantity and size; large boulders can be bounced along in water that is in spate. It is at these times when most of the scouring of rock occurs, producing large quantities of debris which is further ground up in the moving water (Figure 10.4).

- **Wind**. In a similar way, the wind carries abrasive particles that 'sandblast' exposed rock; typically, this happens in hot dry climates where dry sand particles are readily picked up.
- **Heat**. In hotter regions, rocks are broken down as they are heated up in the sun. The rock surface expands whilst being attached to cooler rock, setting up strains within the material. This leads to surfaces flaking off; the 'onion skin effect'.
- **Frost**. In temperate areas, it is the action of frost that does much of the weathering. Water that gets into any cracks or taken up in porous rocks such as chalks and limestones expands when it freezes. The pressures set up to accommodate the expanded water bursts the rock and, like the same effect on frozen pipes, the thaw reveals the damage done.
- **Glaciers**. Ice in a glacier tends to stick to the adjacent rock, but as the glacier moves downhill the enormous forces involved pluck rock away and gradually the glacier becomes a huge mix of rock debris and ice. As this moves downhill, the large rocks embedded in the ice scour away the rock with which it comes into contact; the 'scrubbing brush' action.

Figure 10.5 **Limestone** being broken up by the roots of a plant that had established itself in a crevice

Table 10.1 Soil particle sizes

Soil particle	Diameter (mm)
Stones	> 2
Coarse sand	0.6–2.0
Medium sand	0.2–0.6
Fine sand	0.06–0.2
Silt	0.002–0.06
Clay	less than 0.002

Chemical weathering

Many of the chemicals that make up rocks react with the elements around them. Some of these minerals are soluble in water and others react with oxygen to form new compounds. Rainfall is one of the main causes of rock weathering because the water combines with carbon dioxide in the air to form **carbonic acid**. This weak acid (see below) dissolves away chalk and limestone and reacts with many other minerals making up rock, resulting in their disintegration, for example granite.

Biological weathering

This is the action of all organisms – plants and animals, large and small – that leads to rocks being broken up, from the patches of lichens growing on rocks (Figure 2.20) to the mining and quarrying that creates rock debris in which soil eventually forms.

It is the respiration process of aerobic organisms that provides carbon dioxide (see p. 119) which, when combined with water, forms the carbonic acid. This is concentrated in the plant root zone. It is particularly significant in the continued breakdown of rock, rock fragments and soil particles that tend to be protected against physical weathering because they are down below the soil surface. Roots also contribute physically by growing into cracks and opening them up; most gardeners come across this sort of damage that can be done to paths and driveways (see Figure 10.5).

The characteristics of soil formed from the weathered rock depends greatly on the proportions of the different sizes of these particles which are familiar to us as clay, silt, sand, stones and boulders. They have been classified according to their diameter (Table 10.1). The weathering process also shapes the particles which are normally angular with sharp edges when first formed but become more rounded as they are weathered further, such as river-washed pebbles and windblown sand.

Distribution of rock fragments

The loose material produced by the weathering of rocks is likely to be eroded, that is, moved to other places. The main agents of erosion are gravity, water, wind and glaciers.

- **Gravity**. Loose rock fragments on slopes move downhill. On steep slopes it falls under gravity to form heaps of rock, **'scree'** (Figure 10.6). Any soil that forms on these steeper slopes tends to be very shallow.
- On less steep slopes material is gradually moved downhill under the influence of rain splash leading to soil creep (Figure 10.7), with

deep layers of debris at the base of slopes (colluvial material).

- **Water**. As the water with its load of rock fragments slows down it is unable to carry as much. The larger particles drop to the bottom first. Typically, it is in the streams that boulders, pebbles and coarse sands can be seen, whereas the smaller particles are carried further and not dropped until the water is in the slower moving rivers. Rivers in flood spread out over the valley bottom or across the plain beyond their banks, so when the water level goes down layers of fine sand and silt are left behind along with much organic material. This gives rise to fertile '**alluvial**' soil (see Figure 10.4) which is much valued in production horticulture. A lot of the silt remains in the very slow-moving river until it is dropped around estuaries. The clay fraction is made up of such small particles that it stays suspended in the water to be carried out into the still water of a lake or out to sea where it gradually settles.
- **Wind**. The wind also carries particles that are small enough, typically sands in hot, dry conditions or in the whirling winds at the end of glaciers. Again, the larger particles are dropped first as the wind slows down (forming dunes) or when the particles touch water (thus filling ponds and lakes). Wind-blown sand leads to large areas of **loess** ('brickearth') soils much valued for production horticulture.
- **Glaciers**. Large areas of Britain and Ireland were covered by glaciers in, geologically, recent times. When glaciers melt, they drop their load which comprises everything from boulders to finely ground material ('boulder-clay'). The soils that develop in it are highly variable, and much of it is not easy to cultivate.

Figure 10.6 Scree. The rock weathered on steep hillsides falls under the influence of gravity. The eroded rock continues to break down to smaller and smaller pieces by continued physical and chemical weathering

Soil formation

A shallow layer of rock fragments on the Earth's surface remains subject to erosion until something starts to grow there. Usually, it is the

Figure 10.7 Soil creep on hillsides. On the gentler slopes, gravity causes the loose rock debris and soil particles disturbed by rain splash to move gradually downhill. This process slows as the particles are covered with grasses but continues creating the characteristic striations on the hillside

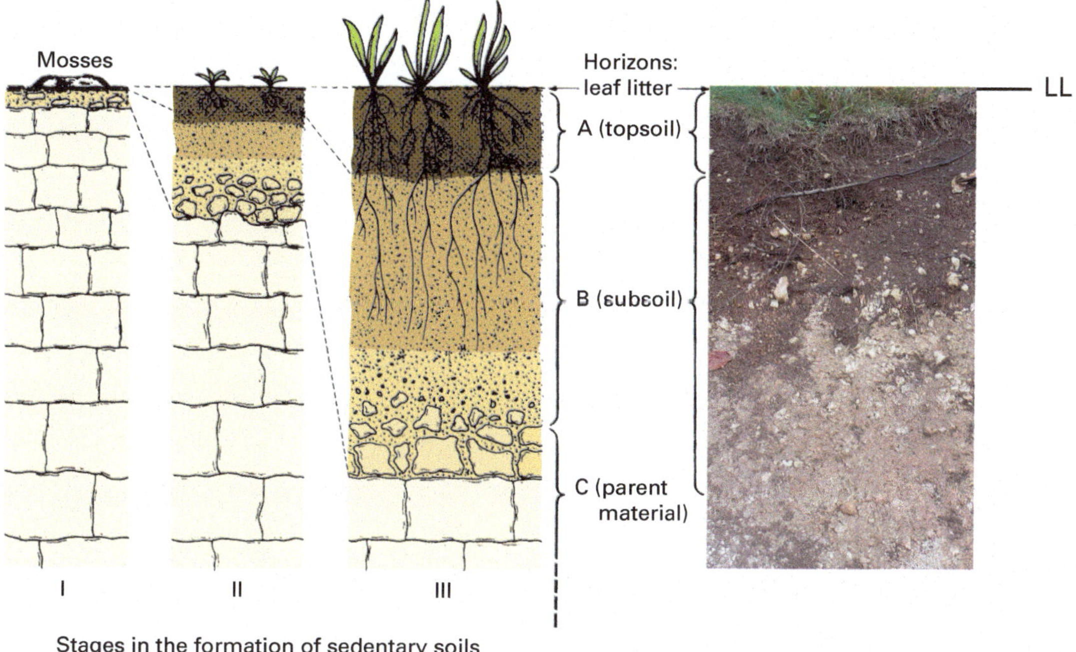

Figure 10.8 The formation of soil. The change from (I) a very young soil comprising a few fragments of rock particles to (III) a deep sedentary soil with a photograph of this final stage on the right-hand side

lower (non-vascular) plants such as lichens, algae and mosses (see p. 14) that initially colonize such inhospitable areas. They bind in the loose material and by slowing wind and water cause more particles to be dropped. Besides this material added at the surface, the soil deepens as the parent rock below continues to be weathered. Furthermore, the plants die, adding organic matter to the soil which decomposes, releasing nutrients.

A hole dug in such a 'sedentary' soil (one that develops from the underlying rock) reveals the characteristic horizons in soils; typically, an organic-rich litter layer at the surface with different horizons down to the unweathered parent rock at the bottom (Figure 10.8). Initially this may be only a few centimetres deep but over time, as the original plants are succeeded by higher plants with roots, eventually shrubs and trees (see succession, p. 140), the soil profile can become more than a metre deep. If this plant cover is removed under cultivation, it becomes vulnerable to erosion again.

> **A soil horizon** is a specific layer in the soil revealed by digging a 'soil pit'.

'Transported' soils form in essentially the same way, but in material that comes from another area, that is, not made from the rock found underneath the site but rather from particles that have been eroded and transported, often from many miles away (Figure 10.8). **Alluvial soils** form in the material brought by rivers and **loess** (brickearth) in the wind-blown deposits. Average agricultural soils develop in much of the variable '**boulder clay**' left behind by the glaciers; very little of which makes good horticultural soil unless substantially improved or if used to grow crops that need very little cultivation such as orchards, but it is the material with which many of those working in domestic and public gardens are familiar.

Organic matter is added at the surface in the form of leaves, dead annuals and perennial plant 'tops' giving rise to the 'litter layer' (also known as the 'organic layer'). This is food for many small animals such as earthworms that start the decomposition process and incorporate much of it into the layers below (see primary decomposers, p. 183). It becomes a source of nutrients for the next generation of plants. The natural development is for there to be a steady decline in the organic matter levels away from the surface until none is found below the root zone. This can be seen because the organic matter gives rise to a black 'jelly' (see humus, p. 185) that coats all the particles.

Soil development

The great variation in the **parent material** goes some way to explain the wide range of soils with all the associated complications for those trying to achieve the best plant performance whether for production, display or to deliver excellent sports surfaces. However, once formed soils undergo

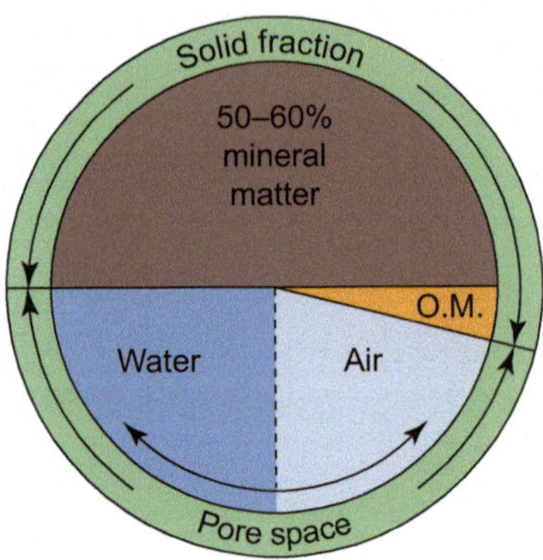

Figure 10.9 Composition of a typical cultivated topsoil; the proportions by volume of solid material (comprising mineral and organic matter), air and water are shown

significant changes over time according to the following factors:

- climate
- vegetation
- topography
- drainage conditions
- time.

Over time, characteristic soil horizons (layers) emerge reflecting the influence of these factors (see Soils of Britain and Ireland in the Support Material).

Under cultivation the soil is turned over so the organic matter concentrated in the surface layers becomes spread out more evenly to spade or plough depth. This creates a distinct boundary between 'topsoil' and 'subsoil' (see Table 10.2).

Composition of soils

Typically, about half of most soils (by volume) is solid material – mineral particles (sand, silt and clay) whose proportions vary little over time and a small amount of organic matter. The gaps between these particles make up the other half, which is filled with air and water in proportions that change significantly with the wetting and drying cycles (see Figure 10.9).

Although we understandably think of soil particles as all just too small to see, there are huge differences between coarse sand, fine sand, silt and clay particles. If a football is used to represent the size of fine sand, then coarse sand would be the size of a room in a house, whereas marbles would represent silt and sugar crystals the largest of the clay particles (most clay is hundreds of times smaller than this). This has a significant effect on root penetration, water-holding properties and aeration (see Figure 10.10) in the root environment.

Soil particles can be classified in several ways. The most commonly used methods of classifying soils by size internationally, in the USA, England and Wales, are illustrated in Figure 10.11.

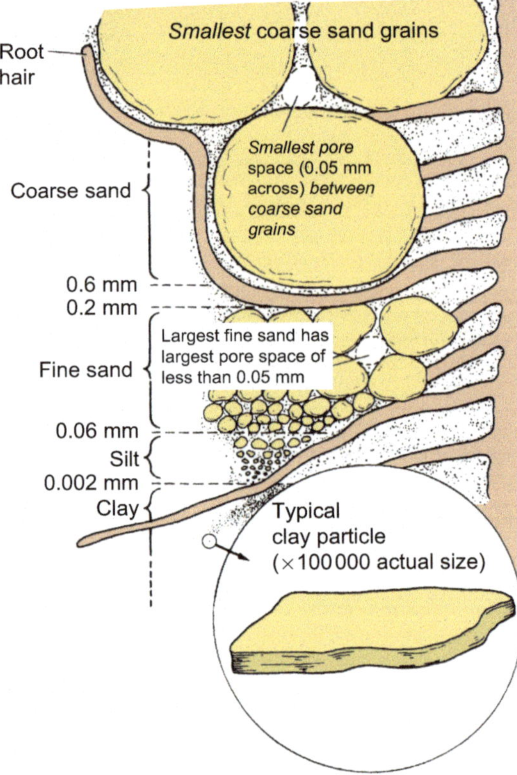

Figure 10.10 Relative sizes of sand, silt and clay based on Soil Survey of England and Wales classification (SSEW), with root hairs drawn alongside for comparison. Note that even the smallest pore spaces between unaggregated spherical coarse sand grains allow water to be drawn out by gravity and so allow some air in at field capacity, whereas most pores between unaggregated fine sand grains remain filled

Particle size classes

There is a continuous range of particle sizes but it is convenient to divide them into classes. Three major classification systems in use today are those of the International Society of Soil Science (ISSS), United States Department of Agriculture (USDA) and the Soil Survey of England and Wales (SSEW). These are illustrated in Figure 10.10. In this text the SSEW scale used by the Agricultural Development and Advisory Service of England and Wales (ADAS) is adopted. In each case, soil is considered to consist of those particles that are less than 2 mm in diameter. The silt and clay particles are sometimes referred to as 'fines'. Those larger than 2 mm are 'stones'.

Sand

Sand grains are soil particles between 0.06 and 2.0 mm in diameter.

The shape of the particles varies from gritty (angular) to the more weathered, rounded soft sand.

Colour varies according to the iron oxide coating from very pale yellow to rich, rusty reddish brown. Silver sand has no such coating.

Sand grains are inert; they neither release nor hold on to plant nutrients. They are not sticky, which makes sandy soils easy to cultivate but vulnerable to being blown when not covered with vegetation.

Soils dominated by coarse sand are usually free draining but have poor water retention, whereas those composed mainly of fine sand can hold much larger quantities of water against gravity.

Water in films clinging on to sand particles is readily removed by roots.

Clay

Clay particles are those less than 0.002 mm in diameter.

Clay particles tend to be platelets that pack together closely.

Their combination of very small size and chemical characteristics makes clay soil sticky when wet and hard when dry. Unless there are cracks between these blocks of packed clay particles, water movement is very restricted.

Many types of clay shrink on drying so cracks are introduced.

Clay soils can be opened up by cultivation or the action of soil organisms such as earthworms (see p. 182) encouraged by the addition of organic matter.

Clay particles are porous so hold water inside as well as on their surface. Consequently, clay-rich soils can hold on to a lot of water, but about 15% of it is too tightly bound to be released to plant roots (see p. 175).

As clay continues to weather it releases plant nutrients, especially potash.

Clay holds on to some nutrients in such a way that they remain available to plants but protected against being leached (washed down the profile). Once below the reach of roots, such nutrients can find their way into watercourses (see leaching of nutrients, p. 178).

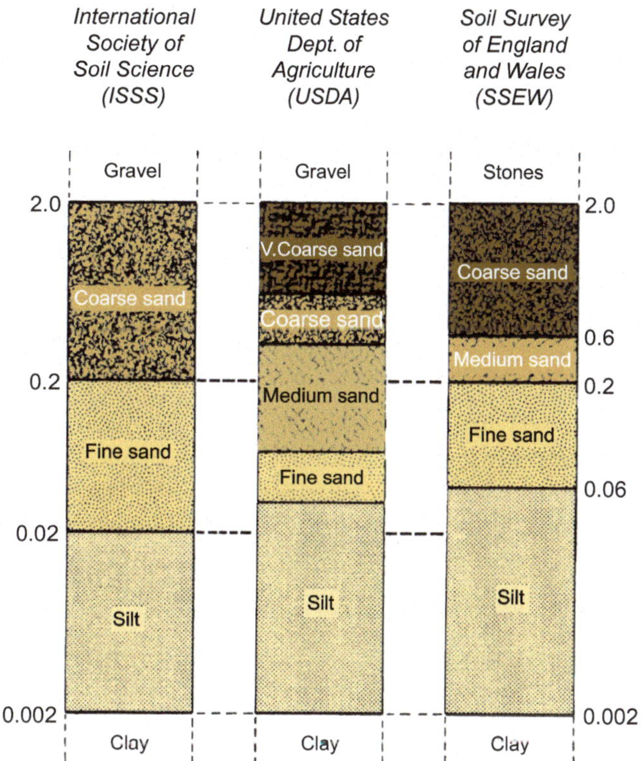

Figure 10.11 Particle size classes as defined by the International Society of Soil Science (ISSS), United States Department of Agriculture (USDA) and the Soil Survey of England and Wales (SSEW)

Further details of the characteristics of clay can be found in the Support Material.

Silt

Silt particles are those between 0.002 and 0.06 mm in diameter.

Most particles in this size range are inert and non-porous like sands, so they behave like very fine sand. They have good water-holding capacity and plants can take up a high proportion of this water (see p. 175). However, the smallest 15% has the properties of clay particles, so soils dominated by silt do yield some nutrients. Similarly, they can hold on to some nutrients like clay, but to a lesser extent.

Stones

Stones are particles larger than 2 mm in diameter.

Particles bigger than sand are stones but more commonly known as grit, gravel, pebbles, cobbles and boulders, according to size and shape. The effect of stones on cultivated areas depends on the type of stone, their size and the proportion in the soil. In general, they make soils difficult to cultivate; digging is harder and the spade is more easily blunted. They have detrimental effects on mechanized work; tines and tyres are worn more quickly especially if the stones are hard and sharp such as broken flint. Stones interfere with drilling of seeds and the harvesting of roots. Close cutting of turf is more hazardous where there are protruding stones.

A high proportion of stone reduces the fertility of the soil as it dilutes the soil components that supply nutrients and hold the water. The amount of soil that roots can explore is reduced according to the volume of stone present.

Soil texture

Soil texture describes the mineral composition of a soil. In most cultivated soils the mineral content forms the framework and exerts a major influence on it. It is normally considered to be a fixed feature, that is, not one that is altered in preparation for a forthcoming planting. As such it provides a useful guide to a soil's potential.

Soil texture can usefully be defined as the relative proportions of the sand, silt and clay particles in the soil.

In simple terms, a soil dominated by

- sand particles is called '**a sand**' and feels 'gritty' or abrasive
- silt particles is called '**a silt soil**' which feels 'silky' (or 'soapy') when wetted
- clay particles is called '**a clay**' which feels 'sticky' when wetted.

A **loam** is an idealized soil for growing because its proportions of sand, silt and clay particles are such that none of their individual properties (grittiness, soapiness and stickiness) is evident. Soils like this are relatively easy soil to manage.

In practice, when a loam is wetted and moulded between the fingers, it is usually possible to detect at least a little

- grittiness, making it a sandy loam
- soapiness, indicating a silt loam
- stickiness, indicating a clay loam.

In the hands of a skilled soil analyst some 36 different soil textures can be determined in the field by feel (see also Soil Texture in the Support Material).

The most accurate method of determining texture is to find the proportions of sand, silt and clay in the laboratory and use a 'textural triangle' to assign the texture (see Figure 10.12 Textural Triangle).

Properties associated with the different textural groups are considered to be as follows:

Soil water-holding properties. In general, fine-textured soil such as clays, clay loams, silts, silty loams and fine sands have good water-holding properties (see p. 178) but poor water movement unless improved by cultivation and/or the addition of organic matter. In contrast, coarse-textured soils (coarse sands and coarse sandy loams) have good water movement but low water-holding capacity unless improved by addition of organic matter.

Soil temperatures correlate closely with soil texture because water has a much higher specific heat value than soil minerals (see properties of water). Consequently, well-drained coarse sands warm up more quickly in the spring compared with other soils.
Darker soils, for example those enriched with organic matter, also warm up quicker than do light-coloured ones. Conversely, plants growing on darker and drier soils are more vulnerable to frost damage.

Nutrient levels. Soils with high clay content continue to release nutrients as they weather and they have good nutrient retention. In contrast, nutrients are not released from sand because most of it is inert, and those that are present in sandy soils are readily lost by leaching. This constant need for nutrients is often met by adding manures (see p. 191) which need frequent replenishment because they decompose rapidly in the well-aerated sandy soils. Hence sandy soils are often referred to as 'hungry soils'.

Ease of cultivation. The differences in the effort required to dig various soil is familiar to most gardeners. The power requirement to cultivate a clay soil is much greater than that for a sandy soil. The expressions 'light' and 'heavy' reflects the working properties rather than the actual weight of the soil, that is, only one horse required to pull a plough in sandy soil (light) but four horses needed on clay (heavy).

The texture of a soil also influences the soil structure and cultivations.

Soil structure

Soil structure is the arrangement of particles in the soil.

To provide a suitable root environment for cultivated plants, the soil must be constructed in such a way as to ensure:

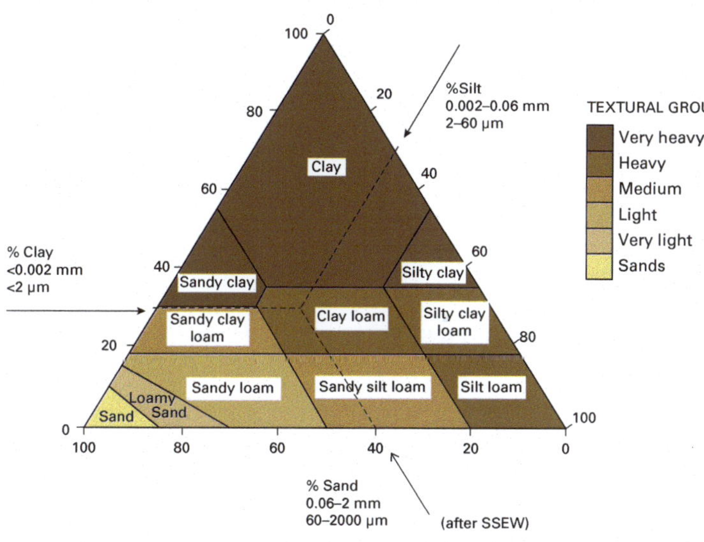

Figure 10.12 Textural triangle. The proportions of sand, silt and clay are plotted on the textural triangle and their intersection pinpoints the texture of the soil, e.g. equal parts sand, silt and clay is a clay loam.

- gaseous exchange (oxygen in, carbon dioxide out)
- adequate reserves of water available to plants.

There should be:

- a high water infiltration rate
- free downward movement of water ('drainage')
- an interconnected network of spaces allowing roots to find water and nutrients without hindrance.

There should be no large cavities that:

- dry out seeds or roots
- prevent good contact between soil and seeds or roots.

Soil crumbs

The plant roots and soil organisms live in the spaces (pores) between the solid components of the growing medium (sand, silt, clay, organic matter). In the same way that a house is mainly judged by the living accommodation created by the solid material (bricks, wood, plaster, mortar), so a soil is evaluated by examining the spaces between the particles. By aggregating the particles correctly, a range of pore sizes can be achieved that allow water holding, free water movement, gaseous exchange and thorough root exploration. This is achieved with a 'crumb soil' which is ideal for most horticultural purposes.

Tilth is the crumb structure of the seedbed.

The interior of a soil crumb is made up of many small pores which hold water against the pull of gravity, whereas the bigger gaps between the crumbs can be large enough to allow some of the water to be pulled out by gravity. This means that after being fully wetted and allowed to drain, there will be mainly water in the crumbs and mainly air between the crumbs; ideal for plant growth and for beneficial organisms living in the soil.

The best arrangement of small and large pores for establishing plants is illustrated in Figure 10.13 alongside a 'dusty' tilth with too few large pores and a 'cloddy' tilth that has too many large pores.

The fineness of a seedbed should be related to the size of seeds, so it usually consists of crumbs between 0.5 and 5 mm in diameter. Cloddy surfaces lead to poor germination of all but the largest seeds and they make weed control difficult. If made too fine, then there will be too few soil pores that hold both air and water. Fine crumbs tend to form a 'soil cap' when wetted (see p. 171) which reduces gaseous exchange and can prevent seedlings emerging.

Compaction. Good soil structure can be damaged in many ways but mainly by anything that compacts the soil thus reducing the proportion of air-holding pores. This includes the collapse of the wetted soil crumbs (soil caps) but also by the pressure of foot or vehicle traffic when the soil is not load bearing ('plastic').

(Further information on Soil Structure can be found in the Support Material.)

Soil conditioners such as manure and compost (see organic matter, p. 171) help the soil to form a good crumb structure. When fresh they can 'open up' the soil (i.e. improve aeration and drainage) and the humus created from it improves crumb formation in very sandy soil and in heavy clays (see p. 201). Lime is added to remedy soil acidity (see soil pH, p. 204) which ensures that the beneficial soil organisms are encouraged, but it also contributes calcium that encourages clay to form crumbs. When clay needs to be improved without raising pH, gypsum (calcium sulphate) can be added to the soil.

There is a great deal of concern about the continuing degradation of our soils, notably erosion and the depletion of organic matter (see Chapter 12). Further information on soil structure and soil management planning are provided in the Support Material.

Topsoil and subsoil

Once a soil has been cultivated the natural concentration of organic matter near the surface is dispersed downwards. This spreads the beneficial effect of organic matter down

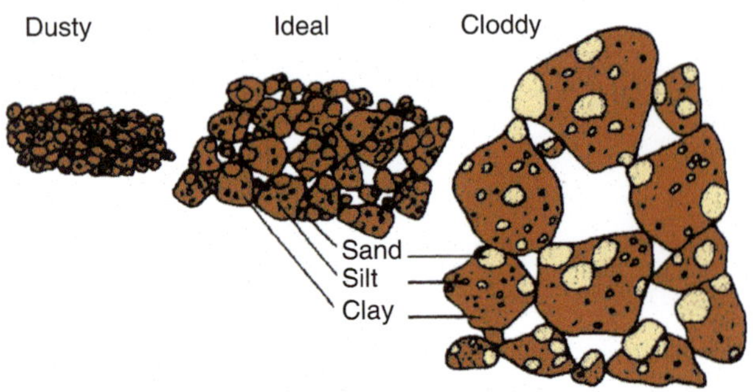

Figure 10.13 Soil crumbs – the 'ideal' arrangement of small and large pores for establishing plants is illustrated alongside a 'dusty' tilth with too few large pores and a 'cloddy' tilth that has too many large pores

to cultivation depth and gives rise to a distinct boundary between a darker **topsoil** and a lighter-coloured **subsoil**. However, the disadvantage is that this reduces the concentration of it at the surface where organic matter provides protection against the detrimental effects of weather (see soil capping, p. 171).

There are important differences between topsoil and subsoil that affect growing plants. The differences are largely because the topsoil is nearer the surface (see Table 10.2).

> **Topsoil** is the uppermost layer of soil normally moved during cultivation. It is typically 10–40 cm deep and darkened by the decomposed organic matter it contains.

> **Subsoil** is the layer below the cultivated zone. It is lighter in colour because of its low organic matter level.

Besides the burying of organic matter by cultivation, the topsoil is naturally richer in organic matter because most of the roots are near the surface, particularly in the top 15–30 cm. Also, the plant remains that fall on the surface are taken down into the top layers by earthworms and many types of insect. The presence of undecomposed organic matter and the burrowing of earthworms tend to keep the topsoil 'open', that is, with a high proportion of air spaces. The topsoil is more exposed to the effects of weather such as freezing and thawing, wetting and drying, whereas the subsoil tends to be protected from it. As it is near the surface, topsoil is within the range of cultivating equipment, but this also makes it vulnerable to compaction by feet or wheeled traffic. All the dead organic matter that accumulates there is a source of food that attracts a vast population of living organisms (see Chapter 12). The topsoil is richer in nutrients because of the decomposition of the organic matter or because fertilizer is added at or near the surface of the soil. The darker colouring is a result of the particles being coated by humus (p. 185).

Subsoil lacks a coating of humus so its colouring depends on the other chemicals on the surface of the particles, commonly iron oxide. In well-drained soils the iron oxide present is in an oxidized form giving the soil a rusty brown colour, but in waterlogged (anaerobic) conditions the iron oxide is reduced to a form that has a grey or bluish tone. Subsoils tend to accumulate the finer soil particles washed down from above, making them heavier', that is, with a higher clay content. Along with the more limited biological activity at this depth (especially of roots and earthworms), water and air movement is reduced.

Cultivations

The conventional preparation of land for planting starts with thorough disturbance of the top 15–30 cm of soil (primary cultivation), usually by digging on small plots or ploughing on larger areas. Then the final surface (the seedbed) is prepared with rakes or harrows (secondary cultivation).

Inverting the soil

Digging or ploughing inverts the soil. This is very energy demanding but has been justified by the need to incorporate crop/plant residues, weeds and bulky manures. It breaks up compacted soil and exposes the clods to weathering leading to further breakdown. Typically, single digging, no more than the depth of the spade (a 'spit') is done on an annual basis. In contrast, double digging or subsoiling is only undertaken to eliminate a deep lying soil structure problem or when a deeper rooting depth is required, such as for long root vegetables.

Forking

Forking can be adopted where loosening or breaking up the soil is the major requirement, and the burying of weeds and trash is not a priority. Forks can also be used to improve the structure of soil that cannot be dug over, such as established flower beds, taking care not to damage shallow feeding roots. The equivalent on a larger scale is the use of tined implements or discs pulled by a tractor. These have the advantage that less energy is required. On lawns a fork can be pushed into the turf to improve aeration and the infiltration of water. An improvement on this is the hand-held hollow

Table 10.2 The main differences between topsoil and subsoils in typical garden soil in Britain and Ireland

Characteristic	Topsoil	Subsoil	Further details:
Colour:	Dark brown/black; lighter if high chalk content	Light browns or grey/blue	see p. 185
Texture:		Higher in the finer soil particles, especially clay	see p. 182
Organic matter content (%):	2–5	> 1	see p. 182
– living:	Enormous numbers, especially near the surface Many roots	Comparatively low numbers Few roots	see pp. 182–185
– dead organisms:	Large quantity	Very little	see p. 184
– humus:	Present	Virtually none	see p. 185
Pore space:	Naturally more 'open' Can be maintained near 50%+ Can be improved by cultivating**	More compressed Too deep for cultivation***	see p. 209
Aeration:	High proportion of large pores so good aeration	Limited large pores so poorer aeration*	see p. 000
Water content:	Depends mainly on soil texture but improved by the organic matter content	Depends on soil texture	see Table 11.1 see p. 185
Nutrient content:	In nature, the site of nutrients Enriched by addition of fertilizers	Low nutrient content* Below main feeding roots	see Table 13.1 see Table 13.1
Suitability for plants	Can be ideal, primary source of nutrients and water	Poor,* but important water reserve	
Effects of weather	Exposed to extremes of freezing and thawing; wetting and drying	Protected from the extremes of weathering by topsoil	

Note:

* Avoid mixing with topsoil

** Cultivation depth normally considered to be a spade depth ('a spit') or plough depth

*** Reached only by subsoilers, see Support Material

tine, which is pushed into the ground to extract a 10 cm core of soil (Figure 10.14); powered versions are available for dealing with larger areas.

Raking

Raking is used on the roughly prepared ground for removing stones and debris, producing a suitable tilth, levelling and incorporating fertilizers.

There are many types of rake available. Suitable ones for meeting objectives should be selected and used appropriately, for example if levelling is the prime requirement, then a wide one should be selected.

The traditional rake is used in two main ways:

Firstly, as a means of reducing soil aggregates to the size of the crumbs required for the seedbed. For this the rake should be kept low to the ground with soil pushed back as much as forward. The impact of the tines breaks the clods which is made easier if the soil is in a friable condition (see p. 171). Weathering over the winter before starting secondary cultivation helps the working of heavier soils.

Secondly, as a means of removing stones and unwanted vegetation by using the rake as a sieve. For this the rake should be pulled through the seedbed in one direction while being held at 45°, leaving the soil crumbs behind.

Rotary cultivating

Rotary cultivators are used to create soil crumbs on uncultivated or roughly prepared ground instead of digging, forking and raking. Small pedestrian types are available for use in gardens and on allotments. There are much larger ones for use in nurseries, market gardens and commercial greenhouses. The type of tilth produced depends not only on the soil's conditions but also on the adjustment of forward speed, rotor speed, blade design and layout, shield angle and depth of working.

Creating a seedbed

Sandy soils are easily broken down to the right size with cultivation equipment but there is a risk of 'overcultivating' resulting in 'dusty' tilth. Heavier soils are more difficult to cultivate. Traditionally, clay soils are dug over (or ploughed) in the autumn and exposed to wetting and drying and especially freezing and thawing to produce

Figure 10.14 Aeration of turf (a) by hand with a simple corer and (b) removing cores with machinery

a 'frost mould'. Weathering over the winter helps the working of clay (heavier) soils in the spring. **Timeliness** (the 'right time to cultivate') is essential because not only does it make cultivation easier when the soil is **friable** but going on to the soil at the wrong time can damage the good soil structures.

Friable is the consistency of the soil when it is easily cultivated, that is, readily forms crumbs.

The right time for soil cultivation (the **'cultivation window'**) depends on the weather, but more specifically the **soil consistency** (sometimes referred to as the 'workability' of the soil). Many soils are too sticky when wet; soil clings to the equipment. As the soil dries it loses it stickiness and becomes 'plastic' (mouldable), and walking or driving on it at this stage damages the soil surface; footprints or tyre tracks are evidence of compaction. As the soil dries out further it becomes **load bearing and friable** – ideal for cultivating. However, if they dry out more the lighter soils can fragment too easily when raked or harrowed, producing a dusty tilth. In heavier soils, the clay can become too hard and tend to form 'clods' which take a great deal of effort/energy to break down to crumbs.

The delay to work on the land can be reduced by using boards to walk on when it is plastic; the pressure on the ground is spread and compaction is minimized. Equipment with very wide tyres (or tracks) does the same thing for mechanized work on a larger scale.

Once the seedbed is created, care is needed to maintain it. Good crumb structure can be destroyed by cultivating at the wrong time, but also by working the soil too much – 'overcultivation'. Soil crumbs are vulnerable to collapse when wetted especially if they are low in organic matter. Rain hitting the surface causes crumbs to break up. Puddles on seedbeds lead to crumbs collapsing. In both cases the particles released fill the gaps between remaining crumbs; on drying these can form a **'soil cap'** which is a crusty layer at the surface that reduces gaseous exchange and trap seedlings (see Figure 10.15).

General, fine tilths should be avoided outdoors until well into spring when conditions are becoming more favourable and seedling emergence through any developing cap is rapid. The surface can be protected by using mulches (see p. 188), and a leaf canopy reduces the problem. Maintaining high organic matter levels helps protect the soil (see p. 182). Increasingly, 'no dig' methods are being adopted.

When irrigating, care should be taken to avoid large droplets hitting the surface and to ensure water is not added faster than the infiltration rate.

'No dig' methods. The traditional preparation of a seedbed, involving as it does the inversion of the soil and creating the appropriate tilth, is very demanding on energy, labour and time. The resulting bare, loose soil is vulnerable to erosion. Furthermore, cultivating tends to interfere with the natural crumb-forming agents such as earthworms (see p. 182) and many other organisms. When

Figure 10.15 Soil 'capping' is caused by the collapse of the soil crumbs at the surface as a result of being hit by raindrops and/or saturated with water for long periods

done at the wrong time compacted zones on the surface ('caps') and in the soil ('pans') are produced. Above all cultivation reduces the organic matter levels near the surface where it is most useful. 'No dig' methods have increasingly been adopted to eliminate these weaknesses, especially as weeds can be controlled with herbicides. In gardens, on allotments and in horticulture generally, much of the damage caused by cultivation is avoided by using 'bed systems' or growing in containers (see Chapter 14).

Bed systems. The compaction problems can be overcome by cultivating in beds, which confines traffic to well-defined paths between the growing areas.

On a garden scale and in much of horticulture, beds are constructed so that all parts of the growing area can be reached from a path. This eliminates the need to step on the growing area. Beds can be laid out in many ways but should be no more than 1.2 m across. The paths should be minimized while allowing access for all activities through the growing season. Beds are often raised above the normal ground level to improve drainage and ease of working. Much of this extra height results from the addition of large quantities of suitable bulky organic matter along with good topsoil taken from these sacrificed areas for paths.

On a field scale the width of the 'beds' is adjusted to the distance between the wheels of the vehicles used so that the majority of the growing area is unaffected by the traffic passing during the life of the crop. The equivalent of this is done in farming and is seen from the roadside as 'tramlines' where cereals are being grown.

Further reading

Adams, C., Early, M., Brook, J. and Bamford, K. (2015) *Principles of Horticulture Level 3*. 7th ed. Routledge.

Ashman, M.R. and Puri, G. (2002) *Essential Soil Science*. Blackwell Science.

Bell, B. (2016) *Farm Machinery*. 6th ed. Old Pond Publishing.

Bell, B. and Cousins, S. (1997) *Machinery for Horticulture*. 2nd ed. Old Pond Publishing.

Bray R. (2023) *Soil Science for Beginners*. Monkey Press.

Brown, L.V. (2002) *Applied Principles of Horticultural Science*. 2nd ed. Butterworth-Heinemann.

Davies, D.G., Eagle, D.T. and Finley, J.B. (1993) *Soil Management*. 5th ed. Farming Press.

Ellis, S. and Mellor, A. (1995) *Soils and Environment*. Routledge.

Holmes, S. and Bragg, N. (2024) *A Gardener's Guide to Soil*. The Crowood Press.

Ingram, D.S. et al. (eds.) (2008) *Science and the Garden*. 2nd ed. Blackwell Science Ltd.

McIntyre, K. and Jakobsen, B. (1998) *Drainage for Sportsturf and Horticulture*. Horticultural Engineering Consultancy.

Prentice Baily, R. (1990) *Irrigated Crops and Their Management*. Farming Press.

White, R.E. (1997) Principles *and Practice of Soil Science*. 3rd ed. Blackwell Science.

www.gov.uk/guidance/create-and-use-a-soil-management-plan

The online material is accessible via the QR code and includes further information on many of the topics in the book, such as

- Soils of Britain and Ireland
- Rocks
- Clay
- Soil texture by feel
- Water is unusual
- Soil structure
- Soil management
- Soil particles

Soil water

Figure 11.1 Land made productive by regional drainage. Water is pumped up from a system of ditches around the fields to lower the water table. This was done in the past by windmills but now the power required comes from turbines or electric pumps (source: Shutterstock, Olena Znak)

This chapter includes the following topics:

- Soil water and growing plants
- The wetting of soils
- Drainage
- The drying of soils
- Irrigation/watering plants
- Water conservation

DOI: 10.4324/9781003581260-11

Soil water and growing plants

The healthy growth of a plant requires a constant supply of water, which is taken up through the roots from the growing medium (see p. 124). If there is too little water for the plants, watering (irrigation) becomes a consideration (see p. 178). However, drainage might be required if there is too much water in the soil. It is particularly important in production horticulture where cultivation can be delayed because the soil is too sticky or plastic (see soil consistency p. 171) or if plants are harmed by 'waterlogging'. Note that if the cause of the problem is in the root zone, then the bad structures need to be dealt with using cultivators or subsoilers (see Support Material).

The wetting of soils

Most rain falling on the soil surface soaks in, but if it exceeds the rate that it can infiltrate then water accumulates on the surface. This standing water (also known as 'ponding' or simply as puddles) leads to soil capping because soil crumbs tend to collapse when wet (see p. 171). Soil surfaces can be protected with mulches (see p. 000) and they become less vulnerable when covered by plants. As the soil becomes saturated (full of water, **waterlogged**) the standing water (puddles) becomes more extensive and long lasting, so damage to soil structure at the surface increases.

A saturated soil is one which has all the soil pores filled with water.

On slopes there is surface run-off. Depending on the steepness of the slope and the intensity of the rain, soil can be carried away (erosion) leaving rills that can turn into gullies with further erosion (Figure 11.2). Seeds, fertilizers and mulches can also be lost which is wasteful but is also bad for the watercourses that receive the fertilizer. Growing on steep slopes is a major problem worldwide and is commonly overcome by terracing to reduce 'run-off' as well as creating flat areas on slopes that are otherwise too steep for cultivation.

Figure 11.2 Soil erosion is a result of water 'run-off'; the moving water carries particles downhill. The faster the water moves; the more soil is carried and larger 'rills' are created eventually enlarging to 'gullies'

As water soaks into the dry soil all the pore spaces are filled with water, that is, they become **waterlogged** (saturated). The roots in this saturated zone can only obtain oxygen from what is dissolved in the water and trapped within soil crumbs.

The saturation point of a soil is when water has filled all the soil pores (i.e. no air in the pores).

Field capacity

When rainfall ceases, the water in the larger soil pores continues to move downwards under the influence of gravity. As this gravitational water (sometimes referred to as 'excess water') is removed, air returns in its place, bringing with it a fresh supply of oxygen.

Gravitational water is the water that can be moved from the soil by the force of gravity.

On sandy soils this may take a matter of hours after the rain has stopped, but far longer on clays where this process may continue for many days. The soil is then said to be at field capacity (FC).

Field capacity is the amount of water the soil can hold against the force of gravity.

The amount of water held at field capacity is known as the water-holding capacity (WHC) or moisture-holding capacity (MHC) of the soil (Table 11.1). Most soils in the lowland areas of Britain and Ireland hold about an average month's rainfall (60 mm) in a topsoil of 30 cm. Coarse sandy and gravelly soils hold far less and peaty soils hold much more.

The amount of water in a given depth of soil at FC can be calculated by simple proportion; similarly with water held at the permanent wilting point (PWP) and the available water (AW).

A more detailed description of water in the soil is given in the Support Material.

Drainage

Continued rainfall eventually brings the entire root zone to field capacity. In many soils the gravitational ('excess') water can drain naturally to

Table 11.1 Water-holding capacity of different soil textures

Soil texture	Water held in 300 mm soil depth (mm)		
	At field capacity (FC) i.e. water-holding capacity (WHC)	At permanent wilting point (PWP)	Available water (AW)
Coarse sand	26	1	25
Fine sand	65	5	60
Coarse sandy loam	42	2	40
Fine sandy loam	65	5	60
Silty loam	65	5	60
Clay loam	65	10	55
Clay	65	15	50
Peat	120	30	90

below rooting depth. Alternatively, some form of artificial drainage can be installed to achieve this. This ensures that saturation in the root zone is temporary.

Drainage is the removal of gravitational (excess) water from the rooting zone.

When more water is added to soils overlying impervious material such as non-porous rocks and many subsoils (e.g. wet clay), it leads to poorly drained (waterlogged) soils. Here oxygen only gets into the top layers of the soil during the drier parts of the year. A water table marks the level below which the soil is saturated with water. This varies over the year with the water table falling to a minimum in the summer. A hole dug in the ground will fill up with water to reveal the water table.

Saturated (waterlogged) soils can be recognized by:

- standing water and surface run-off
- much reduced working days for cultivation (see timeliness p. 171)
- grey or mottled soil colours (Figure 11.3)
- smell of hydrogen sulphide – 'bad eggs'
- indicator plants (such as rushes, mosses)
- restricted rooting
- some disease problems such as clubroot (see p. 293).

Figure 11.3 Poorly drained soils. Soils that have developed in aerobic conditions have bright orange, yellow or brown colours (top), whereas those developed in waterlogged conditions have grey, green or blue tones (bottom). Fluctuating waterlogged and aerobic conditions as the water table rises and falls leads to mottled soil (middle sample)

Wet soil problems should be tackled according to the cause or causes. Many of the symptoms of poor drainage occur when there are compacted layers in the cultivation zone. This is essentially a soil structure problem that should be dealt with by appropriate cultivation to remove the obstruction, for example digging, ploughing, forking/coring (see p. 170) or subsoiling (see Support Material). If the problem is as a result of water being held back by the conditions below cultivation depth, then artificial drainage may be installed to advantage.

Improving poor drainage by removal of gravitational ('excess') water trapped in the root zone should be undertaken before any other remedial work on a soil. It is normally achieved on large areas by laying drainage pipes such as 'clays' (Figure 11.4) or perforated plastic pipes in parallel lines under the ground about a metre deep and taking the drainage water to a ditch.

Mole drainage. An alternative method is to mole drain, which is achieved by pulling a 'bullet' through the soil when the soil is plastic at the depth of the proposed tunnel. A powerful tractor unit is required (see Figure 11.5) to pull the mole plough across the field from a ditch leaving behind a tunnel. If the topsoil is friable or dry, cracks are created by the massive tine ('ripper') which facilitate water movement from the root zone to the drain. The resultant 'soil heave' makes it unsuitable for draining areas which are meant to be level such as lawns or sports fields. This is not a suitable method for light soils (tunnels rapidly collapse) or stony soils.

Sand slitting can be used where a level surface is required (see Figure 11.6); a series of very narrow trenches backfilled with graded sand can move water rapidly away from the surface. Increasingly sports turf for golf greens, football and hockey pitches is established on graded sand with free

Figure 11.4 Drainage pipes. A range of 'clays' (earthenware pipes) is shown. Most commonly they are 30 cm long and either 75mm or 100mm diameter; these are butted up close to each other to form straight lines to intercept and carry away 'excess' water. An example of a plastic join at the top with old-fashioned pipes in the foreground

Figure 11.5 Mole plough showing the robust tine that is lowered into the ditch then drawn across the field. An expander is attached to the foot of the tine which produces a tunnel as it is pulled through the clay soil

Figure 11.6 Sandslitting. Very narrow trenches cut through the turf and backfilled with free-draining sand. This removes surface water very quickly and leaves a level surface

Figure 11.7 Soakaway back-filled with plastic crates which maximize the volume left to store water after it has been covered again. Also note the plastic drainage pipe entering the void (source: Shutterstock, theapflueger)

Figure 11.8 Swales. These wide shallow ditches intercept water coming off higher ground and allow it to soak away

drainage often combined with an underground irrigation system.

Interceptor drains. In domestic situations, the most common problem is the large amount of water running off hard areas such as terraces/patios and driveways on to borders and lawns. This can be reduced or even eliminated by using permeable material to construct them. French drains can be placed to intercept the run-off. A trench is dug to collect the water and lead it away. Ideally, it is lined with woven fibre and has clay (tile) drains, perforated plastic pipe or even 'rubble' laid in it with a slope to the discharge point. The trench is backfilled with gravel. It can be planted up if the trench is part filled with gravel and 'blinded' with sand or woven fibre before putting on a layer of soil. A ditch or a soakaway is required to receive the water (see Figure 11.7).

Note that 'byelaws' normally prevent water being discharged into any pipes servicing a house. Finding an outlet below the level of the drainage pipe is a problem in a garden surrounded by other gardens. Most people have to resort to a soakaway which is a large hole in-filled with rubble or plastic crates that can store a large volume of water while it soaks into the surrounding soil. They work very well if the bottom of the hole is permeable; much less so if it isn't when it just becomes a large water reservoir, useful until it is full. Again, these should be protected with woven fibre to prevent silting up.

Swales can be used to protect urban areas by intercepting water coming off higher ground. They are large, shallow ditches, often with vegetation in them or lined with cobbles, that slow run-off speeds and allow the water to soak away over their length (Figure 11.8).

Regional drainage. Problems occur where the water table is too high. Drains set at the desired depth are of no use because there is nowhere low enough into which to discharge the water. This can be dealt with on a regional basis by creating an artificially low water table; water collecting in the ditches around the fields is pumped up into a network of waterways that take the water away from the low-lying area. A line of windmills that provided the required energy was a familiar sight in lowland areas such as the Fens in East Anglia or in the Netherlands where such land reclamation has been undertaken (see Figure 11.1). Now water is lifted by modern pumping equipment.

The drying of soils

Water drains from the soil profile but also leaves from the surface directly (evaporation) or indirectly through plants (transpiration). The rate of drying depends on the drying capacity of the air; increased water loss will be dependent on the same factors that make up a 'good drying day' for clothes on the washing line or good haymaking weather, that is, sun (or warm air temperature) and wind.

If it is a good drying day, a wet bare soil surface gives up water very quickly. This water is slowly replaced from below and evaporation from the surface continues. The replacement water has to come from deeper in the soil and gradually the surface layers begin to dry out and give up water much more slowly. There comes a point when virtually no water can reach the surface. This happens for most soils when a layer of about 10 cm becomes dried out. This is sometimes referred to as a 'dry mulch'. So long as moist soil from below is not brought to the surface such as by forking over the soil, further water loss does not occur except through roots. Once there are plants growing in the soil, water can be taken from below the surface. Although water loss from an area covered with leaves is not as rapid as from a wet bare soil surface, it is now being lost from the whole root zone. The losses from ground covered by foliage for different lowland areas are shown in Table 11.2. Soil water deficits can be found by

Table 11.2 Potential transpiration rates. The calculated water loss (mm) from a crop grown in moist soil with a full leaf canopy based on weather data collected in different parts of Britain and Ireland.

Area	April	May	June	July	Aug	Sept
Ayr	50	80	90	80	65	40
Cheshire	50	75	80	90	75	45
Channel Isles	50	85	90	100	85	45
Glamorgan	50	80	85	85	75	45
Hertfordshire	50	80	90	95	80	45
Northumberland	45	65	80	75	60	35
N Ireland (coast)	45	70	80	80	70	40

comparing local rainfall figures with the losses of water from full crop cover that occur in that area; for example, much of south-east England loses more water from the soil than it gains from rainfall over the summer (i.e. **semi-arid**). In contrast much of the rest of England, Scotland, Wales and Ireland receives more water than is lost in each of the summer months.

As the drying of the soil in the root zone continues it becomes more and more difficult for plants to extract, and the plants' cells begin to lose turgor (see p. 124). The leaves begin to look 'stressed' and they wilt. Growth is affected because flaccid (wilted) leaves are less effective at intercepting sunlight. It also leads to the stomata closing to reduce water loss (see p. 128) which, in turn, affects photosynthesis (see p. 113). The inability of the plant to keep up with water loss may occur even in ideal soil conditions with the roots taking water up efficiently. A temporary wilt of a plant commonly occurs when there are very strong drying conditions; typically, in Britain and Ireland this occurs on summer afternoons. However, the plants recover as the temperature drops and the rate of transpiration falls. As more water is lost from the root zone it becomes more difficult for the plant to extract so temporary wilt becomes more frequent. If water is not added to the soil, there comes a point when no more water can be taken up by the plant so there is no recovery of turgor overnight. This is when the PWP of the soil has been reached. At this point most sandy soils have virtually no water left in them, whereas clay loams may have 15% left inside the clay particles and in the very smallest pores but is too tightly held to be extracted (see Table 11.1).

The permanent wilting point (PWP) is the water content of the soil when a wilted plant does not recover overnight.

Available water. This is the water that plants are able to take from a soil and that is held between FC and PWP (see Table 11.1). The amount of available water for plants can be increased by:

- greater rooting depth
- improving root exploration by eliminating poor structure
- increasing organic matter levels (see p. 185).

Roots can remove the water at field capacity very easily. However, any restriction of rooting such as poor soil structures makes wilting more likely. Water uptake is also reduced by high soluble salt concentrations, for example from excess use of fertilizers (see osmosis, p. 124), and by the effect of some pests and diseases (see vascular wilt diseases, p. 293).

Temporary wilting becomes significantly more frequent when about half the available water content has been removed. Watering/irrigating the soil before it reaches this point helps to maintain growth rates.

Available water content (AWC) is the water held in the soil between field capacity and the permanent wilting point.

Irrigation/watering plants

For most deep-rooted plants (trees and shrubs) there is usually no need for water after they are established. Ideally, once seeds and plants have been 'watered in', there should be no need for more unless significant amounts of water are lost from the soil. Plants with roots down to 50 cm in loamy soils will have access to an available water content of about 100 mm (see Table 11.1). When half of this has been lost (i.e. about 50 mm of water), it has a soil moisture deficit of 50 mm, that is, the amount of water needed to return the soil to field capacity.

The soil moisture deficit (SMD) is the amount of water needed to return the soil to field capacity.

For most lowland parts of Britain and Ireland, this is roughly a month without rain in April, 20 days in May, 15 days in June and July, and 20 days in August. If this water loss occurs at a time when water shortage affects the performance of the plants involved (the 'response periods'), the soil should be returned to field capacity in one go (so long as rain is not forecast), that is, add 50 mm of water. If a sprinkler is being used the quantity of water added can be checked as it is delivered by having a straight-sided container receiving the same amount of water as the plants: switch off as the water level comes up to 50 mm.

'Response periods' are when the plant/crop benefits from the addition of water

The general rule about adding water is to avoid 'little and often', which encourages shallow rooting and maximizes water loss (evaporation is rapid from wet surfaces and droplets in the air). Instead, hold off watering and then apply large quantities to return the soil to field capacity, that is, add the soil moisture deficit but no more because that leads to water being wasted. An alternative approach is to drip water close to those plants that benefit from continuous water supplies. This wets only a small proportion of the soil surface so losses by evaporation are minimized. The distribution of water as it soaks in depends mainly on the soil texture with very little spread in sandy soils, so delivery points need to be closer together (Figure 11.9) than for loams and clay.

Methods of applying water

Applying water should be carefully related to plant requirements, soil texture and weather conditions. The following methods are commonly used.

- Watering cans are a sound choice in many gardens. The use of 'dipping ponds' can save time waiting for cans to be filled (almost as quick as using hoses when water is being added to small areas or a limited number of plants). Care should be taken to avoid damaging the crumb structure of the soil (see p. 168) or disturbing the growing medium in containers. Consideration should be given to the use of fine roses on the cans. For watering seeds and cuttings, a fine rose turned upwards is recommended to reduce the impact of droplets.

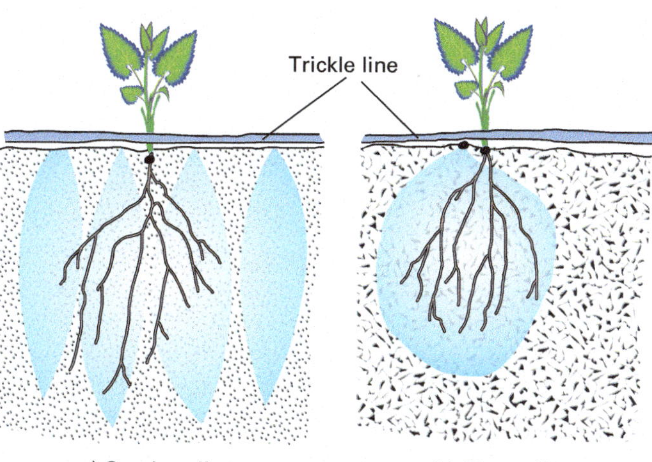

Figure 11.9 Infiltration patterns from trickle or seep irrigation – how water spreads below the surface is shown in a sandy soil (a) and a clay soil (b). Note the dry soil at the surface between the delivery points

- Hoses fitted with trigger lances with means to adjust flow rate and fineness of spray are an alternative to using watering cans. Care is needed because poor adjustment of flow can lead to damage to soil, growing media and plants, especially if a powerful jet of water is selected.
- Sprinklers become an obvious choice, when allowed, to deliver water to a large area rather than to individual plants or containers. Bare soils are vulnerable to droplet damage; large and fast-moving droplets damage soil crumbs in seedbeds as well as disturb seeds. Standing water develops when the delivery exceeds the infiltration rate, leading to soil cap formation (see p. 171). On slopes, run-off can cause erosion of soil and losses of seeds and fertilizer. There is high water loss by evaporation associated with water in the air and on surfaces. It also creates a humid atmosphere.
- Trickle lines or seep hoses (Figure 11.9) deliver water very slowly to the soil, leaving plant foliage and most of the soil surface dry, which in turn ensures a drier atmosphere and reduced water loss. However, care is needed because there is very little sideways spread of water into coarse sand, loose soil or a growing medium that has completely dried out.
- Drip irrigation is a variation on the trickle method, but the water is applied through pegged-down thin, flexible 'spaghetti' tubes to exactly where it is needed such as the base of each plant (see Figure 14.9). Typically, this is used in protected culture and minimizes water loss because no water is put in the air or on plant leaves, and only a small proportion of the growing media surface is wetted.

Water conservation

The need to manage water efficiently is a major concern in the use of this limited resource. Responsible action is increasingly supported by legislation and encouraged by the higher price of water.

There are many ways by which water use can be reduced if certain principles are kept in mind and acted upon appropriately:

- Plant selection; choose drought-tolerant rather than water-intensive plantings.
- Use recycled water. Whenever possible, the direct use of mains water should be avoided. Much grey water from horticultural units can be cleaned by passing through reedbeds.
- Rainwater capture (from the roof of houses, sheds and greenhouses) is an important

consideration in the choice of water source. If it is to be used for seedlings, then care needs to be taken to ensure that it is clean (sterilized).

- Minimize evaporation of water. This is best achieved by not spraying water into the air and by minimizing the time when the soil surface is moist. When water does have to be applied overhead, this should be undertaken in cool periods of the day.
- Increase the water reservoir of the soil. The application of water can be reduced by increasing the available water for the plants by increasing rooting depth, improving soil structure, adding organic matter and/or adding water-holding gel supplements.
- Improve soil structure. Help roots explore the maximum amount of soil. Compacted layers in the soil (soil pans) should be eliminated and good soil structure maintained to increase the effective rooting depth.
- Plant root system. Plants should be encouraged to establish as quickly as possible but, after the initial watering-in, infrequent applications will encourage the plant to put down deeper roots by searching for water.
- Water loss is much reduced by reducing wind ideally by the use of semi-permeable barriers such hedges.
- Minimize water lost by adding no more than the soil can hold. Avoid losses of water from the root zone which wastes water but also leads to leaching of soluble nutrients which then pollutes watercourses. Adding water to an outdoor soil (irrigation) should be to field capacity and no more. Avoid adding the full soil moisture deficit when rain is imminent.

Further reading

Adams, C., Early, M., Brook, J. and Bamford, K. (2015) *Principles of Horticulture Level 3*. 7th ed. Routledge.

Ashman, M.R. and Puri, G. (2002) *Essential Soil Science*. Blackwell Science Ltd.

Baily, R. (1990) *Irrigated Crops and Their Management*. Farming Press.

Bell, B. and Cousins, S. (1997) *Machinery for Horticulture*. 2nd ed. Old Pond Publishing.

Brown, L.V. (2008) *Applied Principles of Horticulture*. 3rd ed. Butterworth Heinemann.

Castle, D.A., McCunnell, J. and Tring, I.M. (1984) *Field Drainage: Principles and Practice*. Batsford Academic.

Culpin, C. (1992) *Farm Machinery*. 2nd ed. Blackwell Science.

Davies, D.G., Eagle, D.T. and Finley, J.B. (1993) *Soil Management*. 5th ed. Farming Press Books and Videos.

Ellis, S. and Mellor, A. (1995) *Soils and Environment*. Routledge.

Hope, F. (1990) *Turf Culture: A Manual for the Groundsman*. Cassell.

Ingram, D.S. et al. (eds.) (2008) *Science and the Garden*. 2nd ed. Blackwell Science Ltd.

McIntyre, K. and Jakobsen, B. (1998) *Drainage for Sportsturf and Horticulture*. Horticultural Engineering Consultancy.

Munns, D.N. and Singer, M.J. (2005) *Soils: An Introduction*. 6th ed. Prentice Hall.

The online material is accessible via the QR code and includes further information on many of the topics in the book, such as

- Water is unusual
- Water in the soil
- Water potential

CHAPTER 12

Organic matter in soils

Figure 12.1 Garden compost featuring the original ingredients and the final product with the brandling worms that play an important part in the process

This chapter includes the following topics:

- Organic matter in the soil
- Role of living OM
- Dead organic matter
- Humus
- Decomposition of organic matter
- The carbon cycle
- The nitrogen cycle

DOI: 10.4324/9781003581260-12

- Mulching
- Composting
- Organic fertilizers and compost teas
- Bulky organic matter including green manures

Organic matter in the soil

Organic matter in its many forms has important effects on soil structure (see p. 167), water availability (see p. 178), soil colour and the cultivation window (see p. 171).

Most mineral soils in temperate areas contain just 2% to 5% organic matter. The main types of organic matter in the soil are:

- living organisms
- dead organic matter
- humus.

Soil organic matter is derived from living organisms such as plants, earthworms, insects, fungi and bacteria. On death these plant and animal remains decompose and are recycled along with any other organic matter that is added. The remains of cultivated plants are incorporated directly by digging them in or indirectly having been used as a mulch or been composted. Green manuring involves growing suitable plants specifically for this purpose. Alternatively, more bulky organic matter can be added in the form of material imported from elsewhere such as composted municipal waste, farmyard manure (FYM), mushroom compost, leaf mould, chipped bark and composted straw.

Living organisms

Most living organisms in the soil contribute to the decomposition of organic matter and the formation of humus, but many also affect soil structure by moving soil and creating a network of interconnected tunnels, thereby improving aeration and water movement.

Saprophytes are organisms that live on dead plant material.

Benefits of living organisms include:

- plant and animal debris is converted to minerals ('plant food/nutrients') and humus
- soil structure is improved by the activity of plant roots, earthworms and other burrowing organisms
- the formation of soil crumbs is helped by the 'stickiness' of bacterial exudates
- the bacteria *Rhizobia* and *Azotobacter* spp. fix gaseous nitrogen
- detoxification of harmful organic materials such as pesticides is undertaken by many bacteria species.

Plant roots move soil as they penetrate the soil and grow. Plants exude a 'slime' containing sugars which makes a thin (2–3 mm) layer that is a very attractive zone for other organisms (see rhizosphere, p. 184). When they die, roots leave large quantities of organic matter in the soil as well as the tunnels they have created which are stabilized by the coating of their decomposed remains. Roots take up water, and as clay and clay loam soils dry out they shrink and crack thereby helping to develop and improve their structure. A good demonstration of the effect of abundant roots on soils is seen under long-term grasslands.

Earthworms affect soil structure directly as they create a network of burrows which significantly improves aeration and water movement. This activity is particularly important in uncultivated areas including 'no-dig' cultivation (see p. 171) and organic growing (see p. 9). Many species feed on soil and digest the organic matter in it before returning the waste into the tunnel they have created. The crumbly nature of this soil, which is now an intimate mix of partially digested organic matter and mineral matter, is evident from the worm casts left by those that cast on the surface. Other species, including those that are important in composting, only eat organic matter. Earthworms play an important part in incorporating leaf litter from the surface, often dragging whole leaves underground (see Figure 12.2). Only two species regularly cast on the surface; unfortunately, these cause problems on turf areas, lawns and sports surfaces. Left unchecked on fertile soil as much as 5 mm soil can be left all over the surface each year. It is this that also leads to the burying of stones and, eventually, even ruined buildings. Further information on earthworms can be found in the Support Material.

Earthworm activity and distribution is largely governed by the following factors:

- **Organic matter**: soils with low organic matter levels support only small populations of worms.

Figure 12.2 Earthworm cast on the surface which is a mix of organic matter and finely divided soil. Most casts are left underground. Those on the surface can be a problem on fine turf. Earthworms play a major role in burying organic matter such as leaves

Figure 12.3 (a) *Lumbricus terrestris* (the common earthworm) that collects organic matter from the surface and takes it down into the soil; (b) *Eisenia fetida* (brandling or tiger earthworms) that live mainly in the litter layer

In contrast, compost heaps and stacks of farmyard manure have high populations. In oak and beech woods where the fallen leaves are palatable to worms, their populations are large and they remove a high proportion of the annual leaf fall.

- **Soil texture**: light and medium loams support a higher total population than clays, peat and gravelly soils. Earthworms tend to avoid coarse sandy soils.
- **Moisture levels**: earthworm numbers decrease in dry conditions, but they can take avoiding action by burrowing down to moister soil.
- **Temperature**: each species has its optimum temperature range; for *L. terrestris* and many others this is about 10°C, which is typical of soil temperatures in the spring and autumn in Britain and Ireland.
- **Soil pH (see p. 201)**: most thrive in soils around neutral (pH 7) and are abundant in soils where there are good reserves of calcium (chalk). Earthworm populations are usually lower in the more acid soils.

There are ten common species of earthworm in Britain, some of which are less than 3 cm long when fully grown. The largest, *Lumbricus terrestris*, which can be in excess of 25 cm (see Figure 12.3), is the organism mainly responsible for the burying of large quantities of litter by dragging plant material down its burrows. Casting species of earthworms are those that take in soil as well as organic matter and their excreta consist of intimately mixed, partially digested, finely divided organic matter and soil. Some species never produce casts; they eat only organic matter. These include *Eisenia fetida* (brandling or tiger earthworms) that are familiar to those with compost heaps.

Slugs and snails move on a single broad foot and have a rasping 'tongue'. They are generally seen as pests in horticulture (see p. 261). However, they are primary decomposers in soils and the process of composting.

Arthropods such as woodlice, millipedes and centipedes as well as insects (notably springtails and beetles) play an important part in decomposing and burying the litter layer as well as similar activity in the soil below it.

Fungi (see p. 29). Most fungi live saprophytically on soil organic matter. There are many that are tolerant of acid soils and they are responsible for much of the decomposition of organic matter in these conditions. Most are well adapted to survive in dry soils but few thrive in very wet conditions. Some fungi species are amongst the few organisms that can digest and break down wood. The white rot fungi are very unusual as they can digest lignin in wood leaving the white cellulose, for example bracket fungi and bootlace or honey fungus (*Armillaria* spp. see p. 300). Soft rot fungi attack and break down the cellulose in wood and along with brown rot fungi (formerly 'dry rot') can damage wooden structures.

Bacteria (see p. 301) are present in soils in vast numbers. More than 1000 million occur in each gram of fertile soil. Consequently, despite their microscopic size, the top 150 cm of fertile topsoil carries about one tonne of bacteria per hectare. There are thousands of different species of bacteria to be found in the soil and most play a part in the decomposition of organic matter and the release of plant nutrients. They are particularly numerous in the 'slime' exuded by plant roots. This 'slime' plays an important part in crumb formation, especially in sandy soils. Bacterial activity eventually leads to the creation of humus that also improves soil crumb development and stability (see p. 168).

Most of the soil bacteria are inactive at temperatures below 6°C, but their activities increase with rising temperature up to a maximum of 35°C. Actively growing bacteria are killed at temperatures above 82°C, but several species can form thick-walled resting spores under adverse conditions. These spores are very resistant to heat and can survive temperatures up to 120°C . Partial sterilization of soil or compost can kill the actively growing bacteria but not the bacterial spores. The growth rate and multiplication depend also upon the food supply. High organic matter levels support high bacteria populations so long as a balanced range of nutrients is present. Bacteria thrive in a range of pH 5.5–7.5.

Actinomycetes (actinobacteria) are bacteria with 'fungi-like' characteristics such as forming a mycelium. They are abundant and widely distributed in soil and compost with a pH range 6.5 to 8.0. In decomposing organic matter many are able to tolerate temperatures up to 65°C. The population of actinomycetes increases with depth of soil. They help decompose a wide range of organic matter, notably some of the more resistant substances, giving rise to many of the dark brown and black colours of humus.

The **rhizosphere** is a zone in the soil that is highly influenced by roots. Living roots change the atmosphere around them by using up oxygen and producing carbon dioxide (see respiration, p. 119). Furthermore, roots exude a 'slime' made up of a variety of organic chemicals that are rich sources of food for many microorganisms. Some of these compounds hold water and form a coating that bridges the gap between root and nearby soil particles. Microorganisms occur here in greatly increased numbers and are more active when in proximity to roots. Some actually invade the root cells, where they live as **symbionts**; *Rhizobium* spp. live symbiotically with many legumes (see nitrogen cycle, p. 186).

Symbionts are organisms that live together either parasitically ('parasitism') or to mutual advantage, 'mutualism'.

A **mycorrhiza** is a fungus living at least partly within the roots of a plant, to mutual advantage.

Within a rhizosphere there are usually mycorrhizae; fungi living intimately with the plant root to mutual advantage (Figure 12.4). There is considerable interest in exploiting the potential of **mycorrhizae**, which appear to be associated with a high proportion of plants, especially in less fertile soils.

In this relationship the fungus obtains its carbohydrate requirements from the plant. In turn, the plant gains greater access to nutrients in the soil, especially phosphates, through the increased surface area for absorption and because the fungus appears to utilize sources not available to higher plants. Most woodland trees have fungi covering their roots and penetrating the epidermis. Orchids and heathers have an even closer association in which the fungi invade the root and coil up within the cells. The association appears to be necessary for the successful development of their seedlings. Mycorrhizal plants generally appear to be more tolerant of transplanting and this is thought to be an important factor for orchard and container-grown ornamentals. Mycorrhizae are available to gardeners as well as growers to establish plants more reliably and are retailed in pelleted form.

Figure 12.4 Mycorrhizal structures in the soil (source: Shutterstock, green scent)

Dead organic matter

This is the source of food for the living organisms. In its undecomposed state it 'opens up' the soil, that is, it creates bigger gaps between the soil aggregates (see p. 168), improving air and water movement. As it decomposes it yields plant nutrients (see p. 196). In general, 'green' (succulent, leafy) organic matter decomposes very rapidly, if conditions are right. Consequently, it tends to have just a short term physical effect but yields nutrients, especially nitrogen compounds. The 'brown' (fibrous/woody) plant material tends to decompose very slowly, so its physical effect persists, but the nutrient contribution is low. The distinction between the 'green' and 'brown' organic matter is a crude but a useful one when selecting materials for composting (see p. 188).

Benefits of dead organic matter in the soil include:

- microbial activity is increased because it is food for soil organisms
- soil is physically opened up and aeration improved by the dead but recognizable organic matter
- water-holding capacity of the soil is improved by fine (unrecognizable) organic matter
- a slow-release source of plant nutrients as the organic matter decomposes and releases minerals.

Humus

Large quantities of organic matter yield tiny amounts of humus, which is one of the end products of decomposition in aerobic soils. This black jelly coats soil particles and darkens the soil where it is present. It is also sticky when wet so it helps in the creation of crumb structure especially in soils where there is little clay to bind particles together. It also forms clay-humus complexes which helps heavy (i.e. high clay) soils to crumble more readily. Like clay particles, it plays an important part in holding on to the nutrients that would otherwise be leached from the soil profile whilst continuing to release them to plants.

Benefits of humus in the soil:

- in sandy and silty soils it helps to form stable crumbs
- in clays the clay-humus complexes formed make them less sticky and more friable
- helps reduce the leaching of some nutrients from the root zone
- soils darkened by the humus absorb more of the sun's radiation
- improves water-holding capacity.

Decomposition of organic matter

As in any other plant and animal community, the organisms that live in the soil form part of a food web (see p. 138). The organic matter derived from dead plants and animals of all kinds is digested by a succession of species: large animals by crows, large trees by bracket fungi, small insects by ants, roots and fallen leaves by earthworms, mammal and bird faeces by dung beetles, and so on. Subsequently, progressively smaller organic particles are consumed by millipedes, springtails, mites, nematodes (eelworms), fungi and bacteria to leave, eventually, just water, carbon dioxide, minerals (including plant nutrients) and humus (humification). This recycling process makes available water, carbon dioxide and minerals to a new generation of plants. The production of minerals from dead organic matter is sometimes referred to as mineralization. The minerals include plant nutrients such as ammonia, nitrates, phosphates, potash and sulphates that are readily taken up by plants from the soil solution (see Plant Nutrition Chapter 13).

The type of living organisms, numbers and their activity depends greatly on:

- food supply (organic matter)
- moisture/water content
- temperature
- pH (see p. 201)
- air (oxygen) levels.

In general, decomposition is favoured by warm, moist and neutral conditions, for example earthworms are most active and abundant in neutral, moist, aerobic, organic matter–rich soils when temperatures are about 10°C (i.e. spring and autumn). In **anaerobic** (oxygenless, e.g. waterlogged), cold and/or very acid areas there are fewer and completely different organisms; decomposition is slower and the organic matter is not fully broken down so it accumulates, leading to the development of peaty soils or peat bogs. More detail is provided in the Support Material.

The carbon cycle

Green plants obtain their carbon from the carbon dioxide in the atmosphere, and during the process of photosynthesis they are able to fix the carbon, converting it into sugar (see p. 113). Sugar is the basic building block for the manufacture in the plant of all its carbon-based compounds, such as starch, cellulose, lignin, proteins, oils and so on. Carbon is locked up in these compounds until released as carbon dioxide. In aerobic conditions, about three-quarters of this carbon is returned to the atmosphere by the green plants themselves during respiration to provide their energy needs (see p. 119). Plant material that has not decomposed (i.e. preserved) acts as a carbon store, and there are huge quantities that have accumulated over time as fossil fuels (coal and oil) and peat. This carbon is released as carbon dioxide if burnt or provided with the conditions for decomposition. In this way carbon is stored in wood (e.g. forests) and peat until burnt as fuel. Peat that has been reclaimed for cropping (drained and limed) releases the carbon that has been stored. Likewise, the vast quantities of carbon stored in soil as organic matter is released when cultivation makes conditions ideal for decomposition; hence the interest in minimal cultivation (e.g. 'no-dig' methods, p. 171) because it keeps more carbon stored. Huge quantities of carbon are released as carbon dioxide when land is brought into cultivation for the first time and even when long-term turf is ploughed up.

As the plant material decomposes, each organism in the food web uses some of the carbon taken in as a source of energy until eventually it is all released as carbon dioxide, as illustrated in Figure 12.5.

The sugars, cellulose, starch and proteins in **succulent** plant tissue, as found in young plants, are rapidly decomposed to yield plant nutrients (minerals containing essential elements), water and carbon dioxide. This 'green' material has only a short-term effect. In contrast, the 'brown' (woody, fibrous, lignin rich) tissue of older plants rots more slowly, but eventually gives rise to **humus**, a complex of organic acids that take a hundred years or more to decompose.

Plants grown in the vicinity of vigorously decomposing vegetation (e.g. cucumbers on straw bales) live in a carbon dioxide-enriched atmosphere. Organic materials such as paraffin or propane that do not produce gases harmful to plants when burned cleanly are used in protected culture for **carbon dioxide enrichment** (see p. 114).

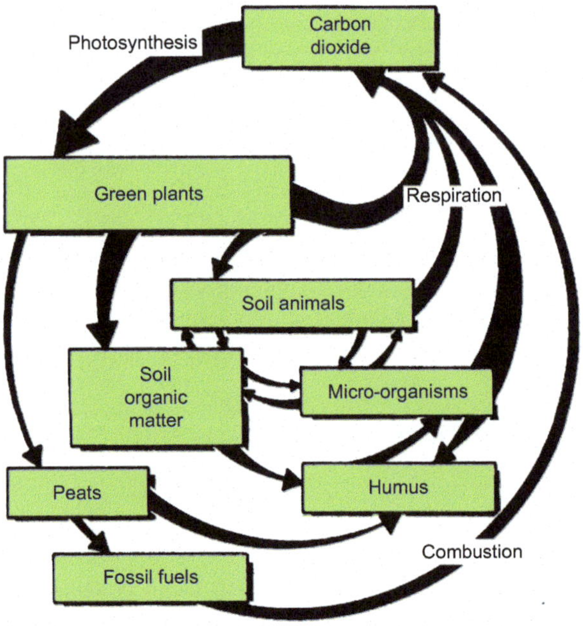

Anaerobic digestion

In anaerobic conditions (no oxygen) there are different organisms present. The decomposition of the organic matter proceeds very differently and the end products usually include methane. This is a feature of marshes or bogs where methane ('marsh gas') appears – seen as 'will o' the wisp'. Using special equipment, it makes an attractive alternative to the normal composting of plant waste if the methane can be collected and utilized.

The end products also include the production of many organic acids that slow down the activity of beneficial bacteria. This is how acid peat bogs (dominated by sphagnum mosses) originate (see p. 210). Unlike the normal decomposition process, it proceeds very slowly and the organic materials in the bog can be preserved for hundreds of years.

Figure 12.5 The carbon cycle. The recycling of the element carbon is illustrated. Note how all the carbon in organic matter is eventually released as carbon dioxide by respiration or combustion. Green plants convert carbon dioxide, by photosynthesis, into sugar which forms the basis of all the organic structures and energy needed by the plant but also animals and microorganisms (source: Dr E.G. Coker)

The nitrogen cycle

The nitrogen cycle follows the fate of nitrogen in its many forms in the plant, the soil and the atmosphere. Although plants live in an atmosphere largely made up of nitrogen, they cannot utilize gaseous nitrogen. They require a source of soluble nitrogen, and it is taken up from the soil water as:

- nitrates
- ammonium.

Both of these are derived from proteins by a chain of bacterial reactions as shown in Figure 12.6.

Ammonifying bacteria (many different species) digest soil organic matter and convert the proteins to ammonia which readily dissolves in soil water. The ammonia from this source or from inorganic fertilizers is toxic to organisms at quite low concentrations. However, it is converted to **nitrates** by **nitrifying bacteria**. This nitrification is accomplished in two stages. Ammonia is first converted by *Nitrosomonas* spp. to nitrites which are even more toxic than ammonia, but, again, this is normally rapidly converted to nitrates by *Nitrobacter* spp. before they reach harmful levels. Ammonifying and nitrifying bacteria thrive in aerobic conditions.

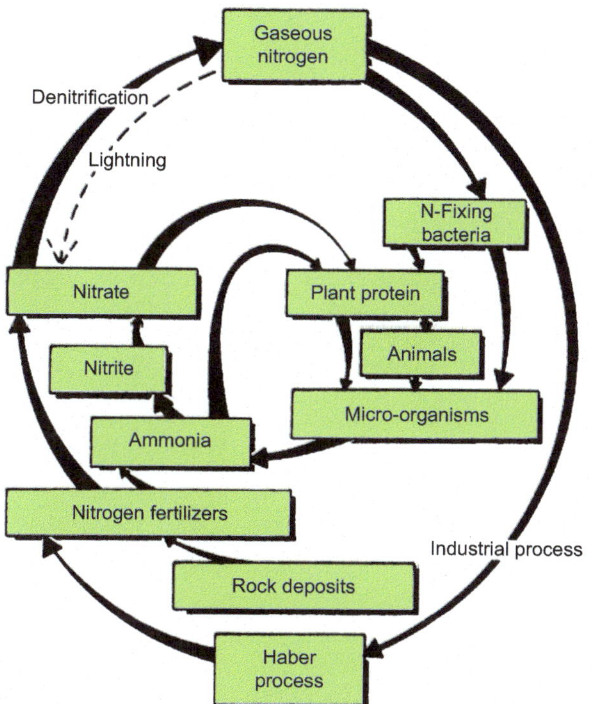

Figure 12.6 The nitrogen cycle. The recycling of the element nitrogen is illustrated. Note the importance of nitrates and ammonia that can be taken up and used by plants to manufacture proteins. Microorganisms also have this ability, but animals require nitrogen supplies in protein form or as amino acids. Gaseous nitrogen only becomes available to organisms after being 'fixed' (captured) by nitrogen-fixing organisms or via nitrogen fertilizers manufactured by humans. In aerobic soil conditions, bacteria convert ammonia to nitrates (nitrification), whereas in anaerobic conditions nitrates are reduced to nitrogen gases (denitrification) (source: Dr E.G. Coker)

Where there is no oxygen, anaerobic organisms dominate. Many anaerobic bacteria utilize nitrates and in doing so convert them to **gaseous nitrogen**. This **denitrification** represents an important loss of nitrate from the soil, which is at its most serious in well-fertilized, warm and waterlogged land.

Nitrogen fixation. Although plants cannot utilize gaseous nitrogen, it can be converted to plant nutrients by some microorganisms. *Azotobacter* spp. are free-living bacteria that obtain their nitrogen requirements from the air. They release ammonia into the soil while active and again when they decompose. The *Rhizobia* spp. that live in root nodules on some legumes (Figure 12.7) also fix nitrogen to the benefit of the host plant.

Finally, nitrogen gas can be converted to ammonia industrially in the **Haber–Bosch process**, which is the basis of the artificial nitrogen fertilizer industry. Besides the associated pollution of the process, it has a very high energy requirement so there is increasing interest in finding alternative approaches to using large applications of artificial nitrogen.

Figure 12.7 Rhizobium nodules on legume

Carbon to nitrogen ratio (C:N)

All nutrients play a part in all nutrient cycles simply because all organisms need a similar range of nutrients to be active and grow. Normally there are adequate quantities of nutrients, with the possible exception of carbon or nitrogen, both of which are needed in relatively large quantities. A shortage of nitrogenous material would lead to a hold-up in the nitrogen cycle, but would also slow down the carbon cycle, that is, the decomposition of organic matter is slowed because the microorganisms concerned suffer a shortage of one of their essential nutrients. A useful way of expressing the relative amounts of the two important plant foods is in the carbon to nitrogen (C:N) ratio.

The most common problem associated with an imbalance of carbon and nitrogen is that a shortage of nitrogen slows the decomposition of organic matter. This occurs in the compost heap when the mixture available has a high C:N ratio, usually because there is too much 'brown' (fibrous or woody) material. The organisms involved use up the nitrogen and then become inactive (**N 'lock up'**) despite there being plenty of carbon material to provide their energy. This can be resolved by adding compost activators/accelerator which are essentially sources of nitrogen. In the soil the same process occurs, but if the nitrogen in the plant material is used up the bacteria can continue by using other sources of nitrogen in the soil. This is known as 'soil robbing' as the nitrogen that would otherwise be used by the growing plants becomes 'locked up' in the soil organisms. Adding any materials with high C:N ratio such as chipped bark or straw, whether dug in or as mulches, can lead to 'locked-up' nitrogen so precautions need to be taken when appropriate.

Nitrogen in soluble form is released during decomposition if the organic material has a C:N ratio narrower than 30:1, such as young ('green') plant material or nitrogen-enriched plant material such as farmyard manure.

Mulching

Mulches are materials added to the soil surface in order to provide one or more of the following:

- maintaining/increasing soil organic matter
- decorative finish (e.g. chipped bark on borders)
- weed suppression
- moisture retention
- stimulating beneficial soil organisms
- modifying soil temperatures (insulation)
- protecting edible crops from soil contact/splash (e.g. straw under strawberries).

Mulches are materials applied to the surface of the soil to suppress weeds, modify soil temperatures, reduce water loss, protect the soil surface and/or reduce erosion.

Many organic materials are used as mulches including farmyard manure, garden compost, mushroom compost, composted municipal waste, leaf mould, chipped bark, composted straw and green manure. There are several non-organic materials used for mulches including minerals (pebbles, slate, stone chippings), tumbled glass and sheets such as woven polypropylene/fibre, polythene, paper and old carpet.

Except for the sheet mulches (woven fibre, polythene, paper, carpet), the materials need to be laid thickly to be effective; organic materials need to be at least 5 cm deep. Most can be applied at any time when the soil is moist in accordance with their function, but those that are insulatory should be applied after the soil has warmed up. Care should be taken to remove perennial weeds before applying and to keep mulch away from the base of woody stems.

Composting

Compost is a dark, soil-like material made of decomposed organic matter (see Figure 12.1). Most gardeners depend on composting as a means of recycling their garden and kitchen waste to maintain organic matter levels in their soils. Many councils now collect 'green waste' and supply composting equipment to encourage householders to recycle garden waste as well as their paper, glass and metals. This is not as environmentally friendly as home composting as there is significant transport involved. Horticulturists are increasingly concerned with the recycling of waste and attention is being given to composting methods to deal with the large quantities of material generated on their units. Increasingly, green waste is used to generate energy to use on site and the surplus exported to the grid. Composting is fundamental to successful organic growing (see p. 9).

Composting is the decomposition of organic matter (including plant residues) in a heap before they are applied to soils.

Conditions must be favourable for the decomposing organisms (see p. 185) for successful composting:

- **Air (oxygen)**. The beneficial organisms are aerobic (see p. 119) so they require well aerated conditions throughout. In practice this means 'turning' the contents to loosen the contents which tend to compact.
- **Water**. The decomposers require moist material to live in and digest – most are inactive when too dry. If too much water is added the air (oxygen) is driven out (in waterlogged conditions the anaerobic organisms take over leading to poor compost). Once the heap is moist enough, a roof/cover is advantageous by keeping out rain (excess water) and retaining heat.
- **Organic matter mix**. To achieve a mix that provides the right balance of nutrients, water and aeration, it is convenient to distinguish between 'green' (leafy/succulent/tender) and 'brown' (fibrous/woody/tough) materials and combine them in approximately equal measure (Table 12.1).
- **Accelerators or activators**. These are essentially nitrogen fertilizers and are not normally needed in garden composting. Also available but even less likely to be needed are materials that make good a shortage of 'brown' material such as sawdust.
- **Shredding**. The rate of decomposition is speeded up by reducing the size of the material put on the compost heap. Shredding increases the surface area so making more of the material accessible to the organisms. The degree of shredding should be such that aeration is maintained and waterlogging is avoided.
- **pH** (see p. 209). The compost mix should not be too acid. Thin layers of lime (see p. 204) can be added as the heap is built.
- **Temperature**. The rate at which the organisms decompose organic matter also depends on the temperature of their environment. The composting process gives off heat (exothermic reaction); under ideal circumstances the

Table 12.1 Compost ingredients

Proposed ingredient	Category
Cardboard	Brown
Farmyard manures	Intermediate
Fibrous prunings	Brown
Haulm (old plants)	Brown
Hedge clippings	Brown
Herbaceous plants (old)	Brown
Grass mowings	Green
Grass – long	Brown
Kitchen (plant) waste	Green
Leaves – young	Green
Autumn leaves	Brown
Nettles – young	Green
Nettles – old	Brown
Paper	Brown
Sawdust	Brown
Seaweed	Green
Straw	Brown
Woody prunings	Brown

temperature can rise to over 70°C within seven days, a high enough temperature to kill harmful organisms and weed seeds. However, most garden composting temperatures do not get near this this level.

- **Heap size**. Heat is generated within the volume of the heap by the exothermic process, but heat is lost from the surface. Consequently, the degree to which the compost heats up depends on the surface area to volume ratio. Small heaps (less than a cubic metre) have a disproportionally large surface area which dissipates the heat generated within the small volume too quickly to allow the heap to heat up ('cold composting'). Insulation reduces heat loss from the surfaces but this also reduces the flow of air (oxygen) into the heap.

Organic waste brought together in large enough quantities under ideal conditions heats up quickly, but the process then tends to slow down because the availability of oxygen becomes reduced. Rather like stirring up the dying embers of a fire (i.e. getting oxygen around unburnt fuel), decomposition rates can the restored by 'turning' the compost heap on a regular basis to improve aeration. This continues until all that is left is a crumbly material with no recognizable plant material and an 'earthy' smell.

Garden (home) composting. Most gardeners are not usually able to obtain enough components at any one time to create the ideal compost heap. It is difficult to be successful with batches less than one cubic metre at a time. This is a 'cold' method that can produce good compost but tends to take many months or even a year or two to complete. When making compost this way, care should be taken to avoid perennial weeds or infected plant material which is unlikely to be rendered harmless.

Figure 12.8 Compost bins: a typical set large for efficient composting; slatted to allow in air and removable to allow easy access to add new material and for regular turning of the contents

Compost bins can be purchased but slatted wooden sided bins can easily be made (see Figure 12.8). An open base over soil allows organisms and air in. It is advantageous to have a second bin alongside so that the compost heap can be turned from one to the other frequently (slats between the two should be removable to make 'turning' easier). A suitable cover (e.g. old carpet) keeps heat in and rain off.

Compost tumblers are containers that can be rotated on an axis to provide an easier method of turning small batches to create compost in a relatively short time (see Figure 12.9). Batches can heat up sufficiently to kill off weeds and diseases and the enclosed nature deters vermin. The compost ingredients should be gathered and the tumbler filled in a short space of time. Nothing further is added until that batch is completed.

Worm composting lends itself to handling small quantities of organic waste which can be added as they arise, such as kitchen waste, especially over the winter period when there is little plant material to accumulate. Brandling or tiger worms (*Dendrobena* spp. and *Eisenia foetida*) are compost worms that feed on organic matter. These can be purchased, but they are readily found in rotting vegetation such as compost heaps (see Figure 12.1) and quickly multiply.

Each kilogram of worms digests 1–2 kg of waste food per day.
The container can be a plastic dustbin ideally equipped with a tap to drain off liquids which can be diluted and used as liquid feed (see p. 201). Smaller containers, wide rather than narrow, can be made of wood, with some insulation to maintain temperatures. The worms digest the vegetation as it starts to rot, which means that once in balance there is no smell.
Typically, a 10 cm layer of sand is used at the base. Bedding material such as well-rotted compost or farmyard manure is needed for the worms to live in until the process begins. Chopped waste is spread to a depth of 5 cm. Around 100 worms are added and covered with wet newspaper to keep out the light and maintain moisture levels. A lid is needed to keep out the rain. Ideally, temperatures should be maintained between 20°C and 25°C and the pH kept between 6 and 8; lime can be sprinkled on if the compost becomes too acid. The compost is removed when ready; the decomposing top layer is separated off and used to start the next run. The rest is spread out to dry in the sun and the worms are recovered by placing a wet newspaper on the compost, under which they will congregate.
On a larger scale, wormeries are used to compost farmyard manure, with continuous systems available that separate the composted material from the worms, which can be recycled with surpluses being available as animal feed.

Composted municipal waste. This is essentially the same as garden compost, but because the composting can be done in bulk, it is heated sufficiently to inactivate harmful weeds and plant diseases. The high wood content gives the final product good stability (see p. 208) but can lead to nitrogen 'lock-up' (see p. 187). The product is usually too acid (see soil pH p. 201) and too high in nutrients to be used alone as compost for container growing unless mixed with composted barks, coir and the like to create good container compost for a wide range of plants. Unless the composting is done in the expensive composting vessels, the most common issues are contamination with plastics and glass.

Figure 12.9 Compost tumbler: turning frequently is made easier by the use of a handle to rotate the load; once filled it is left until the batch is completed

When very large quantities of green waste from households or parks departments are available the ingredients can be collected and heaped up on a concrete base. This makes it easy to use large, powered equipment to turn the ingredients to maintain good aeration. The cooler outer layers are mixed in to ensure all parts are heated up and decomposed rapidly. There are specialist composting vessels with automatic turning equipment and biofilter beds that ensure that the exhaust vapour is cleaned ('scrubbed'). Windrow turners are used in big plants to turn and loosen the green waste. This tends to dry out the material and so are often housed to confine the dust created.

Hot beds. Using the composting process to generate heat has been used since at least Roman times. The hot bed process was much exploited by Victorian gardeners to produce exotic fruits such as melons and pineapples and to ensure a continuity of supply of fruit and vegetables for the kitchen even through the coldest months of the year. The Victorians had access to large quantities of fresh horse manure that remains the best material for a hot bed system.

The emphasis of this decomposition is the generation of heat so there must be enough fresh manure to complete the process in a short period of time, days rather than weeks. Compare this with the cold process where the build-up of a typical garden compost heap is over months. The process is started by building a loose heap, that is, with plenty of oxygen circulating. Within two to three days the pile heats up and generates steam. At this point the pile is inverted and 'fluffed up'. This is repeated three days later sprinkling on water when necessary to keep the mix moist.

After nine to ten days the decomposing organic matter is made into the hot bed at the site where plants are to be grown. The material in the heap is forked into a layer about 10 cm deep and tamped down. This is repeated with further layers to a depth of 20 cm or more. This can all be done within a brick construction on which 'lights' (glass) can be placed like a 'cold frame'. Alternatively, a free-standing frame can be put on top of the block created. The size of the frame is usually made to be a multiple of the 'lights' being used. Traditionally heat was retained by earthing up the sides, but modern materials such as polystyrene can be used to line the frame. The 'lights' are put on but kept open for ventilation.

In less than a day the material starts to heat up significantly. This can be tested with a thermometer or, more crudely, by hand. Temperature should be monitored over the next three days when the maximum is usually achieved. As soon as the temperature begins to drop, a 15 cm layer of moist soil is put on top and the compost is ready to use. Once seeds or plants are in place, the temperature is controlled by opening and closing the 'lights' in the same way as growing in cold frames. The heat provided from the decomposition process will last until the days warm up in the spring and growing in the frame can continue until the autumn when the well-rotted manure can be used as a mulch or dug into the soil.

Organic fertilizers and compost teas

Most organic fertilizers are of animal origin, for example blood meal, hoof and horn, bone meal (see summary of nutrients provided, Table 13.2). However, concentrated sources of nutrients can be produced from plant material to make 'compost teas'. A wide range of plants including comfrey, nettles, borage, clover and bracken can be used. Using comfrey is particularly attractive to organic growers as it is deep-rooted and able to extract nutrients below the reach of most plants. However, these deep roots also make it difficult to get rid of so choosing to put it in a garden needs to be thought through. The leaves are harvested to produce nutrient-rich plant food. There are many different methods of making liquid feed (see p. 000) from comfrey and other plants which are steeped in water to produce a stock tea which is diluted before use.

Bulky organic matter

Maintaining and improving organic matter and humus levels in the soil is usually achieved with one or more of the following bulky materials:

- garden compost (see p. 189)
- composted municipal waste (see p. 190)
- farmyard manure
- shoddy
- spent mushroom compost
- leaf mould
- pine needles
- chipped bark
- biochar
- green manures

These materials also 'open up' the soil, that is, improve aeration (see soil structure, p. 167). The main problem is obtaining cheap enough sources locally because their bulk makes transport and handling a major part of the cost. They can be evaluated on the basis of their effect on the physical properties of soil and their (usually small) nutrient content.

Farmyard manure. This is the traditional material used to maintain and improve soil fertility. It consists of straw, or other bedding, mixed with animal faeces (dung) and urine that provides the nitrogen that offsets the problem of using bedding alone. The exact value of this material in nutrient terms depends on the proportions of the ingredients, the degree of decomposition and the method of storage. Samples vary considerably. Much of the manure is rotted down in the first growing season but almost half survives for another year, and half of that goes on to a third season, and so on. A full range of nutrients is released into the soil and this should be allowed for when calculating fertilizer requirements. The continued release of large quantities of nitrogen can be a problem, especially on unplanted ground in the autumn, when the nitrates formed are leached deep into the soil over the winter beyond the root zone and can end up polluting watercourses.

Farmyard manure is most valued for its ability to provide the organic matter and humus needed to maintain or improve soil structure. It must be worked into soils where conditions are favourable for its continued decomposition. If fresh organic matter is worked into wet and compacted soils or deep into clay, the need for oxygen outstrips supply and anaerobic conditions prevail to the detriment of any plants present. These soils develop grey colouring and a foul ('bad eggs') smell.

Spent mushroom compost

This compost is a by-product of the mushroom industry – it is the 'spent' (i.e. used) material that becomes available direct from the growers or some garden centres. It used to be made from well-rotted horse manure but now it is almost all composted straw capped with chalk which gives its characteristically high pH, that is, basic reaction (see p. 202). This makes it useful for raising soil pH as an alternative to liming (see p. 204). It works well with the growing of calcicoles (see p. 202), especially when growing brassicas (cabbages, cauliflowers, Brussels sprouts).

It is an excellent source of organic matter with added nutrients left over from mushroom growing with which to mulch or to incorporate in the soil,

Table 12.2 Plants used for green manuring (see also Figure 12.10)

Legumes	Non-legumes
Lupinus angustifolius **(Bitter blue lupin)**	*Fagopyrum esculentum* **(Buckwheat)**
Medicago lupulina **(Trefoil)**	*Phacelia tanacetifolia* **(Phacelia)**
Trifolium hybridum **(Alsike clover)**	*Secale cereale* **(Grazing rye)**
Trifolium incarnatum (Crimson clover)	*Sinapis alba* **(Mustard)**
Trifolium pratense (Essex red clover)	
Trigonella foenum-graecum (Fenugreek)	
Vicia faba (Winter field bean)	
Vicia sativa (Winter tares)	

although its use is limited. It must not be used in plants that prefer acid conditions (see calcifuges, p. 202). If used too freely nutrient deficiencies in plants can be induced (see iron induced chlorosis, p. 199) leading to poor performance. Leftover fertilizer levels in the sample can make it unsuitable for young plants, particularly when growing in containers.

Leaf mould is made of the rotted leaves of deciduous trees and makes a highly prized compost. The leaves are often composted separately from other organic matter and much valued in ornamental horticulture for a variety of uses such as an attractive mulch or, when well rotted down, as a compost ingredient. They are commonly composted in mesh cages, but many achieve success by putting them in polythene bags well punched with holes. The leaves alone have a high C:N ratio so decomposition is slow (see p. 187). Usually, it is not until the second year that the dark brown crumbly material is produced, although the process can be speeded up by shredding the leaves first.

It is low in nutrients because nitrogen and phosphate are withdrawn from the leaves before they fall and potassium is readily leached from the ageing leaf. Unless they are from trees growing in acidic conditions, the leaves are rich in calcium and the leaf mould made from them should not be used with calcifuge plants (see p. 202).

Pine needles are covered with a protective layer that slows down decomposition. They are low in calcium and the resins present are converted to acids. This extremely acid litter is almost resistant to decomposition. It is valued in propagation, for growing calcifuge plants such as rhododendrons and heathers and as a material for constructing decorative pathways.

Chipped bark. This is mainly used as a mulch either alone on the soil surface or on one of the non-organic sheet mulches such as woven fibre to make an attractive finish. It is valued in many situations, but there are several characteristics that need to be noted. Bark is nitrogen deficient so it does not decompose readily and can last a long time on the surface. However, once in the soil it tends to rob plants of nitrogen (see p. 187). It is light when dry so tends to be blown around and floats on water. It is difficult to manage on slopes and birds throw it around when looking for food underneath making adjoining paths untidy.

Biochar is organic material that has been carbonized under high temperatures in the presence of little or no oxygen ('pyrolysis'; like charcoal burning). It varies according to the plants from which it is made but all are carbon rich and resistant to decomposition. The honeycombed structure improves soils and composts by providing greater aeration, drainage, water-holding capacity and, especially in sandy soils, nutrient retention. Furthermore, this porous structure in larger particles (rather than dusty material) is ideal for beneficial soil microorganisms.

Some biochars on the market are 'nutrient enriched' to offset the tendency of the product to 'lock-up' nitrogen (see p. 187) and/or biologically enriched (see mycorrhiza, p. 184). The variability of biochar can bring problems and it is recommended using a reliable source such as FSC certified. Many types of biochar can improve plant health by neutralizing acidity, but care needs to be taken as others with a high pH can cause problems.

Green manuring is the practice of growing plants to:

- add organic matter
- develop and maintain soil structure and fertility
- increase microorganism activity in the soil
- cover bare ground to protect the surface
- reduce soil erosion
- compete out weeds
- capture soluble nutrients that would otherwise be leached.

It is an important part of organic growing systems, but also to prevent pollution by the leaching of nutrients from bare ground, especially in the autumn and over the winter. The plants used are typically agricultural crops that cover the ground quickly and yield a large amount of leaf that can cover the ground (see Table 12.2

Figure 12.10 Plants used for green manuring: (a) mustard, (b) *Phacelia*, (c) clover, (d) lupins

and Figure 12.10). Several are legumes which have the advantage of adding extra nitrogen to the soil. The seeds are normally broadcast sown in the autumn when there is no overwintering crop. The green manure is then dug in or cut, left to wilt then dug into soil. Whilst it can be left as a mulch, this can make it difficult to put the next crop in.

Further reading

Adams, C., Early, M., Brook, J. and Bamford, K. (2015) *Principles of Horticulture Level 3*. 7th ed. Routledge.

Brinton, W.F. (1990) *Green Manuring: Principles and Practice of Natural Soil Improvement*. Woods End Agricultural Institute.

Brown, L.V. (2002) *Applied Principles of Horticulture*. Butterworth-Heinemann.

Caplan, B. (1992) *The Complete Manual of Organic Gardening*. Headline Book Publishing.

Dixon, G. (2019) *Garden Practices and Their Science*. Routledge.

Dowding, C. (2012) *Vegetable Course*. Frances Lincoln Ltd.

Dowding, C. (2013) *Organic Gardening*. Green Books.

HDRA. (2005) *Encyclopaedia of Organic Gardening. Henry Doubleday Research Association (Garden Organic)*. Dorling Kindersley.

Killham, K. (1994) *Soil Ecology*. Cambridge University Press.

Littlewood, M. (2007) *Organic Gardeners Handbook*. The Crowood Press.

Lowenfels, J. and Lewis, W. (2006) *Teaming with Microbes: A Gardener's Guide to the Soil Food Web*. Timber Press.

Pavlis, R. (2021) *Soil Science for Gardeners*. New Society.

Pears, P. and Sticklands, S. (2007) *The RHS Organic Gardening*. Bounty Books.

Readman, J. (2004) *Managing Soil without Using Chemicals*. Dorling Kindersley.

SUPPORT MATERIAL

The online material is accessible via the QR code and includes further information on many of the topics in the book, such as

- Earthworms, slugs and snails
- Peat development

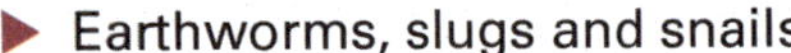

CHAPTER 13

Plant nutrition

Figure 13.1 Plant nutrition has a big effect on the growing of vegetables whether in the home, on the allotment or commercially (source: Shutterstock, marcin jucha)

This chapter includes the following topics:

- The plant's nutrient requirements
- Nitrogen
- Phosphorus (phosphates)
- Potassium (potash)
- Magnesium
- Providing plant nutrients
- Fertilizers
- The importance of soil pH
- Soil testing
- Adjusting soil pH/liming

DOI: 10.4324/9781003581260-13

The plant's nutrient requirements

Gardeners are familiar with the need to provide fertile soil to grow most of their plants, especially if they are intending to grow productive vegetables and fruit. They have probably also seen the consequences of having an impoverished soil where a shortage of one or more nutrients has led to disappointing plant growth. However, it can be puzzling to find that in some circumstances gardeners are advised to reduce the fertility of their soil to get better flowering. Those who have considered establishing a wildflower meadow (Figure 13.2) might well have wondered why it usually involves stripping the good topsoil away. In the feeding of plants so much depends not just on the plants involved but also on the soil available and the intentions of the gardener. Objectives can be as varied as high production from vegetables and fruit, prolific flowering, attractive wildflower meadows and organic gardening, all leading to very different approaches to nutrition. Rather like the use of medicines or herbicides, twice the recommended dose does not produce better results; nor should it be assumed that poor plant performance would be solved by adding more nutrients.

Figure 13.2 Wildflower meadows are often best established on impoverished soils (source: Shutterstock, kritskaya)

Plants are mainly made up of water from the soil and carbon dioxide from the air (see photosynthesis, p. 113). However, a small but significant amount of what is needed is made up of minerals commonly referred to as 'plant nutrients' or 'plant food'. These are normally taken up in soluble form through the roots from the soil solution but can also be taken in through other parts of the plant, for example foliar feeding.

Essential minerals are inorganic substances necessary for the plant to grow and develop and are often referred to as nutrients.

Those using fertilizers will have noted that the main nutrient contents listed on the containers are nitrogen, phosphorus (also known as phosphate) and potassium (also known as potash). These are the major nutrients because they are needed in relatively large quantities. Also needed in quite large quantities are magnesium, calcium (usually supplied in lime) and sulphur. Iron is an example of a nutrient that is essential but needed in much smaller quantities. It is a minor nutrient (micro- or trace element) along with the lesser-known manganese, boron, zinc, copper and molybdenum (see Table 13.1).

The essential elements have very specific functions in plant cell processes. When deficient (in short supply) the plant shows certain characteristic symptoms usually related to the role of the element in the plant.

Table 13.1 Major and minor elements in healthy plants

Element	Chemical symbol	
Major nutrients		**Grams per kilogram of dry matter**
Nitrogen	N	20–50
Phosphorus	P	2–5
Potassium	K	20–50
Magnesium	Mg	2–10
Calcium	Ca	2–5
Minor nutrients		**milligrams per kilogram of dry matter**
Iron	Fe	100–300
Manganese	Mn	50–300
Zinc	Zn	20–200
Copper	Cu	10–20
Boron	B	20–200
Molybdenum	Mo	1–10

Nitrogen (N)

Nitrogen is needed by plants to form chlorophyll (see p. 116) and is associated with leafy (vegetative) growth. Consequently, large dressings of nitrogen are given to plants grown for their leaves such as cabbages, lawn grasses and sports turf.

Deficiency causes slow, spindly growth in all plants. There is usually a general yellowing (chlorosis) of the leaves due to lack of chlorophyll (see p. 116), often preceded by a bluing of the foliage appearing first on the older, lower leaves. When nitrogen shortages occur in the plant the chlorophyll in the oldest leaves is broken down to release nitrogen for use in the new young efficient leaves (Figure 13.3).

Excess nitrogen produces soft, lush leafy growth, making the plant vulnerable to pest attack and more likely to be damaged by cold. Very large quantities of nitrogen are undesirable because they can harm the plant by producing high salt concentrations at the roots (see osmosis, p. 124) and can easily be lost by being leached and thus polluting watercourses.

Nitrogen fertilizers commonly used and their nutrient content are given in Table 13.2.

Figure 13.3 Nitrogen deficiency: note that it is the lower, older leaves that have been affected before the younger ones at the top of the plant

Phosphorus (P)

This element is important in the production of the major chemical required for energy transfer in the plant (adenosine triphosphate, see ATP, p. 119). Consequently, large amounts are concentrated in seeds and the growing points (meristems) of roots and shoots. The growing root has a high requirement, and the plant's ability to establish itself depends on the roots being able to tap into supplies in the soil before the reserves in the seed are used up. Phosphorus in the plant is constantly being recycled from the older parts to the new growing points. This means that most plants have a low phosphate requirement compared with quick-growing plants that are harvested young.

Deficiency symptoms are related to poor root development, which leads to reduced growth of stem and root. There can also be bluey or purplish stem and leaf colourings with or without speckling associated this deficiency (Figure 13.4).

Phosphate fertilizers (containing phosphorus) commonly used and their nutrient content are given in Table 13.2.

Excess. The presence of more than it needs can be detrimental as this interferes with the development of mycorrhiza (see p. 184).

The availability of phosphate from soil is mainly influenced by soil pH (see p. 201). Mycorrhiza play an important part improving many plants access to phosphorus.

Phosphorus is a limiting factor in aquatic plants so any addition of it (e.g. surface run-off including fertilizers) leads to algae blooms (eutrophication) of rivers, ponds and lakes. This leads to oxygen levels being depleted, and reduced fish numbers follow.

Potassium (K)

This is associated with successful flowering and fruiting. The element is present in relatively large amounts in plant cells in solution where it acts as

Figure 13.4 Phosphate deficiency; note the purple colouring of the leaf caused by severe deficiency of phosphorus (source: Shutterstock, Rupinder singh 0071)

Table 13.2 Nutrient content of commonly used fertilizers

	N	P_2O_5 (P)	K_2O (K)	Mg	Ca	S
	%	%	%		%	%
Ammonium nitrate	33–35					
Ammonium sulphate	20–21					24
Bone meal*	3	20 (9)				
Dried blood*	12–14					
Hoof and horn*	12–14					
Keiserite				16		
Meat and bone meal*	5–10	18 (8)				
Potassium chloride			59 (49)			
Potassium sulphate			50 (42)			17
Shoddy*	3–15					
Superphosphate		18–20 (8–9)			20	12–14
Triple superphosphate		47 (20)			14	
Urea	46					

*an organic fertilizer

an osmotic regulator, for example in stomata (see p. 128). It also has a role in providing hardiness, resistance to chilling injury, drought and disease in plants.

Deficiency results in brown, scorched areas on leaf tips and margins of eudicots (dicotyledonous) plants (Figure 13.5). Monocotyledons show similar brown markings at the growing tip of the leaves which spread down the leaf as the deficiency continues. Low potassium levels are usually associated with poor performance of flowers and fruit made worse if large quantities of nitrogen are added.

Potash fertilizers (those containing potassium) commonly used and their nutrient content are given in Table 13.2.

Figure 13.5 Signs of potash deficiency on tomato – typical of a eudicot leaf; note the desiccation of the leaf margin

Magnesium (Mg)

Magnesium has many roles in the plant including being required to make chlorophyll (see p. 116).

Deficiency symptoms appear as a characteristic yellowing between the leaf veins (interveinal chlorosis). This appears on the lower, older leaves (Figure 13.6) because the inefficient old leaves release magnesium from their chlorophyll to enable it to be used to the build new young leaves. Consequently, other than affecting the look of a plant, the deficiency has little or no effect on the performance of the plant initially. However, continued shortage leads to more of the leaves

Figure 13.6 Signs of magnesium deficiency in a eudicot leaf; note the interveinal chlorosis (source: Shutterstock, walkerone)

being affected and plant growth starts to be affected. Gradually the affected areas become reddened or brown before dying (necrosis).

Magnesium fertilizers commonly used and their nutrient content are given in Table 13.2.

Other nutrients

Calcium is a major constituent of plant cell walls as calcium pectate which holds the cells together after cell division before the addition of cellulose (see p. 47). It also influences the activity of meristems, especially in root tips.

Deficiency symptoms tend to appear in the younger tissues first because calcium is immobile in the plant. It causes weakened cell walls, resulting in inward curling, pale young leaves and sometimes death of the growing point. Specific disorders include 'topple' in tulips, when the flower head cannot be supported by the top of the stem, 'blossom end rot' in tomato fruit (see Figure 18.30) and 'bitter pit' in apple fruit (see Figure 18.31).

In general, growing plants within their recommended pH range ensures adequate calcium is available to them (see p. 202).

Sulphur is required by plants in large quantities but is rarely in short supply. It is another nutrient needed for the synthesis of proteins including chlorophyll (see p. 116).

The sulphur in organic matter becomes available to plants when aerobic microorganisms mineralize the organic form to produce soluble **sulphates**. Some plankton in the sea release dimethylsulfoniopropionate (DMSP) which gives rise to the distinctive 'smell of the sea' and deposits small amounts of sulphur in coastal areas. The burning of fossil fuels releases sulphur that pollutes the atmosphere, but also replenishes the soil sulphur reserves along with volcanic fallout and the weathering of pyrite rocks. As air pollution and the use of fertilizers such as ammonium sulphate and potassium sulphate are reduced, there may become a deficiency to address.

A **deficiency** produces a yellowing (chlorosis) of leaves appearing in the younger leaves first.

Iron. Although it does not form part of the chlorophyll molecule, it is a component of some enzymes required to synthesize it.

Deficiency of iron results in yellow ('chlorosis') or even white leaves appearing in the younger leaves first. The deficiency is commonly caused by the presence of large quantities of lime (see p. 204). This **'lime-induced' chlorosis** (Figure 13.7) occurs on overlimed soils and calcareous soils (see p. 203). Calcicoles are adapted for the chalk and limestone areas, but most other plants grown in such conditions do not fare well and have a typically yellow appearance. Good drainage and soil structure should be maintained because a waterlogged root zone can contribute to the problem of lime-induced chlorosis.

Iron supplements such as ferrous sulphate are commonly added in small quantities to amenity turf

Figure 13.7 Lime-induced deficiency on *Elaeagnusx ebbingei*; note that it is the younger leaves that have chlorosis (yellowing) whilst the older leaves are still green

to 'green up' its appearance without the need to add more nitrogen. Chelated iron is a supplement that can be used as a precaution or to treat affected plants being grown on an unsuitable growing medium.

Copper, boron, zinc, manganese and molybdenum are also essential minerals that are required in minute quantities but are rarely deficient in cultivated soils in Britain and Ireland. They are added as supplements to some growing media made up of materials with little or no mineral trace elements present.

Providing plant nutrients

In nature the recycling of nutrients ensures the continued growth of plant communities (see p. 134) unless there are net losses through leaching. This can also be the situation in some gardens, especially where plantings are like natural ones, predominantly trees and shrubs. In practice, even in decorative gardens there is a need to import some nutrients. Soil becomes depleted of nutrients by the removal of crops. There are significant differences in maintaining a lawn where mowings are allowed to stay on the lawn (recycling) compared with one where the arisings are 'boxed off' (i.e. collected and not put back). Most gardeners make good the nutrient losses with their garden compost (see p. 189), brought in as composted municipal waste (see p. 189), bulky organic matter (see p. 191) and/or fertilizers (see Table 13.2).

Fertilizers are concentrated sources of plant nutrients that are added to growing media.

Manure is a source bulky organic matter comprising animal faeces ('dung') and bedding.

Organic growers put the emphasis on plant nutrition by 'feeding the soil', that is, the recycling of organic matter, preventing losses by leaching and making good on nutrient loss by crop removal through the importation of manures (see also Organic Growing in the Support Material).

Fertilizers

Organic fertilizers include those derived from organic sources such as bone meal, hoof and horn, dried blood, fish blood and bone, comfrey feed and nettle tea.

Inorganic fertilizers are those commonly referred to as 'synthetic', 'artificial' or 'bag' fertilizer such as ammonium sulphate, triple superphosphate, potassium chloride and National Growmore. Some are mined/quarried materials such as rock phosphate.

The quantity of nutrient supplied by a fertilizer is expressed in terms of a percentage of the contents.

- Nitrogen fertilizers are given terms of percentage of the element nitrogen in the fertilizer: %N
- Phosphate fertilizers provide the element phosphorus, given as %P, but are sometimes described in terms of the 'equivalent amount of phosphoric oxide': $\%P_2O_5$
- Potash fertilizers provide the element potassium, given as % K (the 'old' name for potassium is kalium), but are sometimes described in terms of the 'equivalent amount of potassium oxide': $\%K_2O$
- Magnesium fertilizers are described in terms of %Mg.

The percentage figures show the quantities of nutrient in each 100 kg of fertilizer:

For example, ammonium sulphate is 20%N, so there is 20 kg of the element nitrogen in every 100 kg of the fertilizer and by proportion there is 5 kg in a 25 kg bag or 1 kg in a 5 kg bag.

Fertilizers are a concentrated source of nutrients as illustrated by the following materials that supply the element nitrogen:

1 kg of nitrogen is supplied in a 5 kg bag of ammonium sulphate

whereas the nutrients in bulky organic matter are more diluted (or non-existent):

1 kg of nitrogen is supplied in 2–5 tonnes of fresh farmyard manure.

There are many different types of fertilizers in terms of

- content (straight/compound)
- formulation (granules/powders, quick/slow/controlled release)
- the ways they are used (base/top dressing, liquid feed, foliar feed).

Straight fertilizers are those that supply only one of the major nutrients: nitrogen, phosphorus, potassium or magnesium, such as ammonium sulphate (supplying N), triple superphosphate (P), potassium chloride (K).

Compound fertilizers are those that supply two or more of the main nutrients – nitrogen, phosphorus and potassium, for example Growmore. The accepted convention for describing the fertilizer content of compounds is to label the content in the order N P K so Growmore is described as 7:7:7, that is, 7%N: 7% P_2O_5 : 7% K_2O . Most of the proprietary fertilizers for gardeners are compounds including those for tomatoes, roses, orchids, cacti and so on.

Besides the major nutrient content, fertilizer regulations require that details of trace elements, pesticide content and phosphorus solubility should appear on the packaging.

Quick release or soluble fertilizers are those that dissolve immediately in water before or after application to soil, for example ammonium sulphate and urea, both of which yield nitrogen that can be taken up by plants within days of application. The downside of quick release materials is that if too much is applied the plants can be harmed or even killed; often referred to as 'root scorch' or 'burnt plants' because water is drawn out of the root by the high salt concentration (see osmosis, p. 124, and disorders, p. 125). Furthermore, care needs to be taken to ensure ground water is not polluted by applying more than can be taken up by plants, otherwise watercourses may be contaminated by leached nutrients or run-off following top dressings.

Organic gardeners avoid the use of quick release fertilizers because the elevated level of solubles in the soil water reduces the effect of beneficial soil organisms.

Slow release fertilizers do not dissolve immediately in water but release nutrients over a long period. Many are organic fertilizers such as shoddy (dirty wool), meat meal, bone meal, and hoof and horn that are decomposed by microorganisms, so the rate at which nutrients are released depends on temperature. This fits well with plant requirements which increase as they grow larger over the late spring and early summer. However, a downside is that there is little or no nutrient to take up when plants start to grow after the winter when temperatures have been too low for microorganism activity.

There are several fertilizers that dissolve very slowly in the soil water. These include urea-formaldehyde (UF) often used on turf because it has the advantage of releasing nitrogen even in cold conditions. Rock phosphate releases phosphate in a form for plant uptake very slowly (over many years).

Controlled release fertilizers are slow release fertilizers that are formulated to release nutrients in a controlled way over a specified long period. One group comprises quick release fertilizers held within a permeable resin coating that lets water in and allows nutrients to diffuse out. The other group is soluble fertilizer coated with sulphur which is broken down by microorganisms so nutrients are released. In both cases the thickness of the coatings can be varied to enable fertilizers to be designed so they release nutrients in line with particular plant needs and for the required period of time, such as a three-month formulation for bedding plants or a nine-month one for container roses.

Base dressings are the fertilizers that are applied to the soil and worked into the seedbed, or incorporated in composts, before sowing/planting.

Top dressings are granular fertilizers applied to the surface because nutrients are needed after plants have been established for ongoing nutrition often in the form of a compound fertilizer such as autumn treatment for lawns and spring dressing for established borders.

In order to reach the root zone, top dressings are usually quick release fertilizers such as applications of nitrogen, often with ferrous sulphate, to 'green up' lawns. This means that care needs to be taken to avoid harming plant leaves ('scorch'). An alternative approach is to use liquid feed.

Liquid feeds are fertilizers dissolved and watered on to soils or other growing media to provide ongoing plant nutrient requirements over the growing season typically for pot plants, hanging baskets, houseplants and bedding and in greenhouse production. When used in conjunction with irrigation systems ('fertigation'), nozzle blockages can occur unless pure materials are used. As with top dressing, care needs to be taken to avoid damaging leaves ('scorch').

Fertigation is supplying nutrients in the irrigation water.

Foliar feed is a liquid feed diluted sufficiently so that it can be applied to leaves without causing 'scorch'. It can be used for routine feeding and has the advantage of immediate uptake to treat deficiency symptoms. Application is usually undertaken first thing in the morning by misting or spraying the leaves until there is 'run-off'.

Fertilizers are provided mainly in the following forms for use in the aforementioned situations:

- **Granules** which have the advantage of being easy to deliver by hand or by fertilizer spreaders when applying base or top dressings.
- **Powders** (or crystals) are provided for those making up liquid or foliar feeds.

More details of Plant Nutrition are provided in the Support Material.

The importance of soil pH

The pH scale is a means of expressing the degree of acidity or alkalinity (see Figure 13.8).

The soil pH (its acidity or alkalinity) has a significant effect on the plant's ability to take up nutrients.

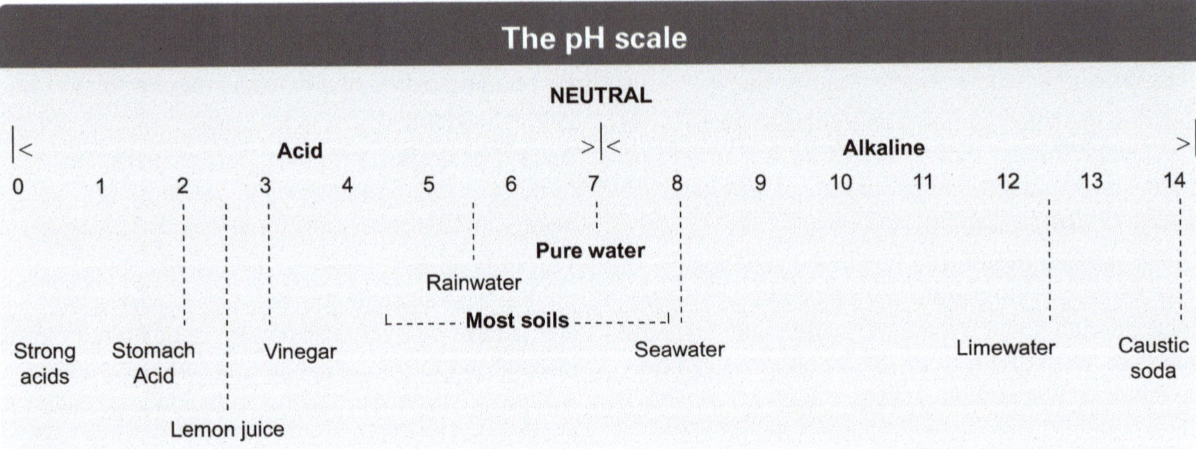

Figure 13.8 A diagram illustrating the pH scale from 0 to 14. Pure water is shown at pH 7 (neutral). Stomach acid is shown at pH 2, lemon juice (2.5) and rainwater (5.5). Caustic soda is shown at pH 14, limewater at 12.5 and seawater at 8. Most soils are between pH 4.5 and 8

Most plants grow in slightly acid, neutral or slightly alkaline conditions where the full range of plant nutrients are most available.

Acids, bases and alkali

Acids are a group of chemicals that have a sour taste and are corrosive, for example vinegar dilute (acetic/ethanoic acid).

Bases are those chemicals that neutralize acids. They include lime (calcium carbonate).

Alkalis are soluble bases. They have a soapy feel and tend to be irritants; strong bases/alkalis are corrosive.

When bases (alkalis) and acids are mixed they neutralize each other and form 'salts', for example common salt which is sodium chloride made from sodium hydroxide and hydrochloric acid. Many fertilizers, such as ammonium sulphate and potassium chloride, are formed this way.

Calcicoles are plants such as *Clematis*, honeysuckle (*Lonicera*) and *Buddleia* that have adapted to calcium-rich (high pH) soils. Cultivated plants that thrive on such soils include *Agapanthus*, beet, *Geranium*, *Echinacea*, *Jasminum*, lilac (*Syringa*), *Parthenocissus*, *Rudbeckia* and *Verbascum*.

Calcicoles are plants that are adapted to grow on calcareous soils (calcium-rich, chalky).

In contrast, the growth of many other plants (calcifuges) is adversely affected in soils greater than pH 5.5 (soils with small amounts of lime or chalk in it). Cultivated plants that grow well in the more acid soils (below pH 5.5) include *Rhododendron* spp., *Camellia* spp., *Pieris* spp., blueberries (*Vaccinium corybosum*) and some heathers such as *Daboecia* spp.

Calcifuges are plants that are adapted to grow on soils below pH 5.5.

Many *Hydrangeas* are blue when grown in soils below pH 5.5 but are pink in soils above pH 5.9. In between they tend to have a transitional mix of neither one nor the other (see Figure 13.9).

The tolerance of a range of plants is given in Table 13.3.

The soils of Britain and Ireland are mainly between pH 4.5 and 8 and the vast majority of cultivated soils are between 5.5 and 7.5. Although there does not appear to be much difference across this range, the significance for organisms living in the soil is considerable. The pH scale is logarithmic (like measuring sound in decibels or earthquake energy release on the Richter scale); each 'unit' is ten times larger or smaller than the one next to it, that is, pH 5 is ten times more acid than pH 6 and pH 4 is one hundred times more. More information

Table 13.3 Soil pH and plant tolerance: **pH below which plant growth may be restricted on mineral soils:**

Celery	6.3	Rose	5.6
Daffodil	6.1	Raspberry	5.5
Bean	6.0	Cabbage	5.4
Lettuce	6.1	Strawberry	5.1
Carnation	6.0	Tomato	5.1
Chrysanthemum	5.7	Apple	5.0
Carrots	5.7	Potato	4.9

Figure 13.9 Hydrangeas: the blue colour of the flowers depends on the availability of aluminium to the plant; this element is readily available in growing media with a pH less than 5.5 and gradually less so at higher pH levels where the colour tends to be pink

on Acidity, Alkalinity and pH is to be found in the Support Material.

It is mainly the presence of calcium or 'lime' (in the form of any calcium carbonate but usually chalk or limestone) that determines the pH of soil. It is the abundance of calcium in soils which interferes with the uptake and utilization of several of the plant nutrients (induced deficiencies). Calcicoles are adapted to these conditions, and these plants are able to thrive in the limey/chalky soils. In contrast, calcifuge plants are adapted to growing in soils with a pH below 5.5; above that the lime (calcium) present interferes with the uptake and utilization of many of the nutrients they require.

So the ideal growing condition for most plants is a soil of pH 6.5 because at this point all the essential plant nutrients are available for uptake by the roots of most plants. Although the majority of plants grow well in soils between pH 6 and 7, there are considerable differences in the tolerance of plants to soil pH conditions (see Table 13.3). Potatoes are considerably more tolerant of soil acidity than most plants and are still productive down to pH 4.9. In contrast, the yield of celery falls significantly in soils below pH 6.3.

Soil pH also has important effects on other organisms besides plants, including the beneficial soil organisms. Like plants, these organisms thrive at their optimum pH creating a more fertile soil. A few soil-borne disease-causing organisms tend to occur more frequently on acid lime-deficient soils (see clubroot, p. 293), whereas others are more prevalent in limey soils, such as common scab of potatoes.

Causes of soil acidity

In Britain and Ireland, where over a year the rainfall exceeds evaporation, many soils tend to become too acid (creating 'sour soils'). This is because rain is weak carbonic acid (see p. 161) which dissolves lime (calcium carbonate), which is then leached from the soil. Lime is very readily leached from free-draining sandy soils in high rainfall areas so they tend to go acid very rapidly. Calcareous soils (i.e. those containing pieces of chalk or limestone) do not become acid until all these base reserves are used up. In addition to the effect of rainfall (carbonic acid), there are several other factors that increase the rate at which soils become more acid:

- **Crop removal**. Some plant nutrients such as calcium, magnesium and potash are bases so when they are taken up by plants but not recycled the soil acidity can increase.
- '**Acid rain**' (polluted rain and snow) is directly harmful to vegetation and also contributes to the fall in soil pH.
- **Organic acids** derived from the microbial breakdown of organic matter, such as humic acids, also lead to an increase in soil acidity.
- **Fertilizers**. Some such as ammonium sulphate increase the rate at which soils become acid.

Soil testing

Soil testing kits are available which can give a useful guide to the nutrient status and soil pH. They usually come with guidance as to how to use the results for liming, fertilizer and/or manure application. Growers of valuable crops would normally use laboratory testing.

Outdoor soils are normally checked in the autumn in readiness for the following year. Nutrient testing is usually limited to determining phosphate and potash levels; 'nitrogen' testing is of limited value at this time because so much of the soluble nitrogen (nitrate) is leached from the soil during the winter.

Gardeners usually use either very simple pH meters (sticks) or colour indicator methods to do their own pH testing (see Support Material), but these methods are not usually better than a half unit either side of the correct value. Soil pH can be measured more accurately in the laboratory with pH meters.

Soil sampling. Above all the usefulness of any soil analysis depends on the degree to which the sample taken is representative of the area from which it is taken (and to be treated). The variability of the soil makes it difficult to obtain a meaningful result on which to base any calculations for nutrient or lime application. It is recommended that several cores of soil are taken (ideally 20 cores down to 15 cm) over the area to be treated, such as 'the lawn' but avoiding any abnormal areas, for example a new area of the lawn that was formerly being used to grow vegetables (this should be tested and treated separately). On a large scale, areas with different textures of soil (sandy loam, clay loam etc.) should be sampled separately. Finding a satisfactory target area in a garden or allotment can be more difficult because of the variable treatment of all the small areas (different history of crops/plants, fertilizer and organic matter applications). See the Support Material for more information on Soil Sampling.

Nevertheless, there is merit in testing the soil when:

- there appears to be a nutritional problem such as deficiency symptoms or general poor growth
- making a new garden or allotment
- establishing a new lawn
- establishing plantations (soft fruit, cane fruit, orchards)
- planting valuable specimens
- planning to grow plants that require an unusual soil pH range to survive, for example calcifuges and calcicoles.

Adjusting soil pH

It's worthwhile to consider selecting plants that will grow in the soil without having to adjust soil pH, especially if it involves soils at either of the extremes of the normal soil pH range. Changing the soil pH has consequences for all organisms, so liming can interfere with the ecology of the area (see p. 134).

Soil pH can be raised ('sweetened') by the addition of **lime** normally as ground chalk or ground limestone (both are calcium carbonate, garden lime). An alternative commonly bought from garden centres is hydrated lime (calcium hydroxide, slaked lime, builder's lime). All should be applied as a fine powder because coarse material takes too long to affect the soil. Care should be taken to protect the eyes as the powders are easily blown around, and gloves should be worn to handle hydrated lime as it can cause skin irritation and chemical burns. Wood ash, the result of burning organic matter, is rich in potash, which has a similar effect as lime so it can also be used to help raise the pH.

If the pH needs to be raised, it has to be done in good time before the new level is required as changes take a few months and can take several years to have its full effect. The amount of lime to add depends on

- final soil pH to be achieved (normally raised to pH 6.5)
- soil pH from which it is to be lifted (as found in the soil test)
- texture of the soil (it is the clay content that resists the effect of lime)
- strength of the lime used (its 'neutralizing value')
- fineness of the lime.

Lime requirement is the quantity of calcium carbonate required to raise the pH of soils to pH 6.5.

A rough guide to how much lime to add can be found in Table 13.4. This table gives the quantities of calcium carbonate required to raise the pH of different soils to pH 6.5 (the recommended level to return mineral soils when their pH is too low for a future planting of 'normal plants'). 'Overliming' a soil must be avoided because this can reduce the availability of plant nutrients, which is easily done on very sandy soils.

Table 13.4 Guidelines for raising soil pH Quantities of calcium carbonate* (g/m^2 of ground chalk or limestone*) required to raise soil to pH 6.5.

Starting pH	Soil texture		
	Light soils	'Loams'	Heavy soils
	(very sandy, low clay)		(very high clay content)
5.0	1,000	1,200	1,400
5.5	700	800	1,000
6.0	400	500	600

* If using hydrated lime these application rates should be reduced by a quarter.

Further reading

Adams, C., Early, M., Brook, J. and Bamford, K. (2015) *Principles of Horticulture Level 3*. 7th ed. Routledge.

Archer, J. (1988) *Crop Nutrition and Fertilizer Uses*. Farming Press.

Brown, L.V. (2002) *Applied Principles of Horticulture*. Butterworth-Heinemann.
Cresser, M.S. (1993) *Soil Chemistry and Its Applications*. Cambridge University Press.
Defra. *Fertilizer Recommendations*. HMSO.
Haylin, J.L. (2004) *Soil Fertility and Fertilizers: An Introduction to Nutrient Management*. Prentice Hall.
Ingram, D.S. (2008) *Science and the Garden*. Blackwell Science.
Marschner, H. (1995) *Mineral Nutrition in Higher Plants*. Academic Press.
Postgate, J. (1998) *Nitrogen Fixation*. Cambridge University Press.
Pratt, M. (2005) *Practical Science for Gardeners*. Timber Press.
Roorda von Eysinga, J.P.N.L. and Smilde, K.W. (1981) *Nutritional Disorders in Glasshouse Tomatoes, Cucumbers and Lettuce*. Centre for Agricultural Publishing and Documentation.

The online material is accessible via the QR code and includes further information on many of the topics in the book, such as

- Plant nutrition
- Basic chemistry
- Organic growing
- Control of pH, acidity and alkalinity
- Lime and liming
- Soil testing
- Soil sampling
- Water is unusual
- Water potential

CHAPTER 14

Growing in containers

Figure 14.1 Hanging baskets on a public house to attract customers

This chapter includes the following topics:

- Growing in a restricted volume
- Loam composts
- Loamless composts
- Compost ingredients
- Plant containers
- Hydroponics
- Green walls
- Advantages and disadvantages of hydroponics

DOI: 10.4324/9781003581260-14

Growing in a restricted volume

Most people have at least some of their plants in containers, usually pots, tubs, troughs or hanging baskets. For some, their plant growing is confined to window boxes. A variety of containers are used for indoor plants, including quite large ones in conservatories. For others whose interest lies in the production of tomatoes, cucumbers and peppers in their greenhouses their containers often include 'grow bags'. Even non-horticultural businesses make good use of the colourful displays in containers that are there to create a pleasant working environment and also to attract customers (see Figure 14.1). Across the horticultural industry there are the equivalents of these methods of growing, and many more, but on a much larger scale.

Growing in containers makes more critical demands on the growing medium for air, water and nutrients because rooting is severely restricted. Compare the volume in containers with the volume that roots explore in soil (see p. 158). Consequently, growing in restricted root volume brings two main challenges:

- meeting the plant's water requirements while maintaining good gaseous exchange to ensure a continuing supply of oxygen and the removal of carbon dioxide
- providing the nutrient requirements in a small amount of water without 'scorching' the roots (see p. 125).

Soil is usually not suitable to use in containers as it tends to collapse when kept wet; most soil types lack the stability needed. If a pot full of soil is kept saturated, it is not long before the container becomes only half full of soil as a result of reduced pore space, that is, much of the air content has been lost. Also soil tends to harbour pests and diseases.

Stability in growing media means the ability to resist collapse in the presence of water.

Compost mixtures need to ensure adequate **air space** after being fully wetted and allowed to drain. Their ingredients must have **stability**. The degree to which the compost should have good **water-holding capacity** depends on the irrigation system to be used. The **nutrient content** of the ingredients needs to be allowed for; most are almost nutrient free, which can be advantageous as fertilizers can be added more precisely. Sometimes there is a need to incorporate 'heavier' (denser) components in order to provide 'pot stability', such as for taller plant specimens. In all cases, the material chosen should also be **partially sterile**.

Partially sterile growing media are those that have been treated to kill pests and diseases but not all organisms.

Consequently, alternative growing media, generally called composts, plant substrates, plant growing media, or just 'mixes' or 'media' are used rather than soil. Increasingly, in intensive production, the preferred alternative to growing in the soil is to grow in water (see hydroponics, p. 213). The limitations of soil for sports grounds leads to its replacement with more appropriate alternatives, such as graded sand for golf greens. Many sports such as tennis and hockey are played on non-grass surfaces that more easily provide a truer and more hardwearing surface (e.g. Astroturf).

Air-filled porosity (AFP)

The importance of supplying water to plants in a restricted root volume is usually understood, but less well appreciated are the difficulties associated with achieving it while maintaining adequate **air-filled porosity** (**AFP**). Roots require oxygen to maintain growth and activity.

Air-filled porosity (AFP) is the percentage of air in a growing medium immediately after it has stopped draining having been saturated with water.

As temperatures rise, the plant requires more oxygen, but the amount that is dissolved in water decreases. Even in cool conditions, the oxygen that can be extracted from the water provides only a fraction of the root requirements. So, unless the plants have special modifications to transport oxygen through their tissues, as in aquatics, there must be good gaseous movement through the growing medium. While large numbers of small pores (micropores) ensure good water-holding properties, it is the relatively large, interconnected pores that allow the rapid entry of oxygen. Creating successful physical conditions depends on the use of components that provide a high proportion of these larger pores (macropores).

It is generally considered that 10%–15% AFP is needed for a wide range of plants. Azaleas, cacti and epiphytic orchids require 20% or more, whereas others, including chrysanthemums, lilies and poinsettia, tolerate 5%–10% AFP. In contrast, carnations, conifers, geraniums, ivies and roses tolerate levels as low as 2%.

Ensuring that a growing medium in a container has adequate air-filled porosity is made difficult because much of the water is held in the container even when it has many drainage holes. Much of the water filling even the larger pores does not readily leave unless the compost has good contact through its holes with similar-sized pore spaces such as on sand or capillary matting. However, when standing out on a hard surface, coarse gravel or wire, the water will cling to the particles in the container. This can be tested by fully watering a pot of compost, holding it until it has finished dripping then touching the compost through a hole; normally a stream of water will run down your finger. Furthermore, unless stood out on appropriate material, the lower layers of the compost remain saturated (i.e. no air) irrespective of the height or width of the container. This makes it particularly difficult to get good aeration in shallow trays and modules. This is demonstrated when a washing sponge is fully wetted then left to drain; after water has left the sponge under the influence of gravity, the lower layers remain saturated.

The sizes of the components used in a compost must be selected to create the larger pores (macropores), but also to ensure that they are not filled in by smaller particles ('fines'). This is most easily achieved by using closely graded particles. The reverse of what is wanted for composts is achieved when combining many different-sized particles, for example making concrete, where the object is to minimize the air spaces, as illustrated in Figure 14.2.

Water-holding capacity

The water-holding capacity of compost ingredients varies enormously. However, the importance of this depends on how the plants in the compost are to be irrigated. It is a major consideration if the plants are in small hanging baskets watered by hand. Attempts have been made to improve water holding by adding additives to the compost. In contrast, if there is to be a constant supply of water provided through one of the many irrigation systems the water-holding capacity is far less significant.

Loam composts

Loam composts, typified by John Innes (JI) composts, are based on loam sterilized to eliminate the insects and soil-borne fungi (see damping off, p. 291) that largely caused the unreliable results from traditional composts. Furthermore, the loam should have sufficient clay and organic matter present to give good structural stability (the original specification identifies 'turfy clay loam' which is not easy to source). Originally peat was added to improve the physical conditions and provide high water-holding capacity but now a peat alternative is used. Finally, coarse sand/grit is included to ensure free drainage and therefore good aeration. The two main John Innes composts are JIP for potting and JI (S) for seed sowing and cuttings, and there is also a 'JI Ericaceous' mix. Further details of the history and formulation of JI composts can be found in the Support Material. Such composts are well proven and are relatively easy to manage because of the

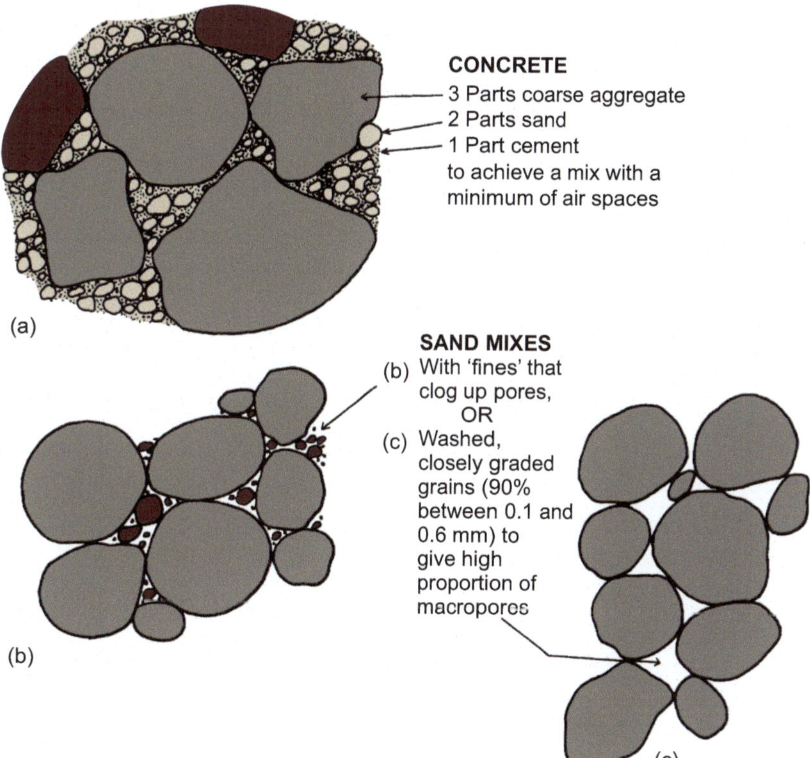

Figure 14.2 Pore spacing in (a) concrete; (b) sand with 'fines' and (c) closely graded clean sand/grit. Note the differences in pore spaces. It is the larger pores that enable the plant roots to have access to air as well as water

water-absorbing and nutrient-retention properties of the clay and organic matter present. They are commonly used by amateurs for valuable specimens and for tall plants where 'pot stability' is important.

The main disadvantage of loam-based compost has always been the difficulty in obtaining suitable quality loam ('turfy clay loam') as well as the high costs associated with steam sterilizing. Furthermore, the loam must be stored dry before use, and these composts are heavy and so difficult to handle in large quantities. Many loam-based composts currently made have relatively low loam content and consequently exhibit few of its advantages. Consequently, they have been superseded in horticulture generally by cheaper, lighter and cleaner alternatives.

Loamless composts

Loamless composts provide the advantages of a uniform growing medium, but with components that are:

- lighter
- cleaner to handle
- do not need to be sterilized (unless being used more than once)
- cheaper to prepare.

However, without the loam content the control of nutrients, including micronutrients, is more critical. Furthermore, many are more vulnerable to leaching of soluble nutrients. Most of the components have low nutrient levels to which manufacturers and growers who make their own compost can add nutrients accurately.

A very wide range of these composts is available but broadly fit into the following main categories:

Multipurpose composts, as their name suggests, have been formulated for a wide range of purposes. The components are sufficiently fine to ensure successful growing from seed or cuttings for at least the first four to eight weeks after which nutrients need to be topped up on a regular basis.

They are made from many different types of sustainable ingredients (traditionally peat was commonly used but now organic alternatives are used). Most contain added nutrients such as nitrogen, phosphorus, and potassium and invariably micronutrients (trace elements).

However, while multipurpose composts are versatile, they may not be suitable for all situations. Some plants, such as the calcifuges, have specific needs.

Seed and cuttings composts are more precisely adjusted for use with seeds, particularly fine seed and cuttings. These benefit from having finer, more closely graded components and lower soluble nutrient content; although phosphate levels are usually similar. Lime is added, so again, these composts are not suitable for calcifuges.

Container composts tend to be coarser and less closely graded than the ones described earlier. They are usually used as potting compost but are appropriate for the larger plants being grown in larger containers. Lime is added so again these composts are not suitable for calcifuges.

Ericaceous mixes are specifically designed for calcifuges ('lime-hating', 'ericaceous plants', see p. 202). They contain essentially the same ingredients as the composts described earlier but no lime is added.

Open composts. Composts with very high air-filled porosity typically made of composted bark are suitable for azaleas, cacti and orchids that require a very open compost.

Compost ingredients

A list of some of the materials commonly used in composts is given in Table 14.1.

Peat has, until recently, been the basis of most loamless composts or used alone but is now replaced by more sustainable, environmentally friendly alternatives. For development of peat see the Support Material. It is now being excluded from all composts because it is formed in areas important because of being;

- a unique natural habitat
- an important carbon sink: they store about a third of the world's soil carbon
- an archive containing archaeological and geochemical historical information going back hundreds of years
- an important part of flood prevention.

Table 14.1 Compost ingredients

Organic materials	Inorganic materials
Coir	Sands and grits
Composted bark	Perlite
Garden compost	Polystyrene
Green waste/compost	Pumice
Heather/bracken	Rockwool
Leaf-mould	Topsoil
Lignite	Vermiculite
Pine needles	
Seaweed	
Sewage sludge	
Spent hops and grains	
Spent mushroom compost	
Straw	
Vermicomposts	
Wood chips/fibre	
Woodwastes	
Wood fibre	

'**Green compost**'. There has been much interest in the use of suitably graded 'green composts' such as that from composted municipal waste (see p. 190). Although the product is too high in nutrients to be used alone, it lends itself to being part of a mix with composted bark.

Coir comprises the husk and dust particles of coconut processing. It has good water-holding capacity, rewetting and AFP characteristics. It has a pH between 5 and 6, which makes it suitable for a wide range of plants but is not suitable for ericaceous mixes. It has a high carbon to nitrogen ratio which means that allowance must be made for its tendency to lock up nitrogen when used in a compost or as a soil conditioner (see p. 187). It is increasingly available in dry compacted forms (bricks, rolls) that expand five to ten times their volume on wetting ready for use. It is a sustainable option but the need for it to be imported increases its carbon footprint.

Composted bark has been used as a mulch and soil conditioner for many years. There are many different types of bark and they have widely different properties. Its problems include the presence of toxins (overcome by composting) and a tendency to lock up nitrogen (see p. 187), which can be offset by adding nitrogen. When composted with sewage sludge or similar material rich in nitrogen, a material suitable as a compost is produced. In its processed form it is increasingly being incorporated into composts especially those needing a high air-filled porosity.

However, the main use of bark used in horticulture is for mulching. **Chipped bark** is popular as a mulch, frequently over a woven fibre to provide a more attractive finish. **Pulverized bark** has been used as a mulch and soil conditioner for many years.

Straw has been used with some success. In general, the main types available, wheat and barley, break down too easily and a practicable method of stabilizing them has not yet been found. Stable, friable material has been derived from bean and oil seed rape straws, although care is needed in mixes because of the high potassium levels and there is a need to add extra nitrogen because of its high carbon to nitrogen ration (see p. 187).

Straw bales were commonly used in the past with crops such as cucumbers planted into a soil layer while the straw decomposed below. It is a method now mainly limited to some keen amateurs.

Lignite is a highly variable soft brown coal formed from compressed vegetation, often found at the base of the larger peat bogs. The dusts ('fines') have been used as carriers for fertilizers and the more granular material can be used to replace grit in mixes usually bringing improved water retention.

Inorganic materials have always been used in composts, but there is now a wider choice available. Note that most of the inorganic alternatives are made from non-renewable resources (sand, loam, pumice) or consume energy in their manufacture (plastic foams, polystyrene) or both (vermiculite, perlite, rockwool).

Sand, grit or gravel are used to change physical properties. They have no effect on the nutrient properties of composts except by diluting other materials. As they are added to lightweight materials the density of the compost is increased which is important for ballast when a 'heavier' compost is preferred such as for tall plants grown in plastic pots. They should be introduced with caution because they tend to reduce the air-filled porosity of the final mix. They can also be used as an inert medium in aggregate culture (see p. 214).

Perlite is a glass-like mineral that is crushed and then expanded by heat to produce an inert white, lightweight aggregate available in grades from fine to coarse (Figure 14.3). It is devoid of nutrients and does not hold on to nutrients. The granules are porous with a rough surface so hold considerably more water than gravel or polystyrene balls. In composts it is used to improve the aeration and the

Figure 14.3 Examples of growing media; from top to bottom: rockwool, perlite, vermiculite, lightweight expanded clay aggregates

uptake of water. Perlite is valued alone or in mixes for propagation and used to cover seeds in seed trays. Graded samples filling 'pillows' can be used in hydroponic systems.

Care should be taken to use dust-free samples and avoid breathing in the dust.

Vermiculite is a mica-like (clay) mineral expanded to 20 times its original size by rapid conversion of its water content to steam. The finished product is available in several grades, all of which produce growing media with good aeration and water-holding properties (Figure 14.3). There is a tendency for the honeycomb structure to break down and go soggy. Consequently, for long-term planting, it tends to be used in mixtures with the more stable perlite. Some vermiculites are alkaline but the slightly acidic samples are preferred in horticulture. Vermiculite has high nutrient-holding properties which makes it particularly useful for propagation mixes; most samples contain some available potassium and magnesium.

Rockwool is derived from a granite-like rock crushed, melted and spun into threads. The resultant lightweight, spongy, absorbent, inert and sterile rockwool provides ideal rooting conditions with high water-holding capacity and good aeration. Shredded rockwool (see Figure 14.3) can be used in compost mixes and cubes, propagation blocks and plugs that can be modularized to create a complete growing system. However, it is most usually supplied as wrapped slabs.

Rockwool is also available in water-absorbent and water-repelling forms. Consequently, mixtures can be used to achieve exactly the right balance between air-filled porosity, water-holding and capillary lift, making an ideal medium for hydroponic growing (see p. 213). Its essentially inert nature means that a complete nutrient feed must be incorporated in the water of this sort of growing system. It is frequently used in tomato, cucumber and pepper production; film-wrapped cubes are available for plant raising and pot plants.

Pumice is a porous volcanic rock that is prepared for use as a growing medium by crushing, washing (to remove salt and fines) and grading. It is mainly used to grow long-term crops such as carnations in troughs or polysacks.

Expanded polystyrene balls or flakes provide a very lightweight inert material that can be added to soils or composts as a physical conditioner. It is non-porous and so reduces the water-holding capacity of the growing medium while increasing its aeration, thus making it less liable to waterlogging when overwatered. This has made it an attractive option for winter propagation mixes. However, it is less popular than it might be because it is easily blown away and sticks to most surfaces.

Expanded clay aggregates (Figure 14.3) such as Leca or Hortag are made from clay balls which are fired and then expanded in a rotary kiln to produce round balls with a smooth surface but honeycomb centre. They come in different grades to suit their many applications. They are lightweight materials with an attractive finish which makes them suitable for interior landscaping planters (containers). Their finish makes them attractive to add to the top of other container plants increasingly. They are increasingly being used in hydroponics (see Figure 14.5).

Plastic foams of several different types are becoming popular for propagation because of their open porous structure. They are available as flakes and balls for addition to composts or as cubes into which the cuttings can be pushed.

Plant containers

The final air-filled porosity (see p. 208) of the mixture that is experienced by the plant roots depends not only on the nature of the contents, but also on the characteristics of the container and the base on which it stands. If containers are stood out on wire mesh or on stones, relatively little water leaves, so the air (oxygen) content remains poor. It is also important to retain contact between the compost and the standing-out material through adequate holes in the base to help drainage, as well as to ensure the uptake of water if irrigated from below.

Plastic containers predominate the wide range of containers available. Most are very functional, but there are many decorative containers available with advantages in terms of lightness and cost. The black ones tend to heat up more than the standard terracotta-coloured ones, and the contents of white plastic pots can be as much as 4°C lower than in other colours. Pots of white or light green plastic can transmit sufficient light to affect root growth adversely and encourage algal growth. Translucent pots are a specialist requirement for growing some plants that have symbionts that need light, such as moth orchids.

Clay pots are porous and water is lost from the walls by evaporation. Consequently, clay pots dry out more rapidly than plastic ones. This is particularly useful in the winter and can help improve air-filled porosity. The higher evaporation rate also keeps the clay pots slightly cooler, which may be beneficial in hot conditions.

Biodegradable containers such as those made from paper have become popular because they can be planted directly into the soil. Some materials decompose more rapidly than others and there can be a temporary lock-up of nitrogen (see p. 187), but most such containers are now manufactured with added available nitrogen. It is essential that these containers are soaked well before planting and the surrounding soil is kept moist afterwards or the roots will fail to escape into the soil.

Modules are made by adding a loose growing medium mix to a tray of cells. Increasingly, traditional seedbed, bare-rooted or block transplant techniques have become replaced by raising a wide variety of plants in these modules. The cells are variously wedge or pyramid shaped, so designed to enable the contents to be easily removed and to make a highly mechanized transplanting process possible. Fine, free-flowing mixes of peat, polystyrene or bark are used to fill the cells, which have large drainage holes and usually no rim to hold 'free' water. Roots in the wedge-shaped cells are 'air-pruned' as they reach the edge of the cell, which encourages secondary root development.

'Plugs' are mini-modules in which each transplant develops in less than 10 cm^3 of growing medium and are used for bedding plants as well as vegetable production. Both are familiar to gardeners buying in garden centres or online. The rate of establishment is largely determined by the water stress experienced by the transplant. Irrigation of the module or plug is found to be more successful than applying water to the surrounding growing medium.

Hydroponics (from Greek hydro = water and ponos = labour/work)

Hydroponics (water culture) involves the growing of plants in water (see Figure14.4). The term often

Figure 14.4 Hydroponic lettuce growing: (a) nutrient film technique (NFT) in shallow troughs on the floor; (b) close-up of roots with no root hairs grown in gullies filled with clay aggregates; (c) gullies stacked vertically to maximize cropping area

includes the growing in an inert solid rooting medium watered with a complete nutrient solution, which is more accurately called '**aggregate culture**'. Plants can be grown in nutrient solutions with no solid material so long as the roots receive oxygen and suitable anchorage or support is provided.

This has been practiced in many forms for thousands of years, but more recently there have been significant developments, which meant that by the beginning of the 21st century a high proportion of the crops grown commercially in protected culture are in hydroponic systems.

These can be divided into two main types which in turn can be subdivided further:

- **solution culture** where there is no medium for the roots, simply nutrient solution, divided into
  – **static solution culture** where the roots are grown in bottles, buckets or tanks of nutrient solution and the air (oxygen) is supplied either by bubbling it through or keeping the water so shallow that sufficient oxygen is taken in from the water surface (rather similar to managing fish tanks),
  – **continuous flow culture**, for example nutrient film technique (see NFT, Figure 14.4a) in which the roots of the plants are held in a long, lightproof tube (to stop algal growth) with a flat bottom. Along this flows a thin film of nutrient solution where a thick mat of roots develop. An ideal rooting environment is created providing adequate water, oxygen and nutrients as long as the solution is correctly adjusted as it is recycled. In practice it has proved ideal for leafy crops such as lettuce rather than for fruits such as tomatoes (further information can be found in the Support Material).
  – **aeroponics** where the root environment is saturated with fine drops of nutrient solution, that is, misting roots suspended in a chamber from which unused liquid can be recirculated. This provides excellent aeration making a very successful propagation method. Fogponics uses a finer droplet size (fogging) that is claimed to be the ideal for nutrient uptake and more energy and water efficient than other systems.
- **Aggregate or medium culture** involves a nutrient solution but the roots are in an inert substrate. Sands and gravels have been used but although initially cheap, they now are considered too heavy and usually need to be cleaned and sterilized. Instead, the usual options now are rockwool culture, perlite culture and lightweight clay aggregate culture (see Figure 14.4b and c). The difficulty in disposing of large quantities from commercial operation has led to a renewed interest in biodegradable alternatives such as coir and wood fibre formulated appropriately. These can be subdivided into:
  - **sub-irrigation**. This can be passive whereby the plants grow in containers/pots filled with inert porous material, typically expanded clay aggregate, that is wetted with a nutrient solution by capillary action from a reservoir in the base (see Figure 14.5). This has proved a particularly attractive method of cultivating orchids as it provides an open root zone and a humid atmosphere. It is commonly used in interior landscaping.
  - **drip irrigation** has the plant roots in an inert growing medium, usually rockwool slabs or lightweight clay aggregate (Figure 14.6), fed by nutrient solution from above. At its simplest, nutrient is added

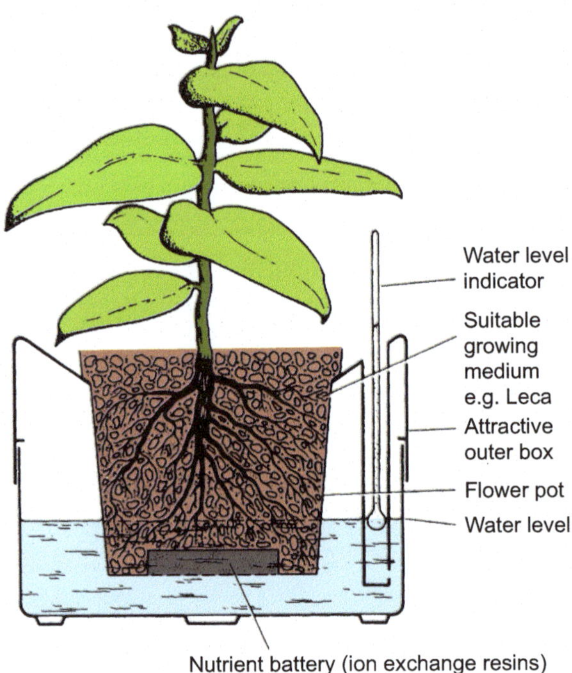

Figure 14.5 Planter with nutrient reservoir with aggregated clay balls; these are a frequent choice for interior landscapers as they are clean, lightweight and attractive

Figure 14.6 Drip feed supplying a lightweight clay aggregate hydroponic system

to the surface once a day. At the other extreme, a carefully adjusted constant drip feed is used although, most commonly in commercial situations, timers and pumps are used to add nutrient solution at appropriate intervals (an example of growing tomatoes in a Rockwool Hydroponics System is to be found in the Support Material).

Green walls

It has been a long-standing part of gardening to cover walls with foliage or flowering plants to add height to the garden display, cover up an unattractive area or increase the areas available for wildlife (nectar for bees, insects for birds, nesting sites). This has been done by planting with climbers that can cling to walls (*Hedera* spp., *Parthenocissus tricuspidata*) or by providing trellis or wire work to tie in wall shrubs or fan-trained fruit trees. The covering can have other benefits including a haven for wildlife, protection of the wall from rain and providing some insulation with some claiming reduction in noise and improved air quality.

The new **green wall** technology increases the possibilities by having a framework that can be secured to a wall or freestanding to provide platforms for growbags for plants and a drip fertigation system (providing nutrient in the water). Alternatively, some have cells that can be filled with compost and others are hydroponic systems. These green walls can be purchased as kits for the domestic garden usually in the form of panels that can be linked to cover the area required. Interior green walls (Figure 14.7) are supplied with appropriate lighting, and there are major installations on large buildings (Figure 14.8).

Figure 14.7 Interior green walls have extended the possibilities for interior landscapers

The plants can be grown from plugs or seed. Plants are selected from a wide range of decorative plants appropriate for the situation (sunny sites, shady/ north facing walls, exposed sites). What appeals to many gardeners is the possibility of growing salads, herbs and fruit (e.g. strawberries) in a similar way to the commercial production of crops such as lettuces (see Figure 14.4).

Managing the nutrient solution is fundamental to the success of the hydroponic systems. The nutrients can be made up on site from basic ingredients or, more often, from proprietary mixes. The mix must take into account not just the plants involved but also the stage of development and water quality. For many of the hydroponic systems a daily check of the water pH and nutrient concentration is needed, especially at critical times. Nutrient and pH control is achieved using, as appropriate, a nutrient mix, nitric acid or phosphoric acid to lower pH and, when the solution is too acid, potassium hydroxide to raise pH. Great care and appropriate personal protective equipment (PPE) is required to handle these chemicals safely.

Advantages of hydroponics

Much of the success of hydroponics is because the plant does not expend much energy producing roots. This is because water, oxygen and nutrients are easily accessed and there are usually not the complications inherent in other growing media such as over- and underwatering, soil-borne pests and diseases.

Hydroponics is suitable for stem, leaf, flower and fruit production. In the wider context, for many crops it can produce higher yields than other methods. Furthermore, one form or another can be used where other growing methods are not possible, including desert areas, roof gardens and walls. More specifically, advantages are:

- efficient water use, not least because in some systems the water is reused
- energy efficient compared with most alternatives
- soil-borne pest and disease problems minimal and easier to deal with than in soil (the main exception is *Verticillium* wilt)
- no nutrient pollution in closed systems (the reverse is true for 'open' systems)
- clean plants including roots (see Figure 14.4b).

Disadvantages

- Vigilance is needed to maintain the correct pH for the plants being grown because in water changes can happen very quickly.

14

Figure 14.8 Green wall in shopping centre

- pH control involves the use of acids and alkalis (bases) which require great care and appropriate PPE to handle safely.
- Some of the systems are vulnerable to equipment failure especially where automatic equipment is involved. This leads to the use of visual and/or audible alarms, back-up pumps and stand-by generators.

Further reading

Adams, C., Brook, J. and Early, M. (2014) *The Principles of Horticulture Level 3*. Routledge.

Bragg, N. (1998) *Grower Handbook 1 – Growing Media*. Grower Books.

Bunt, A.C. (1988) *Media and Mixes for Container Grown Plants*. Unwin Hyman.

Cooper, A. (1996) *The ABC of NFT*. 2nd ed. Casper Publications.

Handreck, K.A. and Black, N.D. (2002) *Growing Media for Ornamentals and Turf*. Revised 3rd ed. New South Wales University Press.

Resh, H.M. (2015) *Hydroponics for the Home Grower*. CRC Press (Taylor & Francis Group).

Smith, D. (1998) *Grower Manual 2 Growing in Rockwool*. Grower Books.

The online material is accessible via the QR code and includes further information on many of the topics in the book, such as

- John Innes composts
- Tomatoes in rockwool hydroponics
- NFT and aggregate culture
- Carbon footprint of growing media
- Peat development

CHAPTER 15

Plant health maintenance

Figure 15.1 Fine mesh netting to protect strawberries from bird damage to the fruits (source: Shutterstock, iva)

This chapter includes the following topics:

- Definitions, benefits and limitations of physical, cultural, biological and chemical control methods
- Integrated pest management
- Biosecurity and phytosanitary requirements in Great Britain and Ireland
- Garden health plans
- Potential impacts of a changing climate on plant health

DOI: 10.4324/9781003581260-15

In the process of maintaining plant health, horticulturists should consider the whole range of prevention and control methods to manage pest, disease and weed problems including physical, cultural, biological and chemical methods and integrated pest management (IPM). For some types of diseases, cultural methods such as managing soil pH or selecting resistant cultivars can be effective, for example in reducing the incidence of clubroot in brassicas. With small pests such as insects, physical controls, such as deterrents and traps, and biological controls are often used to reduce the use of chemical controls. This is particularly appropriate in protected environments where the environmental conditions can be manipulated to provide the necessary conditions for the effective use of biological control organisms. Physical methods of weed control such as hoeing of annual and seedling weeds on dry days is a highly effective method with few of the risks of chemicals.

While the more environmentally preferable, non-chemical methods may appeal to many horticulturists, particularly organic growers, there will be occasions when chemical control might be necessary; control of blight on potato by professional growers in warm, humid weather in summer would be one example. IPM utilizes a range of compatible physical, cultural, biological and chemical methods and is often used in commercial production of greenhouse crops such as tomatoes, where pests such as glasshouse whitefly can be controlled effectively. Garden health plans consider all potential impacts on plant health, not just biotic factors such as pests, diseases and weeds, and are considered to be a more holistic approach to plant health management in the garden than a conventional IPM programme.

Physical control

Physical control is the use of hand or mechanical means to remove and destroy weeds, pests or diseases, or the use of materials to block or trap them.

Benefits of physical control:

- once established, they often remain for a long period of time, for example wire or wooden fencing used against rabbits and deer
- avoids the use of pesticides which reduces the risk of development of resistance in the target organism.

Limitations of physical control:

- some physical methods are expensive to establish on a large scale, for example fencing used against rabbits
- can be labour intensive on a large scale.

Safe practice and environmental effects. In physical control, some hazards are:

- use of hazardous equipment such as ploughs, rotary cultivator and steam sterilization equipment
- hazardous removal of large plants such as trees
- burning of infected plant material.

Natural balances:

- Of the physical controls described next, the most significant method that may affect natural balances is **partial soil sterilization**. If a recently sterilized soil or growing medium contains plants infected with a disease such as 'damping off' or *Fusarium* wilt, the spread of the disease may be more rapid than in a non-sterilized soil. This is because naturally occurring bacterial and fungal competitors to the disease-causing pathogen have been destroyed by the heat process.

Examples of physical controls

Mulches and barriers

Mulches. Bulky organic matter such as leaf mould and composted garden waste have numerous benefits for the soil and garden when used as mulches (see p. 188), including their effectiveness as weed suppressants. A layer of organic mulch 5–10 cm deep can be applied on the soil surface of beds and borders in late winter and early spring before weed seed germination commences; the mulch blocks the light and the germinating seedlings perish before reaching the surface. This method is not effective at controlling established perennial weeds. Care should be taken to avoid covering low-growing or emerging plants, and the lower stems of woody plants should have a mulch-free gap around them. **Weed suppressant fabrics** can be laid on the ground as sheet mulches with planting taking place through slits in the material. In ornamentals and edible crop plantings such as strawberry, they are an effective way of reducing weed problems. Polythene and woven polypropylene sheeting have been used for many years but biodegradable forms of sheeting such as those based on **wool** are also available and can be used for this purpose.

Risks: when the synthetic fabric sheets are no longer needed, they should be disposed of at a local authority recycling centre or collected by an Environment Agency registered waste disposal contractor. They should not be burnt as they release hazardous fumes.

Horticultural fleece placed over growing crops helps to prevent the access on to crop plants of flying pests such as carrot fly adults and the large cabbage white butterfly, and **fine mesh netting**

Figure 15.2 (a) Glue bands which trap female winter moth on fruit trees; (b) yellow sticky trap for monitoring and controlling flying pests in greenhouses; (c) lettuces planted to attract slugs away from tomatoes

protects fruit plants from bird damage (Figure 15.1). **Fences** use sturdy wire-mesh, no larger than 30 cm mesh, approximately 120 cm high and sunk into the ground to a depth of about 30 cm with the lower part bent outwards to deter rabbits and small deer from digging under the fence. Micro-mesh **screens** placed over ventilation fans help to prevent the entry of pests such as sciarid flies or aphids from outside a greenhouse.

Traps

Pheromone traps containing synthetic substances similar to the attractant odour of the female insect are commonly used by growers to lure male codling and tortrix moths which are then trapped on a sticky surface inside the trap. This enables an accurate assessment of their numbers and a more informed decision on effective control methods and timing. It also prevents the trapped males from mating with females, thus reducing offspring. Similar traps are available for other pests such as plum moth, pea moth, leek moth and raspberry beetle.

Risks: minimal.

Insect glue bands (Figure 15.2a) are wrapped around the trunk of apple trees in autumn to prevent the wingless winter moth females from crawling from the soil up into the tree to mate and lay their eggs. This method should be used sparingly as some beneficial organisms can be trapped by the glue band.

Coloured sticky traps (Figure 15.2b) are used in greenhouses to detect the presence of pests before they cause significant damage to the crop: they also help to monitor the effectiveness of control measures and offer some level of control.

Figure 15.3 Marigold placed in a greenhouse to deter glasshouse whitefly (source: Shutterstock, thoughtsofjoyce)

The trap colour attracts the winged adult stages of certain insect pests such as **yellow** traps for thrips, aphids, leaf miner, sciarid fly and whitefly and **blue** traps for thrips.

Deterrents

Marigold (*Tagetes* spp.) can be grown in greenhouses adjacent to crops such as tomato (Figure 15.3) to deter pests, such as glasshouse whitefly, which are repelled by marigold from attacking the crop. **Citronella** extracts and aluminium ammonium sulphate are present in commercially available deterrent products and can be sprayed on to susceptible plants to deter foxes, badgers and rabbits.

Cultural control

Cultural control is a process in which manipulation of the growing environment results in conditions which prevents or controls weeds, pests and diseases.

Benefits of cultural controls:

- avoids the use of pesticides, thus reducing risks to the crop, humans, non-target organisms and the environment
- avoiding the use of pesticides reduces the risk of development of resistance in the target organism.

Limitations of cultural control:

- may be time-consuming to carry out on a large scale
- may lack the rapid control that is seen with some chemical controls.

Safe practice and environmental effects. While most controls described next need no detailed comment in this section, **vegetatively propagated** material such as bulbs and tubers, which are commonly used by horticulturists, bring with them the danger of spreading pests and diseases such as white rot of onions. Such material needs careful examination before being planted.

Natural balances. Two aspects of cultural control may be highlighted in terms of natural balances:

- **Rotation** of crops can effect a beneficial change in the fungal and bacterial population in the soil. Soil pests such as **potato cyst nematodes** may be reduced by soil fungi, and roots containing diseases such as **club root** may be rotted by soil bacteria and fungi at a time when the host crop (potato and cabbage in these two instances) is absent.

Examples of cultural control

Soil fertility

The correct level and balance of major and minor nutrients in the soil is recognized as vitally important for optimum crop yield and quality; it should be remembered that **plant resistance** to pests and diseases is also affected by nutrient levels in the plant. **Excessive nitrogen** levels, causing soft leafy growth, encourages the increase of peach-potato aphid or grey mould on many species of cultivated plants and 'Fusarium' patch on lawn grasses. Adequate levels of **potassium** can help the plant to resist the severity of fungal diseases – for example, *Fusarium* wilt on plants such as peas, tomatoes and carnation. Fertility provided by well-composted organic matter can provide essential nutrients to plants.

Club root disease of brassicas is less damaging in soils with a **pH** (see p. 201) greater than 7, and garden lime in the form of calcium carbonate can be incorporated before planting these crops to achieve this aim.

Risks: the use of increased levels of fertilizers to grow bigger crops and vigorous, dense turf may (particularly in **sandy** soils) lead to nutrients leaching into drains and streams that may cause an increase in algal bloom with poor water quality and death of water animals in streams and ponds.

Soil cultivation

Timely and appropriate management of soils with structural problems such as a soil pan can be carried out by double digging (or subsoiling on a large scale) to rapidly alleviate the soil pan and enable an immediate physical improvement in **soil structure** (see p. 167) in preparation for the growing of crops. The improved structure and drainage and can help reduce diseases such as club root in brassicas and disorders caused by waterlogged soils (see p. 308). Ephemeral and annual weeds such as chickweed (*Stellaria media*) can be buried and soil pests such as leather jackets and wireworms can be exposed to natural predators such as birds. It also allows for the incorporation of bulky organic matter at depth. Minimal or no-dig soil management methods should be considered where possible, as excessive cultivation can disrupt beneficial soil fauna, bring weed seeds to the surface and accelerate the decomposition of soil organic matter. Cultivation during periods of high soil water content can damage soil structure.

Crop rotation

Some important soil-borne pests and diseases affect specific crops. For example, potato cyst nematodes (see p. 283) affects potatoes and tomatoes and clubroot (see p. 293) infects members of the Brassicaceae family. As these problems are soil borne, they are naturally relatively slow in their spread but are difficult to control, largely because they have a life-cycle stage that survives for many years in the ground, even when the host plant is absent, although the simple method of rotation by planting crops in different plots each year can see the pest or disease numbers slowly decline.

A grower can create **four beds** or plots, sometimes bounded by wooden boarding to isolate each plot, and thus create a system to achieve effective rotation (Figure 15.4). Groups of plants, often belonging to the same family, can fit into the rotation system, because they often have susceptibility to the same pests and diseases. The main groups are the **Solanaceae** family, including potato, tomato, pepper and aubergine; **Brassicaceae** family, including cabbages, cauliflowers, Brussels sprouts, swede, turnip and radish; **root crops**, including a diverse range such as beetroot, carrot, parsnip, celeriac and celery; **legumes** (species which were formerly members of 'Leguminosae' family) including peas and beans; and **Alliaceae** family, including onion, leek, shallot and garlic. When considering a rotation plan, it is recommended that the grower tries to restrict species from the same family to the same plot in the same growing season. A typical **four-year rotation** on a four bed plot would be as follows: year 1, Legumes; year 2, Brassicas; year 3, Solanaceae; year 4, Alliaceae and root crops. Over a four-year period, each group will be grown on a different bed.

Rotation is not so effective against pests and diseases which are not host specific such as grey mould (*Botrytis cinerea*), which may infect a wide range of species from all groups and is also not likely to be effective against common and rapidly spreading airborne pests such as peach-potato aphid.

Risks: the **sclerotia** stage (see p. 289) of **white rot of onion** (*Stromatinia cepivora*) and related crop such as leeks, garlic, chives and shallots is an

Figure 15.4 A bed system of growing vegetables facilitates effective rotation of crops to help reduce problems with pests and diseases (source: Shutterstock, AleksB59)

especially difficult disease to manage. The sclerotia can survive in the soil for 15 years or more. A very long rotation period would be necessary to remove this disease completely. A four-year rotation would not normally be sufficient.

Crop spacing

Crop spacing can be an effective cultural method in reducing competition from weeds. Each species or crop has an optimum spacing based on the plant's growth rate, height and spread and the grower's production objectives. However, weeds can compete effectively with cultivated plants when there are big spaces between the crop plants. Consideration can be given to growing plants as closely as possible to minimize the opportunity for weeds to establish in the spaces between them.

Planting and harvesting times

Some pests emerge from their overwintering stage at about the same time each year (e.g. **cabbage root fly** in late April). By planting early to establish resilient brassica plants before the pest emerges, a useful cultural control can be achieved. Similarly, the deliberate planting of early potato cultivars enables harvesting before the maturation of **potato cyst nematode** cysts (see p. 283), so that damage to the crop and the release of the nematode eggs are avoided. Annual weeds may be allowed to germinate in a prepared seedbed. Once they have been controlled by hoeing shallowly on a dry, warm day or applying a non-residual, contact herbicide, such as **fatty acids**, **acetic acid** and **pelargonic acid**, a crop may then be sown into the undisturbed bed or **stale seedbed**, with less chance of further weed competition as the crop seed germinates and establishes.

Companion planting

A common practice in the garden is the cultivation of two or more plant species growing closely together to provide benefits to both or one of the species (Figure 15.2c). Such a situation may seem at first sight to encourage competition rather than mutual benefit, and many of the commonly used **companion planting** groupings are based on anecdotal rather than scientific evidence. However, some growers find certain groupings do work and they routinely use companion planting in their garden.

Two of the many recommended companion-planting groupings are:

- nasturtium (*Tropaeolum majus*) will attract the large cabbage white butterfly to lay eggs on them rather than on brassica vegetables
- poached egg plant (*Limnanthes douglasii*) will attract hoverfly adults to the vicinity, and they in turn lay their eggs on a range of species where aphids are pests; the larvae of hoverflies are aphid predators.

Vegetative propagation material

Such material is used in many areas of horticulture as bulbs (e.g. tulips and onions), tubers (e.g. dahlias and potatoes), runners (e.g. strawberries), cuttings (e.g. chrysanthemums and many woody species) and graft scions of some woody plants. The increase in nematodes, viruses, fungi and bacteria by vegetative propagation is a particular problem, since the organisms can be on or inside the plant tissues, and the vegetative plant tissues are sensitive to any drastic control measures. Thorough **inspection** of introduced material upon receipt can reduce the risk of this problem. Soft, puffy narcissus bulbs, chrysanthemum cuttings with an internal rot, glasshouse whitefly or two-spotted spider mite on stock plants, and virus on nursery stock are all symptoms that would suggest either careful treatment or rejection of the plants.

Healthy stock schemes

Quality of fruit planting material is assured in the UK by the **Fruit Propagation Certification Scheme**. Healthy planting stock of soft-fruit, top-fruit, rootstocks and ornamental cultivars associated with top fruit and micropropagated material of any of these crops is available to growers from government certified and inspected commercial plant producers. The UK government's **Seed Potato Classification Scheme** aims to determine that all seed potatoes delivered to buyers and growers are healthy, true to cultivar and free of mixtures. In the UK, all classified seed potatoes are derived from nuclear stock which is tested to establish they are free of quarantine (notifiable) organisms and certain other pathogens.

Hygienic growing

Some examples of the horticultural practices which reduce pest and disease problems in cultivated plants are included next. During the plant's life, the grower should aim to provide optimum conditions for growth. Soil irrigation should be appropriate to achieve healthy growth (see **field capacity**, p. 174), but not so excessive that root diseases such as damping off in container-grown seedlings and club root of brassica are actively encouraged.

Water sources

Covering of water in tanks in greenhouses helps to prevent the spores of fungal-like organisms such as damping off infecting irrigation water.

Removal of alternative hosts

Alternative hosts are plant species that some pests or pathogens use as a host in addition to their preferred host species and should be removed where possible. Common examples of weeds which could become alternative hosts are the following:

- clubroot, which infects shepherd's purse (*Capsella bursa-pastoris*)
- black bean aphid, which infects fat hen (*Chenopodium album*)
- in greenhouses, two-spotted spider mite can be found on chickweed (*Stellaria media*).

Removal of infected plant material

In fruit tree plants such as apple, routine **pruning** operations may remove serious pests, such as fruit tree red spider mite eggs, and diseases such as apple canker and powdery mildew. Pruning should also aim to reduce the density of shoots in the centre of the tree. The resulting reduction in humidity provides a microclimate that is less favourable to these diseases. **Tree stumps** harbouring diseases such as honey fungus (*Armillaria* spp.) should be removed promptly. Leaving infected stumps is not recommended in a garden where honey fungus is present as the fungus can continue to live on the dead wood for many years (Figure 15.5).

Garden biodiversity

Removal of dead plant material from gardens can be considered from the opposite point of view and the subsequent effect on garden biodiversity. Many **beneficial species**, such as ladybirds, ground beetles, centipedes and ichneumon wasps, use the **dead hollow stems** of garden perennials as overwintering refuges. Hedgehogs may spend the winter within piles of dead branches. Horticulturists may wish to consider a **balance** between tidiness and their desire to achieve a diversity of organisms which in turn leads to natural pest control.

Figure 15.5 Dead tree stumps can be a long-term source of honey fungus infections (source: Shutterstock, berni0004)

Biological control

Biological control is the use of natural predators, parasites and pathogens to control pests and diseases. The process involves the horticulturist directly introducing biological control agents into the growing environment.
Predators such as *Phytoseiulus persimilis*, a mite used to control two-spotted spider mite, eat their prey. **Parasites** such as *Encarsia formosa*, a wasp used to control glasshouse whitefly, lays an egg inside a whitefly larva and the developing wasp larva consumes and destroys the whitefly. **Pathogens** such as *Lecanicillium muscarium* Ve6, a fungus used to control glasshouse whitefly and aphids, develops on and enters the pest's body, ultimately killing it.

Benefits of biological control:

- it avoids the build-up of chemical residue in crops and in the environment
- the risks to humans from pesticides are reduced
- it reduces the development of pesticide resistance in pests and diseases.

Limitations of biological control:

- highly specific environmental conditions are often necessary for it to be effective
- limits the use of other control measures such as chemical control
- requires some of the pests to be present before introduction can take place.

Safe practice and environmental impact. The main problems with biological control are the following:

- unsuccessful application of biological control organisms can lead to a severe pest problem (e.g. late introduction of *Encarsia formosa* wasp used against glasshouse whitefly)
- the potential for non-native organisms which are used as biological control agents subsequently becoming an environmental problem themselves is always present. However, the release of all biological control agents is regulated, and organisms being used for this purpose must first receive government approval.

Risks can be minimized by the following:

- understanding both the pest's and predator's or parasite's life cycles in order to achieve reliable control

- choosing the most suitable biological control agent for the pest or disease concerned
- taking care that non-target organisms are not subject to the activities of the biological control agents.

Natural balances. In most horticultural situations, there are examples of **natural balance** between species:

- With pests, their naturally occurring **predators and parasites** are an important form of plant protection.
- With diseases, naturally occurring parasites are less well understood, but the **nutritional** condition of the plant and the resulting naturally occurring bacterial and fungal populations on leaf, stem and root surfaces (see **organic growing**, p. 151) often help to slow a disease's progress.
- The garden represents a complex and diverse environment. There may be plant species present from numerous continents and any of these plant species may be accompanied by a specific pest from its country of origin. Plant species that have been **established** in Britain and Ireland for many years (e.g. apple) often have beneficial predators and parasites (e.g. *Aphelinus mali*, a wasp parasite on apple woolly aphid), which were introduced accidentally or deliberately from their country of origin, that reduce pest numbers. It is quite likely, however, that for the more **recently introduced** plant species, there may not be appropriate predators or parasites to control a recently introduced pest occurring on the plant species in Britain and Ireland.
- Some horticultural practices can **disturb natural balances**. In a natural habitat such as a woodland, a **climax population** of plants and animals develops. Here, a complex balance exists between indigenous pests and their **predators/parasites**. The **food webs** include several types of predators/parasites found on each plant species that control (but do not eliminate) the pests. This development of food webs is not achieved to such an extent in many gardens, since a natural succession of wild plant species mentioned earlier is not desirable to horticulturists who are aiming for optimum production of edible crops or for a pleasing aesthetic layout of decorative plants free from weeds.
- Regular **movement or removal** of cultivated plants without thought to the natural balance between predator/parasites and pests will make pest attacks more likely in the garden/nursery situation. For example, if a horticulturist stops growing the poached egg plant (*Limnanthes douglasii*) the number of useful hoverfly (feeding on its pollen) may be reduced.
- The **removal** of the dead stems of herbaceous perennials and grasses in autumn and branches of decaying wood, which are common winter sheltering sites for parasitic wasps, predatory beetles, ladybirds and centipedes, may reduce the potential for control of pests.
- Poor **soil structure** (see p. 167) resulting from inappropriate soil management methods in a garden may hinder the movement of useful predatory animals such as centipedes in their search for underground soil pests.
- A poor physical preparation of soil and lack of attention to **pH and nutrient** levels in soil may result in poor soil microbial action (see p. 182).
- The **repeated planting** of edible crops or ornamentals into the same area of soil often leads to serious attacks of persistent soil-borne pests or diseases. Notable examples are club root disease on brassicas (see p. 293) and potato cyst nematode pest on potatoes (see p. 283). A comparable unbalanced situation is found when roses are planted into soil previously occupied by old specimens of roses with the resulting problem called '**replant disease**' thought to be caused high levels of nematodes (see p. 283) and fungal root pathogens.
- The inappropriate or excessive use of **pesticides** may result in a rapid decrease in natural predators and parasites and may considerably delay their appearance and build-up in the following growing seasons.

Examples of biological control species

There are two sources of 'natural enemies' to pests (and occasionally diseases): the **indigenous** species (they are locally present naturally in wild plant communities) and the **exotic** ones which have been introduced from other countries. Garden pests may be controlled by **predators** which eat the pest, or by **parasites** which inhabit the body of the pest at some stage in its life cycle; both methods lead to death of the pest. These beneficial organisms can be encouraged by certain gardening practices and in some cases deliberately introduced by the grower as biological control agents. A range of beneficial organisms useful in horticulture is described next.

Indigenous predators and parasites

Wild birds can contribute greatly to the control of garden plant pests. **Blue tit** (*Cyanistes caeruleus*) consumes significant quantities of aphids and small caterpillars such as the larva of the winter moth. They will also eat scale insects, which are otherwise notoriously difficult to control.

Figure 15.6 (a) A blue tit box; (b) blue tits at a bird feeder (source: Shutterstock, Jgade)

The installation of tit boxes (Figure 15.6a) and feeders (Figure 15.6b) are effective ways to encourage them into a garden.

Hedgehog (*Erinaceus europaeus*). Although their preferred diet is beetles, caterpillars and earthworms, they will also eat slugs. Hedgehogs can be encouraged to enter gardens by means of small holes cut into the base of a fence panel. Wooden hedgehog shelters are commercially available for placing in quiet areas of gardens. Heaps of logs and piles of leaf litter in a quiet location are suitable for their daytime and overwintering retreat. Care should be taken in winter that hibernating hedgehogs are not burnt in bonfires. If the garden contains a pond, it should be remembered that hedgehogs can fall into the water at night, and will avoid drowning if a part of the pond wall has a gentle slope or a small ramp provided for their escape.

Frogs and toads commonly leave their ponds in damp weather and may contribute greatly to the control of slugs and ground-living insect pests. The **common frog** (*Rana temporaria*) is smooth-skinned, about 7 cm in length, and greenish-brown or yellow in colour. It is seen most commonly from March to October. The frog's egg mass is laid in spring as a large round clump, usually in shallow water. The species' numbers have decreased in the countryside in recent years, and it is most commonly seen nowadays in garden ponds. The introduced green and brown **marsh frog** (*Pelophylax ridibundus*), found mainly in Kent and East Sussex, is large (up to 17 cm in length), has dark blotches on its body and often has a yellow stripe down its back. The **common toad** (*Bufo bufo*) has a grey or brown, warty appearance. The female may reach 9 cm in length, while the smaller male is more commonly 6 cm. The toad's egg mass is laid in spring in long strings. This species may be active all year round in Britain when the weather is mild. It prefers deep ponds. The introduced **midwife toad**, which can reach 5 cm in length, is found most commonly in Bedfordshire, South Yorkshire and Devon. The male holds on to the egg mass on his back (hence the common name). This species makes a characteristic high-pitched piping sound. These amphibians commonly leave their ponds in damp weather and may contribute greatly to the control of slugs and ground-living insect pests such as vine weevil adults.

Lacewings (e.g. *Chrysoperla carnea*) (Figure 15.7a) are pale green insects, 1.5 cm, which fold their transparent wings over their bodies when at rest. Several hundred eggs are laid per year, each on the end of fine stalks, on the underside of leaves. They are useful pest predators, their larvae (Figure 15.7b) eat aphids and mite pests, often reaching the prey in leaf folds where ladybirds cannot reach. They also are used commercially in greenhouses.

Ladybirds. There are approximately 40 British species of **ladybirds** and many are predators of plant pests. The red **seven-spot ladybird** (*Coccinella septempunctata*) (Figures 15.8a,b,c,d) emerges from the soil in spring, mates and lays about 1,000 elongated yellow eggs on the leaves of a range of plants, such as nettles, and crops such as beans throughout the growing season. Both the emerging slate-grey and yellow larvae and the adults feed on a range of aphid species. Wooden ladybird shelters and hotels are now available to encourage the overwintering of these useful predators. A development in the recent years has been the rapid spread and increase of the **harlequin ladybird** from South East Asia. This species is larger (6–8 mm long) and rounder than

Figure 15.7 (a) Lacewing adult (source: Shutterstock, Roger Meerts); (b) lacewing larvae control aphids (source: Shutterstock, Muddy knees)

Figure 15.8 (a) Ladybird adult; (b) ladybird eggs; (c) ladybird larva (source: Shutterstock, Tomasz Klejdysz); (d) ladybird pupa (source: Shutterstock, Martin Fowler); (e) ground beetles are predators on soil-borne pests (source: Shutterstock, Marek R. Swadzba)

the seven-spot species (4–5 mm). It has a wider food range than other ladybird species, consuming other ladybirds' eggs and larvae, and eggs and larvae of moths.

Hoverflies (Figure 15.9a), superficially resembling common wasps, are commonly seen darting or hovering above flowers in summer. Many of the 280 plus British species (e.g. *Syrphus ribesii*) lay eggs in the vicinity of aphid colonies, and their legless light green-coloured larvae consume large numbers of aphids. The flowers of some garden plants are especially useful in providing pollen for the hoverfly adults and therefore encouraging aphid control in the garden; examples include poached egg plant (*Limnanthes douglasii*) and Californian poppy (*Eschscholzia californica*).

Other beneficial insects include the common wasp (Figure 15.9b), parasitic wasps, anthocorid bugs and ground beetles (Figure 15.8e). Predatory mites and spiders (Figure 15.9d), naturally occurring parasitic nematode eelworms are also important in controlling pest numbers.

Insect pests may also be parasitized by parasitic fungi. The numerous species of web-forming and hunting spiders (Figure 15.9d) are very important in the reduction of many types of insect pests.

Occasionally, weeds are controlled biologically. The **cinnabar moth** larva (*Tyria jacobaeae*) eats the foliage of ragwort (*Jacobaea vulgaris*). White blister (*Pustula* spp.) is commonly seen infecting ragwort and groundsel (*Senecio vulgaris*) but unfortunately it also infects florist's cineraria (*Pericallis* × *hybrida*). Attention should always be given by horticulturists to the careful selection and use of chemical controls (if they are needed) to avoid harm to the predators and parasites described earlier (see p. 228).

Exotic (introduced) predators, parasites and pathogens

In greenhouses and polythene tunnels, **high temperatures** (often all year round) and exotic species of plants bring with them pests and diseases from other countries; they increase very rapidly and may quickly become resistant to pesticides.

Biological control of exotic pests usually requires exotic **predators and parasites,** and so the health of the major greenhouse crops in Britain and Ireland is due in part to these introduced predators and parasites. Two organisms, a South American mite that eats all stages of the two-spotted spider mite and a tiny chalcidoid wasp that parasitizes the glasshouse whitefly, have been used for many years and are briefly described next.

Phytoseiulus persimilis (Figure 15.10) is a 1 mm globular, orange, predatory tropical mite used in greenhouse production to control two-spotted spider mite (see p. 281). The predator's short egg–adult development period (seven days), laying potentially (50 eggs per life cycle) and appetite (five pest adults eaten per day) explain its extremely efficient action.

Encarsia formosa is a small (2 mm) wasp, which lays an egg into the glasshouse whitefly third and fourth **larval stages** (see p. 268), causing it to turn black and eventually die before the wasp, which has developed inside it, appears. It requires temperatures above 17°C to be effective.

An understanding of each pest's and each biological control organism's life cycle is required to ensure success in control. There are now many commercially available biological control agents, which are available from specialist companies, that are used to control garden and greenhouse pests (Figure 15.9c) such as aphids, caterpillars, glasshouse whitefly, leaf miners, mealy bugs, two-spotted spider mite, sciarid fly, thrips, vine weevil and slugs. Biological control of diseases is limited but developing. Biopesticides such as the parasitic fungus *Trichoderma harzianum* **T-22** is commercially available as a granular formulation which can be mixed with or spread over the growing medium or dispersed in water and applied via an irrigation system to protect many species of plant against a range of soil-borne root diseases such as *Pythium* spp. (damping off) and *Fusarium* spp. These biological control agents have been selected for their effectiveness against their chosen pest or disease species. Also, they have been tested to ensure they do not interfere with the natural balances in the garden/local environment.

Chemical control

Chemical control is the use of a pesticide to prevent or control a weed, pest or disease.

The number of pesticides available to horticulturists has decreased in recent years, reflecting the need to use only products that present minimal risks to human and non-target animals, plants and the environment. All professional users are legally required to be appropriately trained and certified to carry out the application of pesticides.

Benefits of chemical control:

- a wide range of products are available for pests, diseases and weeds
- can be cost-effective on a large scale.

Limitations of chemical control:

- can be hazardous to humans, animals and plants and the environment if not used correctly
- chemical control can cause resistant strains of pests, diseases and weeds to develop.

Figure 15.9 (a) Hoverfly feeding on pollen and nectar; (b) common wasp (source: Shutterstock, Ger Bosma Photos); (c) *Cryptolaemus* ladybird predator on mealy bug (source: Shutterstock, Protasov AN); (d) garden spider (source: Shutterstock, novama)

Figure 15.10 *Phytoseiulus persimilis* feeding on a two-spotted spider mite

Safe practice and environmental impact. In chemical control, the **hazards** include the following possible outcomes:

> A **hazard** is something with the potential to cause injury or illness.

- acute poisoning of humans, domestic animals, farm animals and wild animals
- accumulation of pesticides that lead to harmful levels in humans, animals, plants and the environment
- damage to cultivated and wild plants.
- contamination of soils and waterways
- development of strains of rodents, insects, mites and fungi resistant to pesticides.

Risks can be controlled and reduced by adhering carefully to the following:

> A **risk** is how likely and how seriously someone could be harmed.

- Accurate identification of the pest, disease or weed problem to ensure any methods of control used have a realistic chance of success.
- Restricting pesticide applications to only those situations that justify such a control measure; in many instances, physical, cultural and biological measures may be more effective and involve fewer risks.
- Choosing the least hazardous pesticide to effectively control the problem organism.

- All pesticide products used in the UK. Ireland must have **government approval**, and the situation regarding approval changes frequently; users of pesticides have a responsibility to ensure that any product they use has current approval. Approval of products for **professional use** can be checked in **The UK Pesticide Guide** which is published annually as print and online versions by the **British Crop Production Council** (BCPC). Approval for products for use by **home gardeners** can be checked on the **Health and Safety Executive**'s website; the Royal Horticultural Society also publishes details of products approved for home gardeners on their website which are updated regularly. Pesticides are regulated in the Republic of Ireland by the Pesticide Registration and Control Divisions (PRCD) of the **Department of Agriculture, Food and the Marine.**
- For professional users, the operator must be appropriately trained, certified and equipped to carry out application of the product safely and legally. Certificates of competence for the use of a range of pesticide application equipment in the UK are awarded by **NPTC/City and Guilds Land Based Services.**
- All professional users in the UK are legally required to comply with the requirements of the Health and Safety Executive's **Code of Practice for Using Plant Protection Products**. This gives specific information on training and certification, planning and preparing for work, working with pesticides, disposing of pesticide waste and keeping records.
- **Home gardeners** should always ensure that they read the product label and adhere fully to the instructions to ensure the product is stored, used and disposed of correctly to ensure effective use and to protect the safety of humans, plants, animals and the environment.

Other information on **safety and environmental impact** and on **natural balances** is given separately next in each of the four main sections: on herbicides (see p. 230), molluscicides (see p. 230), insecticides (see p. 231) and fungicides (see p. 232).

The word **'pesticide'** is used to cover all **plant protection products** (PPP), including herbicides (for weeds), molluscicides (for slugs and snails), insecticides (for insects), acaricides (for mites), nematicides (for nematodes) and fungicides (for fungi). Plant growth regulators such as rooting hormones are also classified as PPPs and are covered by the same legislation.

Legislation

There is a range of legislation covering the use of PPPs in the UK, and all professional users of PPPs should be familiar with the following.

The Health and Safety at Work Act 1974. This is a primary piece of legislation covering occupational health and safety in the UK. It sets out the general duties that employers have towards employees and members of the public, that employees have to themselves and others and that self-employed people have towards themselves and others.

Control of Substances Hazardous to Health Regulations 2002 (COSHH) covers substances which are hazardous to health, including PPPs and other more general hazardous substances; it requires that all hazardous substances in the workplace are adequately assessed and controlled to prevent or reduce human exposure.

Plant Protection Products (Sustainable Use) Regulations 2012 is legislation that is specific to PPPs and requires professional users, advisers and suppliers to be in possession of a valid certificate of competence, issued by a **government designated awarding body**, for the work being undertaken. It also requires anyone who uses PPPs to take reasonable precautions to protect human health and the environment. Guidance pertaining to these obligations can be found in two separate Codes of Practice.

The **Code of Practice for Using Plant Protection Products**, for all professional users of PPPs and the **Code of Practice for Suppliers of Pesticides to Agriculture, Horticulture and Forestry** on the storage and transport of pesticides for those who sell, supply and store PPPs. Following this guidance will help in meeting the legal obligations referred to in it.

The **Official Controls (Plant Protection Products Regulations) 2020** supplements the Plant Protection Products (Sustainable Use) Regulations 2012 and requires all professional users of PPPs and adjuvants (substances which affect the way plant protection products behave and the effects they have, but are not PPPs themselves) to register with the **Department For Environment, Food and Rural Affairs** (**DEFRA**) if they are a business, organization or sole trader which uses PPPs and any adjuvants professionally in the UK. Details on whom this applies to and how to register are given on the DEFRA website. Registration is not required for people who use non-professional PPPs in their own garden. Inspection visits by the Health and Safety Executive will take place under the powers given by the above-mentioned regulations to ensure compliance with all relevant pesticide regulations.

15

Herbicides (used on weeds)

Benefits and limitations of herbicide control: see general points for 'chemical control' (see p. 227)

Safe practice and environmental impact of herbicidal control: in many garden situations, the removal of weeds such as those described in Chapter 16 is considered a useful or necessary activity.

- Care needs to be taken when transporting, storing, handling, applying and disposing of herbicides. The active ingredient and other substances in a product may present risks if adequate precautions are not in place. Products for home gardeners are generally chosen for their lower levels of risk, but users still need to adhere to the label instructions.

Natural balances:

- Some weeds have **beneficial effects** in the garden and can be viewed as important constituents of a healthy ecosystem and can be important in sustaining some non-pest invertebrate species. For example, the leaves of perennial stinging nettle (*Urtica dioica*) are the **main food source** for caterpillars of the comma, peacock, red admiral and painted lady butterflies. Removal of nettle plants in an area is likely to reduce these insects.
- Care should be taken when applying herbicides near water; some products have approval for application near or in water but other herbicides may be harmful to **aquatic organisms**.
- Some herbicides may be harmful to beneficial organisms if used incorrectly.
- Careless application of the herbicides may cause damage to cultivated plants. For example, **2,4-D**, a selective lawn weedkiller, will harm non-target broadleaved (**eudicot**) plants if it comes into contact with them through spray drift or other means.

Restoring and maintaining natural balances:

- avoid destroying plants which maintain beneficial predators, parasites and pollinating species such as hoverflies, butterflies (Figure 15.11) and bees
- avoid spray drift onto non-target plants
- avoid spraying flowering weeds with herbicide as they are likely to be of interest to pollinating insects such as bees which may be harmed.

Figure 15.11 Comma butterfly; the caterpillars feed on nettle leaves

Examples of herbicide groups

Herbicides can be classified into the following groups based on their **mode of action** on weeds:

- **Total herbicides** will kill all eudicot and monocotyledon plants they are applied to. Total **contact herbicides** are used on ephemeral, annual and seedling weeds; examples include **acetic acid**, **fatty acids** and **pelargonic acid** which scorch the weed's foliage and leads to its death. Total **translocated herbicides** enter the leaves, stems or roots and then move via the vascular system (see p. 50) to reach all parts of the plant. This property is particularly useful in controlling **perennial** weeds with extensive root system or underground perennating organs such as taproots, rhizomes, tubers and bulbs, an example being **glyphosate**.
- **Selective herbicides** have an ability to control certain groups of weeds, leaving other species unharmed; they usually enter the weed through the leaves but surrounding garden plants are unaffected. The most common examples are active ingredients such as **2,4-D**, **MCPA** and **mecoprop-p** which are used as lawn herbicides to control a range of broadleaved (**eudicot**) weeds whilst leaving the lawn grasses (**monocotyledons**) unaffected. Selective herbicides, if used incorrectly, can be harmful to lawn grasses, too.
- **Residual** (**soil acting**) herbicides, containing active ingredients such as **diflufenican**, are applied to the soil where they remain active over a period of weeks to months and prevent weeds from emerging. They are usually mixed in a formulation containing a total herbicide such as glyphosate. These products are often used for total control, usually in non-crop situations such as gravel paths, driveways and pavements, away from direct contact with cultivated plants.

Molluscicides (used on slug and snails)

Benefits and limitations of molluscicide control: see general points for 'chemical control' (see p. 227).

Safe practice and environmental impact of slug and snail control: although some harmful products are no longer approved and **ferric phosphate** is the only **molluscicide** with approval for garden use in the UK, there are still important points to be remembered:

- Although some people consider slugs and snails to be plant pests, many species are not pests and are an important part of the ecosystem; they are a **food source** for many animals and are active in contributing to recycling vegetation and increasing soil organic matter content when they die. The use of biological control such as parasitic nematodes (see p. 262) and physical control with barriers (see p. 262) or handpicking provide effective alternatives to chemical control.
- Cultivation of **resistant** or less susceptible species and cultivars of plants, for example '**Pentland Dell**' potato cultivar is one of the least affected by slugs.
- Encouragement of **natural predators** such as hedgehogs, frogs and ground beetles by providing suitable habitats in the garden may considerably reduce slug and snail damage to a tolerable level.
- Place chemical control pellets inside containers that allow access to slugs and snails, but prevent the entry of mammals and birds.

Natural balances: Britain and Ireland have a climate that may bring damp conditions to the garden at any time of year. Many plant cultivars (such as those of lettuces, potatoes and *Hosta* spp.) have succulent tissues that favour slugs' and snails' feeding habits.

Examples of molluscicide active ingredients

- **Ferric phosphate**, when used in accordance with manufacturer instructions, will ultimately break down into phosphate and iron in the soil. It is relatively non-toxic to mammals and birds. This molluscicide can be used by organic growers but permission from the relevant **certification body** is required prior to use by commercial organic growers.

Insecticides and acaricides (used on insect and mites)

Benefits and limitations of insecticidal control: (see general points for chemical control p. 227).

Safe practice and environmental impact of insecticide and acaricide control: while some harmful and persistent insecticides are no longer approved for garden use in the UK, there are still important points to be remembered:

- Care needs to be taken when transporting, storing, handling, using and disposing pesticides. The active ingredient within the product represents a particular risk if adequate steps are not taken (see p. 229). Products approved for home use are chosen for their relatively low level of risk to humans, but home gardeners still need to be careful.
- Some of the insecticide ingredients available to home gardeners can be harmful to beneficial organisms in the garden. Alternative methods (see physical, cultural and biological control in this chapter) such as deterrents, barriers, traps and encouraging beneficial predators and parasites should be considered before taking the option to use an insecticide.

Natural balances:

- The use of insecticides/acaricides can significantly affect natural balances in the garden. While the increase in insect and mite numbers is balanced in nature by **beneficial** predators and parasites such as ladybirds and parasitic wasps (see biological control, p. 223), inappropriate application of a product containing the synthetic pyrethroid insecticide **deltamethrin** may harm these non-pest species.
- When using biological control agents such as *Encarsia formosa* against glasshouse whitefly (see p. 268) in **greenhouses**, the selection of appropriate compatible crop pesticides is critically important (see integrated pest management p. 233) to avoid harm to *Encarsia formosa*.
- Some insecticide products are harmful to aquatic organisms in **water** (fish, snails and insects), and care is needed not to apply insecticide near to waterways unless it is approved for that use.

Restoring and maintaining natural balances:

- Read the pesticide product label carefully to check whether the product is harmful to biological control and non-pest species. Manufacturer information sources will provide more detailed information.

Entry point for insecticides into insects

Insects and mites have **three** main points of weakness for entry by pesticides: their **waxy exoskeletons** may be penetrated by wax-dissolving contact chemicals; their abdominal **spiracles** allow chemicals to enter tracheae or are blocked by pesticide formulations; and their **digestive systems**, in coping with the large food quantities required for growth, may take in stomach poisons.

15

Examples of insecticide groups

Insecticides can be classified into the following groups

- **Natural insecticides** – such as **pyrethrins** (pyrethrum) which is derived from the flowers of *Tanacetum cinerariifolium* and used to control a range of pests by contact action including aphids, whitefly, thrips and caterpillars. Although pyrethrins have short persistence and can be used on a wide range of plants, they can be harmful to beneficial organisms. Other natural insecticide/acaricides include **fatty acids** which contains potassium salts of fatty acids and is derived from vegetable oils. It works by contact action and penetrates the external layers of the insect's body causing death of soft-bodied pests such as aphids, glasshouse whitefly and two-spotted spider mite.
- **Synthetic contact insecticides** – only organisms which are touched by these insecticides will be killed, an example being the **synthetic pyrethroid** group of insecticides. They are effective against a wide range of insect pests. They have low toxicity to mammals but are active for longer than natural pyrethrins. **Deltamethrin** is used to control aphids, glasshouse whitefly, caterpillars, codling moth, raspberry beetle and others pests of a range of ornamental and edible crops. It can be residual for up to three weeks but can reduce the effectiveness of biological control by harming beneficial predators and parasites.
- **Synthetic systemic insecticides** – such as **flupyradifurone** are absorbed into the plant through leaves or roots and are effective against **sap-sucking** insects such as aphids, whitefly, scale insects and mealybug which ingest the active ingredient when sucking the plant sap. Some systemic insecticides such as **acetamiprid** also have contact action against biting and chewing pests such as lily beetle and caterpillars.
- **Plant invigorators** – are products which are not formulated or approved as pesticides but have become popular because of their useful effects on plant pests and diseases. These products include surfactants, plant nutrients and seaweed extracts which have a physical action on soft-bodied pests such as whitefly, aphids, mealybugs and two-spotted spider mite, they also act as a bio-stimulant to plants to help them resist and recover from the damaging effects of pests and diseases. They are suitable for use in integrated pest management programmes as they can be compatible with some beneficial biological control agents. They are not subject to the same legal controls as pesticides.

Nematicides (used on nematodes)

No products are currently available to home gardeners for nematode control. **Fluopyram** is an active ingredient in a nematicide product currently approved for potato cyst nematode but its use is only permitted for appropriately trained and certified professional growers.

Fungicides (used on fungal diseases)

Benefits and limitations of fungicidal control: see general points for chemical control (p. 227).

Safe practice and environmental impact of fungicidal control: fungicides used in horticulture do not generally represent such acute risks of human and environmental harm that may be found in insecticides.

- However, as with other pesticides, care needs to be taken when transporting, storing, using and disposing fungicides. Products for home use are chosen for their low level of risk to humans, but home gardeners still need to be careful.

Natural balances:

- Most fungi in the garden ecosystem are beneficial in numerous ways; decomposers help to break down organic matter, especially woody material, to form humus, and soil-dwelling mycorrhizal fungi form beneficial symbiotic relationships with plant roots. Disease controls with broad-spectrum fungicides such as **myclobutanil** can reduce the levels of these beneficial fungi.

Restoring natural balances:

- Provide plants with optimal conditions of soil fertility (see p. 196) and microclimate that will reduce the likelihood of fungal and bacterial infections but will encourage beneficial microorganism activity.
- Provide plants with pathogen-free soil and growing media. Avoid the introduction of infected plants.
- Choose cultivars with a proven record of **resistance** where possible (see p. 233).
- Read the pesticide product **label** carefully to check whether the chosen active ingredient harms biological control agents and beneficial species.
- Be particularly careful when spraying in greenhouses with introduced **biological control** agents present or near **waterways**.

Action of fungicides

Fungicides must act against the disease but not seriously interfere with plant activity. They can act either in a **protectant** way on the plant surface or in a **systemic** way inside the plant. Some **bio-fungicides** which contain spores of beneficial fungi are added to the growing medium to help protect cultivated plants from soil-borne pathogenic fungi.

Examples of fungicide active ingredients

- **Sulphur** is a protectant fungicide for use on diseases of ornamentals and edible plants such as powdery mildew, rose black spot and rust.
- **Tebuconazole** is a systemic fungicide to control box blight, powdery mildew, rust and blackspot on roses and other ornamental plants.
- *Trichoderma harzianum* **T-22** is a bio-fungicide which is applied to the growing medium; it releases a beneficial fungus to protect plant roots against soil-borne diseases such as *Pythium* spp., *Rhizoctonia* spp. and *Fusarium* spp. in protected ornamental and edible crops.

Formulations

Active ingredients are mixed with other ingredients to increase the efficiency and ease of application, prolong the period of effectiveness, or reduce the damaging effects on plants and humans. The whole product, a **formulation**, in its container is given a product name, which often differs from the name of the active ingredient. The main formulations are liquids, granules, wettable powders and dusts.

Product label

The **statutory conditions of use** are clearly stated on the label of every product. It is a legal offence not to follow the statutory conditions of use. This can include the following information.

- field of use restrictions, e.g. horticulture or home use and so on
- user restrictions
- the crop or situation which may be treated
- maximum individual dose/application rate
- maximum number of treatments or maximum total dose
- maximum area or quantity which may be treated
- latest time of application, harvest or re-entry interval
- operator protection and training requirements
- environmental protection requirements.

Other advisory information is also included. The product label should be adhered to at all times to ensure the product is being used legally and effectively and to minimize the risks to humans, non-target organisms and the environment.

Integrated pest management (IPM)

IPM is the use of a range of compatible cultural, physical, biological and chemical methods of pest and disease control. The key stages of an IPM programme and the order in which they are followed are:

Scouting – inspecting the crop regularly to determine if any pests, diseases, weeds or harmful abiotic conditions are present. **Identifying** – accurately identifying any potential problems which are discovered. **Monitoring** – regular observation of the development of any previously identified pests, diseases and weeds. **Determining actions and economic thresholds** – deciding when damage to the crop has reached a predetermined threshold and then deciding what interventions are required. **Implementing cultural and physical controls –** some cultural and physical controls are likely to have been in place from the start of the crop's establishment as prevention methods; any further control interventions can be implemented at this point. **Implementing biological controls** – application of appropriate biological control agents where appropriate; some pests, diseases and weeds cannot be controlled biologically. **Implementing chemical controls** – the use of chemicals within an effective IPM programme is the final intervention to address problems which have not been effectively contained by the other methods. Careful consideration should be given to ensure that the chemicals are compatible with biological control agents and what their implications are for the crop, humans, non-target organisms and the environment. The potential for resistance development in the pest, disease or weed should be also considered.

Selection of plants

The horticulturist has two important considerations when selecting the most suitable species/cultivar to grow. The first is the choice of a plant species which is appropriate for the location, an concept popularized as '**Right Plant, Right Place**' by the esteemed plantswoman, the late **Beth Chatto**. The second is the use of species and cultivars which have known **resistance or tolerance** to certain pests and diseases.

The garden can be seen as a pattern of small habitats, each presenting particular soil and microclimate conditions that are favourable to particular **plant species**. Among the many decisions a horticulturist has to make, choosing the most suitable plant species for each location is one of the most important. Failure to choose appropriately may result in poor plant growth, susceptibility to pests and diseases and long-term

15

implications in terms of time and money being spent on interventions aimed at helping the plant to thrive. Adaptations of plants to specific habitats is discussed in Chapter 5.

Benefits of correct species choice:

- Optimum plant growth is achieved without the need for repeated, time-consuming and expensive horticultural interventions in the future. Avoiding plant replacements reduces the garden`s carbon footprint.
- Healthy plants are likely to be more attractive in the case of ornamental species and more productive in the case of edible plants.

Locations in the garden

Plant selection for resistance

Benefit of resistance control:

- the selected resistant plant has natural characteristics which help the plant to resist a pest or disease with minimal intervention being necessary from the horticulturist
- In this way, resistance may greatly reduce the need for other control methods.

Table 15.1 Suitable garden plants for different locations

Garden location	Suitable plant species
Sunny	*Lavandula angustifolia* (shrub), *Lonicera japonica* (climber), *Verbena bonariensis* (herbaceous perennial)
Dry shade	*Aucuba japonica* (shrub), *Hedera hibernica* (climber), *Liriope muscari* (herbaceous perennial)
Moist shade	*Fatsia japonica* (shrub), *Parthenocissus henryana* (climber), *Eriocapitella japonica* (herbaceous perennial)
Sandy soil	*Cytisus scoparius* (shrub), *Trachelospermum asiaticum* (climber), *Eryngium planum* (herbaceous perennial)
Heavy soil	*Cornus alba* (shrub), *Rosa filipes* (climber), *Filipendula ulmaria* (herbaceous perennial)
Alkaline soil	*Salvia rosmarinus* (shrub), *Lonicera periclymenum* (climber), *Valeriana rubra* (herbaceous perennial)
Acid soil	*Pieris japonica* (shrub), *Berberidopsis corallina* (climber), *Uvularia grandiflora* (herbaceous perennial)
Protected site (by a wall)	*Garrya elliptica* (shrub), *Hydrangea petiolaris* (climber), *Kirengshoma palmata* (herbaceous perennial)
Windy site	*Olearia macrodonta* (shrub), *Schisandra rubriflora* (climber), *Geranium sanguineum* (herbaceous perennial)

Limitations of resistance control:

- new strains of disease or pest may develop which overcomes the plant's resistance.

Natural balances: since plant resistance works against the pest or disease from within the plant, there are not likely to be any changes in the balance of food chain species (see p. 138) when resistant cultivars are chosen. The indirect effect will be that, in the case of pest resistance, a pest's predator and parasite numbers may decrease in that locality.

Examples of resistant cultivars include potato '**Pentland Dell**' which is less prone to **slug attack** than most potato cultivars. Carrot '**Flyaway**' **F1** was developed for resistance to **carrot fly**, Tomato '**Crimson Blush**' **F1** shows resistance to **tomato blight** and the lettuce '**Brighton**' has resistance to **downy mildew**.

Biosecurity and phytosanitary requirements in Britain and Ireland

The term 'biosecurity' refers to precautions which aim to prevent the introduction and spread of harmful organisms which includes **non-native pests**, such as insects, and disease-causing pathogens, such as some bacteria, fungi and viruses. The UK government's strategy on biosecurity is set out in '**The Plant Biosecurity Strategy for Great Britain 2023–2028**'. This strategy recognizes that introduced, non-native pests and pathogens threaten the health of British plants. These increased threats have been attributed to a number of factors, including **climate change** and the increase in **global trade and travel**, which provides more opportunities for the introduction of pests and diseases such as live plants or plant products in trade or passenger baggage from abroad, timber and wood packaging such as pallets, dirty tools, machinery and vehicles and contaminated soil and organic matter. In addition to government initiatives on biosecurity, some public gardens and plant collections have implemented their own policies aimed at ensuring biosecurity protocols are in place. For example, in London, **The Royal Parks**' policy on biosecurity is published on their website. The general process aimed at ensuring biosecurity in professional horticulture includes:

- Plants should be purchased from known, **reliable growers**.
- Whenever possible, purchase plants which have been produced as **locally** as possible.
- Ensure all necessary statutory **phytosanitary** requirements are implemented.
- Ensure that **inspection checks** are carried out when plants are delivered and that **quarantine facilities** are available if required.

- **Regular scouting** for plant health problems should take place.
- Where possible, avoid the use of wood-based materials such a pallets which may house pests such as the **Asian longhorn beetle**.
- Use **best practice** in plant cultivation. This involves the implementation of plant cultivation concepts and practices which have been developed, researched, trialled and analysed for their effectiveness and sustainability and then put into practice and evaluated at horticultural establishments of repute or research institutions before being accepted as 'best practice' by professional horticultural bodies.

Phytosanitary requirements

Imported plants and plant products are a significant route for non-native, invasive pests and diseases to enter Great Britain (GB). To import regulated plants and plant products into GB (England, Wales and Scotland), a **phytosanitary certificate** must be issued by the government's plant health authority in the country where the plants are being imported from to ensure that each consignment is officially inspected, complies with legal requirements for entry into GB and is free from quarantine pests and diseases. Since the UK left the EU, Northern Ireland (NI) has remained part of the EU **single market** and continues to be covered by EU plant health regulations for plant imports. Therefore, movement of plant material from NI to GB requires a phytosanitary certificate. These requirements apply principally to professional importers but the plant health regulations prevent the public from bringing plants and plant products for personal and home use into GB, unless the plant material is accompanied by a phytosanitary certificate. Exporters of plants from GB to other countries must also provide a phytosanitary certificate for the plant material they intend to export so the plants can legally enter all countries, including NI, outside GB. The relevant plant health authority with responsibility for issuing phytosanitary certificates in England and Wales is the **Animal and Plant Health Agency** (**APHA**); in Scotland, **Science and Advice for Scottish Agriculture** (**SASA**); in Northern Ireland, the **Department of Agriculture, Environment and Rural Affairs (DAERA**); and in the Republic of Ireland, the **Horticulture and Plant Health Division of the Department of Agriculture, Food and the Marine**.

GB also has a **plant passport** scheme. A UK plant passport is an official document required to move plants and certain regulated wood within Great Britain and Crown Dependencies such as the Channel Islands and the Isle of Man and includes all plants for planting, some seeds, seed potatoes and wood, specific details are published on the **Department for Environment, Food and Rural Affairs** website. A relevant plant passport must accompany the plants at all stages of movement around GB and Crown Dependencies, and it ensures that plants and plant products can be traced throughout the supply chain and declares compliance with plant health requirements such as freedom from pests, which is essential for maintaining biosecurity. Prior to the UK's exit from the EU, plants were moved around GB on EU plant passports, but since the UK left the EU, a UK plant passport for movement of plants in GB and Crown Dependencies is required. Movement of plants within NI and from NI to other EU countries still requires an EU plant passport (but this could change). The Republic of Ireland also issues EU plant passports. The issuing of plant passports is generally only applicable to professional operators (e.g. nurseries and garden centres) who are producing, moving and selling plants to other professional operators; all such professionals can issue plant passports themselves but must be authorized by APHA in England and Wales (or the **Forestry Commission** for the forestry sector) and SASA in Scotland. DAERA is responsible for authorizing EU passport-issuing operators in NI and the Department of Agriculture, Food and the Marine in the Irish Republic. In Britain, a retailer does not need to pass on the plant passport to the customer if the plants are being purchased from a retail nursey, garden centre or shop. An exception to this rule is if a customer is purchasing plants from a distance retailer such as phone, mail or online sales; in this instance, the retailer must supply the customer with the plant passport for traceability purposes.

Some serious non-native plant pests and diseases are classified by DEFRA as **quarantine** organisms which are **notifiable**, and they are subject to additional statutory measures which help to keep introduced pests and diseases out of areas where they could damage crops, trees, wild plants and ecosystems, and there are controls in place to help combat these. If a quarantine organism is identified or suspected, it must be reported immediately to an APHA Plant Health and Seeds Inspector. Tree-related quarantine pests and diseases should be reported to the Forestry Commission. Full details on quarantine pests and diseases are displayed on the DEFRA website including reporting arrangements for England and Wales, Scotland and Northern Ireland. Examples of quarantine pests include the red-necked longhorn beetle (*Aromia bungii*) and tobacco whitefly (*Bemisia tabaci*), and quarantine diseases include sudden oak death (*Phytophthora ramorum* and *P. kernoviae*) and Xylella (*Xylella fastidiosa*).

Garden health plans

A **garden health plan** is a method of managing plant health which considers all potential impacts on plant health, not just biotic factors such as pests, diseases and weeds. This is generally considered to be a more holistic approach than a conventional IPM programme, which tends to be narrower and more focussed on pest management. A plan will aim to identify all potential plant health risks including both biotic and abiotic factors and then identify what actions can be taken to mitigate the identified risk factors. A garden health plan will take account of the same biotic factors as IPM but also considers abiotic factors such as **site conditions** which could create microclimates that create threats to plant health, including prevailing winds, light intensity and duration, temperature, humidity and atmospheric pollution. Soil factors such as soil texture and structure, water and air content, nutrient status and pH are also considered. **Local factors** which can also influence plant health problems should be identified; this might include a grower observing what crops or plants local growers or neighbours are cultivating and identify their potential impact and future plant health problems. Local wild species or features such as woodlands could also have an impact on the presence of some potential pests and diseases which might threaten cultivated plants. **Historical factors** can also be considered, for example how past vegetation on a particular site could affect future plant selection; a site with a long-established history of woody planting might reasonably be expected to have the presence of the honey fungus pathogen (*Armillaria* spp.) in the soil – this would certainly help inform future plant selection. However, a new garden on land which historically had no woody vegetation might not have the honey fungus pathogen present in the soil.

Potential impacts of a changing climate on plant health

Recent years have seen the appearance and establishment of plant pathogens, pests and weeds from other regions of the world which have not previously been a problem in these islands. It is very likely that climate change will lead to an increase in all three. International trade in plants means that pests and diseases have spread with those plants beyond their normal geographical range.

Diseases. Environmental conditions due to climate change can enable pathogens to survive and adapt to their new homes, and this is problematic, not just for cultivated plants, but also for wild plant communities. It has long been known that outbreaks of disease can occur when conditions become favourable and that variations in humidity and soil moisture are the main drivers of disease (e.g. potato blight). These may happen more frequently with climate change which can indirectly affect plant pathogen interactions through alterations in plant and pathogen physiological processes. A warming climate can affect overwintering survival, population growth and number of generations of pathogens and can shorten their incubation periods. It can also give rise to new strains of pathogens which are more virulent and better adapted. As well as increased temperature, climate-induced changes in humidity and water availability will also significantly impact future disease outbreaks. The effects of climate change on plant diseases will vary from crop to crop (and in some cases it may be beneficial), but there is increasing evidence that, overall, the development and severity of plant diseases will increase. The soil-borne pathogen *Phytophthora cinnamomi* which thrives in soils that are excessively wet in winter and affects woody species such as yew (*Taxus baccata*) could become even more prevalent. Pear rust (*Gymnosporangium sabinae*) from continental Europe was previously rare in Britain and Ireland but over the last 20 years has become more common and widespread. Summer droughts are likely to increase plant stress and lead to a higher incidence of diseases such as powdery mildews. Milder, wetter winters are likely to increase the incidence of fungal diseases such as apple canker (*Neonectria ditissima*). Chalara ash dieback (*Hymenoscyphus fraxineus*) was initially discovered in Poland in 1992, with the first cases reported in both Britain and Ireland 2012, and is now widespread. Many of the same issues described for plant diseases could equally be applied to plant pests too.

Pests. Outdoor pest species which have multiple generations each year such as box tree caterpillar (*Cydalima perspectalis*) could potentially produce more generations each season which would lead to a greater number of individual insects and result in even more direct damage to plants. However, natural predators of pests could also potentially produce more generations each year. There is also the possibility that opportunities will increase to apply a wider range of biological controls outdoors over longer periods of time. Pest species which have in the past have survived only under protection, such as glasshouse whitefly (*Trialeurodes vaporariorum*), might become problematic outdoors. Invasive exotic pest species such as rosemary beetle (*Chrysolina americana*), horse chestnut leaf miner (*Cameraria ohridella*) and oak processionary moth (*Thaumetopoea processionea*) have all arrived and spread in Britain and Ireland in recent years; the arrival and establishment of these pests is possibly linked to the trend for increasingly milder winters which can facilitate their survival. Potentially, many more exotic pests could arrive and become established as well as the possibility that indigenous pests have a greater chance of surviving winter outdoors.

Weeds. Milder winters and earlier springs could lead to weed seeds germinating earlier in the year; consequently weed management practices would need to start earlier, too. Summer droughts can limit the amount of seed germination and weed establishment later in the season, and drought conditions can affect the efficacy of translocated herbicides which require the weeds to be in active growth to work efficiently. Milder winters could enable a wider range of exotic weed species to survive winter and establish in Britain and Ireland. Weed species whose seeds are wind dispersed could potentially be disseminated more widely. Winds can limit the opportunities to the horticulturist to spray pesticides because of an increased risk of spray drift.

Further reading

Adams, C., Brook, J. and Early, M. (2015) *Principles of Horticulture Level 3*. Routledge.

British Crop Production Council. (2025) *UK Pesticide Guide*. BCPC. https://ukpesticideguide.co.uk/

Brown, L.V. (2008) *Applied Principles of Horticultural Science*. Routledge.

Buczacki, S. and Harris, K. (2014) *Pests, Diseases and Disorders of Garden Plants*. HarperCollins.

Greenwood, P. and Halstead, H. (2018) *Pests and Diseases*. Dorling Kindersley.

Helyer, N., Cattlin, N. and Brown, K. (2003) *A Colour Handbook of Biological Control in Plant Protection*. Timber Press.

Ives, J. (2022) *A Gardener's Guide to Biological Pest Control*. The Crowood Press.

RHS. *Fungicides for Home Gardeners*. Online: www.rhs.org.uk/advice/pdfs/fungicides-for-home-gardeners.pdf

RHS. *Pesticides for Home Gardeners*. Online: www.rhs.org.uk/advice/pdfs/pesticides-for-home-gardeners.pdf

RHS. *Weed Killers for Home Gardeners*. Online: www.rhs.org.uk/media/pdfs/advice/WeedkillersForGardeners

Stewart, D. (2023) *A Gardener's Guide to Sustainable Gardening*. The Crowood Press.

www.daera-ni.gov.uk/topics/plant-and-tree-health

https://forestrycommission.blog.gov.uk/category/tree-health/

www.hse.gov.uk/pesticides/

https://planthealthportal.defra.gov.uk/

www.royalparks.org.uk/sites/default/files/2023-07/Biosecurity%20Policy.pdf

www.sasa.gov.uk/plant-health

www.soilassociation.org/causes-campaigns/reducing-pesticides/

www.teagasc.ie/crops/horticulture/plant-health/

The online material is accessible via the QR code and includes further information on many of the topics in the book, such as

- Garden health plans
- Biological control case study

CHAPTER 16

Weeds

Figure 16.1 Ground elder growing in a garden border; it is commonly considered to be a pernicious weed but the flowers attract pollinators such as bees and hoverflies (source: Shutterstock, weha)

This chapter includes the following topics:

- Definition of a weed
- Negative impacts of weeds
- Beneficial effects of weeds
- Weeds in different locations
- Weed identification and biology
- Ephemeral, annual and perennial weeds and their control methods

DOI: 10.4324/9781003581260-16

Weeds are usually wild or uncultivated plants which occur spontaneously, are unwanted by the grower and have the ability to reproduce prolifically and spread by seed or vegetative means such as rhizomes, runners, tubers and bulbs. They can be detrimental to the growth and development of cultivated plants and are often considered to be unsightly in the garden. For many years, weeds have been viewed by gardeners and growers as an unwanted and undesirable feature. However, it is now being recognized they are not wholly problematic and although they can have negative impacts, they also have characteristics which can be both horticulturally and ecologically beneficial. In the immortal words of the 19th-century American philosopher **Ralph Waldo Emerson**, '*What is a weed? A plant whose virtues have not yet been discovered*'. In this chapter, a range of weeds are described, together with their impacts and measures which may be used to control them.

A **weed** is an unwanted plant growing in a cultivated area

Negative impacts of weeds

- **Reduction of plant or crop productivity** occurs because of competition between the weeds and cultivated plants for water, nutrients and light. Cultivated plants are deprived of these essential requirements and poor growth results. The effects are seen when **light** is excluded from the plants as they are crowded out by the weeds. Similarly, the availability of **nutrients** and **water** from the soil is reduced when the weeds compete with cultivated plants.
- **Reduced visual appearance of an ornamental garden**. A horticulturist may consider that any plant spoiling the visual appearance of plants in pots, beds and borders, paths or lawns should be removed, even though the growth of cultivated plants themselves may not be significantly affected. Hedge bindweed growing up established shrubs and ground elder emerging from beneath concrete garden paths are two examples of this kind of effect.
- **Alternative hosts of pests and pathogens**. Plant pests and pathogens are commonly found on weeds. Perennial sow thistle (*Sonchus arvensis*) and chickweed (*Stellaria media*) (Figure 16.2) can host glasshouse whitefly and the two-spotted spider mite in greenhouses. Groundsel (*Senecio vulgaris*) can be infected by white blister disease which affects florist's cineraria (*Pericallis* × *hybrida*).
- **Hazardous properties**. Some weed species have natural properties which create **risks** for humans and animals. The **Weeds Act 1959** lists five species of harmful weeds which occupiers of land have a legal duty to control including spear thistle and creeping thistle (Figure 16.15) (*Cirsium vulgare* and *C. arvense*) which have spiny leaves and ragwort (*Jacobea vulgaris*) which is poisonous to horses and other animals such as cattle. Other species such as giant hogweed (*Heracleum mantegazzianum*) has sap which will cause serious irritation and blistering to skin and damage to eyes when exposed to direct sunlight. The fruits and foliage of deadly nightshade (*Atropa belladonna*) are highly toxic to humans and animals.
- **Other impacts of weeds**. The quality of sports turf surfaces are reduced as a consequence of weed presence. They can affect **drainage** of soft and hard surfaces by preventing the flow of water along ditches (e.g. chickweed). Weeds such as redshank (*Persicaria maculosa*) have strong stringy stems which can **clog machinery** such as the blades of lawn mowers (Figure 16.3). Seeds of weeds growing in crops being grown to be harvested as a **seed crop** can contaminate the purity of crop seed when harvested. Weeds in **hard landscape** situations can destabilize vertical and horizontal structures.

Figure 16.2 Chickweed, a common ephemeral weed of gardens (source: Shutterstock, avoferten)

Figure 16.3 Redshank can clog lawn mower blades (source: Shutterstock, crystaldream)

Roadside weeds can be hazardous and create risks for motorists, cyclists and pedestrians by blocking sight lines and obscuring road signs.

Beneficial effects of weeds

Although weeds can in some circumstances be regarded as being undesirable, it is important to recognize the **beneficial effects** they can provide in a garden. They can be ecologically important by providing **food and habitats** for wildlife such as invertebrates and other animals. Pollinating insects such as bees and butterflies feed on the **nectar and pollen** of dandelion (*Taraxacum* spp.) which flowers from March onwards, and birds such as finches feed on the dandelion seeds when they subsequently develop. Natural predators of pests, such as carabid beetles, feed on the **seed** of chickweed (*Stellaria media*) and will consequently be attracted to the garden and help to reduce pest populations. The leaves of the perennial stinging nettle (*Urtica dioica*) are a **food source** for caterpillars of the red admiral, small tortoiseshell, painted lady and comma butterflies. Weeds can also help in the development of **healthy soil** by adding to soil organic matter content when they die, which subsequently **improves soil structure** and increases **nutrient content** (see p. 182). Weed coverage of the soil surface helps to protect the soil from **capping and erosion** during periods of high rainfall and, therefore, protects soil structure. Surface coverage can also reduce water evaporation from the soil which can be beneficial by conserving water, which is particularly helpful in the establishment of young plants. Some deep-rooted weeds can access nutrients from deeper in the soil which, when the plant subsequently dies and decays, will be recycled into the topsoil where cultivated plants can access the nutrients.

Leguminous species of weed such as white clover (*Trifolium repens*) can improve nitrogen levels in the soil as a consequence of their symbiotic relationship with nitrogen-fixing *Rhizobium* spp., bacteria which live in the root nodules of clover and other closely related 'legume' species (see p. 187).

Weeds in different locations

A range of different situations are listed next, together with examples of weeds which are commonly found in them.

- In **recently cultivated soil**, the seeds of ephemeral and annual weeds such as the common poppy (*Papaver rhoeas*) and chickweed, which have been brought to the soil surface, often germinate in large numbers. A common perennial weed problem in this location is couch grass (*Elymus repens*). The seedlings of these weeds can be controlled by **hoeing** on dry, warm days or by application of a non-residual, **contact** herbicide containing fatty acids, pelargonic acid or acetic acid.
- In an **herbaceous perennial border**, annual weeds can often be relatively few in the growing season as they are crowded out by the cultivated plants and are relatively easily removed by regular hoeing. However, perennial weed species such as couch grass, Japanese knotweed (*Reynoutria japonica*), creeping cinquefoil (*Potentilla reptans*) (Figure 16.4) and field horsetail (*Equisetum arvense*) (Figure 16.21a and b) are more effective at growing amongst the cultivated plants. These must be carefully removed with a **garden fork**. Application of herbicide is difficult and likely to damage cultivated plants if care is not taken. Careful **spot treatment** may be the most suitable method of control in this situation using a translocated herbicide such as **glyphosate**.
- In **woody perennial plantings** (shrubs and trees), perennial weeds range from the species such as broad-leaved dock (*Rumex obtusifolius*) and ground elder (*Aegopodium podagraria*) and to climbing species such as hedge bindweed (*Calystegia sepium*). **Woody** species such as sycamore (*Acer pseudoplatanus*) and ash (*Fraxinus excelsior*) can also establish as tree weeds in woody plantings. Control by **translocated** herbicide (see p. 230) includes spot weeding or use of **residual** herbicide (see p. 230) approved for use in woody plantings.
- In **lawns**, common weed species are usually perennials such as daisy (*Bellis perennis*), creeping buttercup (*Ranunculus repens*), yarrow (*Achillea millefolium*) and dandelion (*Taraxacum* spp.). Most species of **annual** weeds are controlled by **regular mowing** which they cannot withstand. One exception is annual meadow grass (*Poa annua*) which can tolerate the mowing regimes of lawns and even golf greens; **cultural methods** of control have to be applied in this case. Control by use of herbicides can include the use of **selective** lawn herbicides which are effective in controlling **broadleaved (eudicot)** weeds in the **monocotyledonous** lawn grasses.
- In **commercial horticulture**, a diverse range of plants are cultivated ranging from fruit and vegetables to herbaceous and woody ornamental species and numerous production methods ranging from **field grown** to **container production**. Consequently, many species of weeds are likely to be problematic. In principle, commercial growers will utilize variations of the same range of physical, cultural and chemical methods as non-professional gardeners. However, as **professional**

16

Figure 16.4 Creeping cinquefoil (*Potentilla reptans*) produces adventitious roots at its stem internodes (source: Shutterstock, dabjola)

operators, they are likely to have access to a wider range of herbicides than home gardeners; this means they will have additional legal responsibilities pertaining to the use of products approved for use in professional horticulture, including the legal requirement to be appropriately trained and certified in the use of pesticides.

Weed identification

As with any plant health problem in horticulture, correct **identification** is essential before any appropriate control measures can be attempted. A weed seedling initially causes little damage to a crop but will quickly grow to be the more damaging mature plant, usually bearing seeds that will spread the species. The **seedling stage** is relatively easy to control, whether by **physical** or by **chemical** methods. Identification of this stage is therefore important and with a little practice the most significant weeds can be recognized using such plant features as cotyledons and leaf shape, colour and the first true leaves (Figure 3.7).

Mature weeds can often be identified using an illustrated **wild flora book**, which shows details of stem, leaf, flower and fruit characteristics. **Smart phone apps** such as 'Google Lens' are an effective way of identifying plants from digital photos taken in the field.

Weed biology

An **ephemeral** weed is a weed that completes several life cycles in a growing season.

An **annual** weed is a weed that completes its life cycle within one growing season.

A **perennial** weed is a weed with a life cycle of more than two growing seasons but may live for many years.

The range of weed species generally includes **flowering plants**, **ferns** and **horsetails**, **mosses**, **liverworts** and **algae**. These species display one or more special features of their life cycle, which enables them to compete successfully against cultivated plants and consequently cause problems for the horticulturist.

- **Ephemeral weeds** such as groundsel and chickweed flower and **produce seeds** throughout much of the year and seeds often germinate quickly and thus emerge from the soil to **compete** with cultivated plants. Their roots are often quite **shallow**.
- **Annual weeds** such as the common field speedwell (*Veronica persica*), annual meadow grass and fat hen (*Chenopodium album*) are similar to ephemerals in their prolific seed production. They sometimes develop **deeper roots** than ephemerals.
- **Perennial weeds** such as ground elder, creeping buttercup, couch grass, yarrow and docks (*Rumex* spp.) have persistent **organs of perennation** (see p. 72) which can be difficult to control. Ground elder, couch grass and yarrow have narrow **rhizomes**, creeping buttercup has **runners** and docks have deep swollen **taproots**. While seed production may also be prolific, it is the underground organs of perennation which presents the main problems to horticulturists, enabling the weed to emerge quickly from the soil in spring, often from considerable depths if they have been dug in. The **chopping-up** of underground organs by spades and rotary cultivators can cause these species to increase in cultivated soils.

Spread of weeds

Weeds can **spread** in a number of ways:

- Fruits such as those in hairy bittercress **dehisce** (split) and discharge seeds explosively up to 1 metre away.
- Seeds of many species from the Asteraceae family, such as groundsel, thistles and dandelion, are carried long distances in the wind by hairy **pappus** 'parachutes'.
- Seeds of chickweed may be spread by the **moving water** in ditches.
- The fruit of cleavers (*Galium aparine*) (Figure 16.5) **sticks** to the hair of animals in a manner similar to the fastener fabric 'velcro'. Chickweed seed is held in a similar way. Annual meadow grass seeds become **sticky** when wet

and can adhere to soles of shoes and machinery such as rollers or wheels on lawn mowers.

- Some of the seeds of groundsel, annual meadow grass, yarrow and dock can survive digestion in the stomachs of birds. Chickweed and annual meadow grass seed are also able to survive a rabbit's digestive system.
- The surface and underground organs such as **rhizomes** of perennial weeds, for example thistle, yarrow and couch grass, slowly spread the weed from its point of origin.
- Ploughs and rotary cultivator blades can cut underground parts of weeds such as thistles, yarrow, dandelion and couch which can then **regenerate** as individual plants.
- Commercial seed stocks can be contaminated during **harvesting** by seeds of weeds growing in the crop such as speedwells and couch grass.

Other aspects of weed biology

- **Soil conditions** may favour certain weeds. Sheep's sorrel (*Rumex acetosella*) prefers acid conditions. Mosses are commonly found in poorly drained soils. Yarrow grows well in dry soils. Common sorrel (*Rumex acetosa*) survives well on phosphate-deficient land whereas annual meadow grass thrives in phosphate-rich soil. Yorkshire fog grass (*Holcus lanatus*) colonizes impoverished turf. Nettles and chickweed prefer highly fertile soils. Understanding the **preferred soil conditions** of a weed species is useful if **cultural control** by modification of soil conditions is planned to reduce a weed problem.
- The **growth habit** of a weed may influence its success. Chickweed, creeping buttercup, slender speedwell and creeping cinquefoil produce **horizontal** (prostrate) stems bearing numerous leaves that prevent light reaching emergent crop seedlings or turf. Groundsel and fat hen have an **upright habit** that competes for light as they develop. Perennial weeds such as hedge bindweed and cleavers can grow alongside and **climb up** woody plants, such as cane fruit and border shrubs, making control difficult.
- **Seed production** can be high in certain species. A single fat hen plant can produce 70,000 seeds. Not all the seeds will be viable but it is an extremely prolific species.
- **Dormancy** of seeds (see p. 32) is seen in many weed species. In this way, weed seed germination commonly continues in a staggered way over multiple years after seed dispersal and so presents the horticulturist with a long-term weed problem. Seed of the common poppy can remain dormant in the soil for decades.
- **Perennial weeds** with swollen **underground organs** provide challenges for the horticulturist. This is especially so in long-term crops such as soft fruit and turf because many **herbicides** have little effect on these underground organs.
- **Fragmentation** of above ground parts may be important. A lawnmower used on turf which has slender speedwell weed growing in it cuts and spreads the stems that establish, like cuttings, in other parts of the lawn in damp conditions.
- **Greenhouse** production generally suffers less from weed problems because many plants are grown in **hydroponic systems** or in containers in **sterile** growing media which are free of weed seeds. Even soil-grown crops such as cut flowers tend to have less weed problems as greenhouse border soils can be **sterilized by heat**, but weeds such as chickweed (Figure 16.2), groundsel and hairy bittercress may become established.

Figure 16.5 Cleavers produce 'velcro-like' fruits to adhere to animals' fur and aid dispersal (source: Shutterstock, Gerry Bishop)

Ephemeral weeds

Hairy bittercress (*Cardamine hirsuta*)

This species is a member of the Brassicaceae family.

Location. It is common throughout Britain and Ireland. It is often seen in gardens and is particularly common on **bare ground**, in greenhouses and at the side of paths. The compost of **container-grown** plants in nurseries and garden centres may also be inhabited by this species, which is one means by which the weed may be introduced into gardens (Figure 16.6).

Life cycle. This is an ephemeral or annual weed. It flowers throughout the year but peaks from March to August. Seed is most commonly released in spring and summer. A large plant may release several thousand seeds. Weeding may encourage seed dispersal. High temperatures dry the dispersed seed and induce the ripening

16

Figure 16.6 Hairy bittercress can quickly colonize bare ground, containers and greenhouses (source: Shutterstock, Martin Fowler)

process that allows germination. The peak time of germination is in autumn. The seedlings, being **frost tolerant**, survive all but the severest of winters. The species is able to complete its life cycle in as little as five weeks. In fertile soils, the life cycle may take longer. The dormant seeds of this species in the soil can lead to a relatively persistent 'seed bank' that emerges over **several years**.

Spread. This is by means of the fruit **dehiscing** as a discharge mechanism which can result in the seeds travelling up to 1 metre away from the mother plant. Also, the seeds become **sticky** when wet and can be spread on tools and clothing.

Control. As with all ephemeral and annual weeds, it is important to control the plant **before flowering** and seed production. Seedlings should be removed by **physical** methods such as hand weeding or practices such as hoeing among ornamentals or edible plants, preferably when soils are dry and the weather is warm and breezy. In this way, flowering and dispersal of seed is prevented. In container plants, weed seedlings should be removed before seed production and dispersal begin. Covering a garden soil with a 5–10 cm deep mulch of bulky organic matter in late winter will be effective in suppressing the emergence of ephemeral weeds. There are several **chemical** control options for this weed. For bare soil and the side of paths, a contact herbicide containing **acetic acid**, **fatty acids** or **pelargonic acid** can be applied. In planted areas, **spot treatment** with these same herbicides is possible. Care should be taken not to allow spray to contact the foliage of any **non-target plants** as these herbicides are not selective and they are capable of damaging green parts of all plant species.

Chickweed (*Stellaria media*) and mouse-ear chickweed (*Cerastium fontanum*)

These species are members of the Caryophyllaceae family.

Location. Chickweed is an ephemeral weed and found in many horticultural situations growing as a weed of bare soil, among herbaceous and woody perennials, and in vegetables, soft fruit and greenhouse plantings. It has a wide distribution throughout Britain and Ireland and is most significant on nutrient-rich, moisture retentive soils.

Life cycle. The adult plant has a characteristic lush appearance and grows in a **prostrate** manner over the surface of the soil, and its leafy stems crowd out young plants as it increases in size. Small white, five-petalled flowers are produced throughout the year, the flowering response being indifferent to daylength. The flowers are **self-fertile**.

An average of 2,500 disc-like seeds (1 mm in diameter) may result from the oblong fruit capsules produced by one plant. Since the first seed may be **dispersed** within six weeks of the plant germinating, and the plant continues to produce seed for several months, it can be seen just how prolific the species is. The large numbers of seed are most commonly found in the top 7 cm of the soil where, under conditions which provide light, fluctuating temperatures and nitrate ions, they may overcome the dormancy mechanism and germinate. Many seeds, however, survive up to the second, third and occasionally fourth years. Germination can occur at any time of the year, with April and September as peak periods. Chickweed is an alternative host for some aphid-transmitted viruses such as cucumber mosaic virus and the stem and bulb nematode.

Spread. The seeds are normally released as the fruit capsule opens during dry weather. They survive digestion by animals and birds and may thus be dispersed over large distances. Irrigation water may carry them along channels and ditches.

Mouse-ear chickweed is a relative of chickweed but is a **perennial**, with oval leaves and many

Figure 16.7 Mouse-eared chickweed in turf (source: Shutterstock, Martin Fowler)

Figure 16.8 Shepherd's purse with inconspicuous white flowers and heart-shaped fruits (source: Shutterstock, IvanaStevanoski)

leaf hairs that give the leaves a slightly fluffy appearance (Figure 16.7). It often grows in turf, lying close to the soil surface and escaping lawnmower blades and thrives in **poorly drained** soil. It may be infected by cucumber mosaic virus carried by peach-potato aphid (see p. 264).

Control. Chickweed is controlled by a combination of methods. **Physical** controls include partial sterilization of soil in greenhouses, while hoeing on dry days in the spring and autumn periods prevents the seedling from developing. A 5–10 cm deep layer of bulky organic matter can be applied as a mulch in late winter and is effective in supressing ephemeral weeds. **Chemical** controls rely on contact herbicides such as **acetic acid**, **fatty acids** and **pelargonic acid** which are highly effective in controlling chickweed. Mouse-ear chickweed in turf is reduced by mowing the grass close to the ground and by the **physical** control method of raking out the weed's horizontal stems with a spring-tined rake in spring, and by seeding over bare areas in spring or autumn with grass seed cultivars that are suitable for the growing situation. In borders and beds it can be hoed or hand weeded. **Herbicides** containing a combination of active ingredients such as **2,4-D** and **MCPA** may be effectively used on mouse-ear chickweed in turf as a spot treatment for isolated weeds or over the whole area when it is widespread.

Shepherds's purse (*Capsella bursa-pastoris*)

This species is a member of the Brassicaceae family.

Location. This is an annual weed (Figure 16.8) that commonly occurs in beds and borders where the soil is regularly cultivated and is found growing on most soil types.

Life cycle. Flowering and seed production occur throughout the year, most commonly in **May to October**. Each plant can produce several thousand seeds. Three generations can occur per year. Many seeds germinate after a ripening period of two years, but this period is much shorter in some instances, especially when seeds are on the soil surface and exposed to daylight. Turning over soil can result in dormant seeds (from a deeper level) germinating at the soil surface. Shepherd's purse is a host of two significant diseases, **clubroot** and **white blister**, and also the pest **mealy cabbage aphid**.

Spread. This species can be introduced as seed into the garden through garden compost, or on new plants purchased from nurseries or garden centres. It can spread from adjacent gardens.

Control. Several **physical** methods are available for control. Young plants can be pulled out by hand. Seedlings can be hoed when the soil surface is dry. Avoid disturbing the soil so that dormant seeds are not brought to the surface. Mulch around garden perennials in late winter using a 5–10 cm deep layer of bulky organic matter to suppress weed growth. Spraying young weeds with a contact herbicide such as **acetic acid**, **fatty acids** and **pelargonic acid** can be effective; but care must be taken to ensure the spray does not contact non-target species.

Groundsel (*Senecio vulgaris*)

This species is a member of the Asteraceae family.

Location. This is an ephemeral weed, particularly on heavy soil but can grow on a **range of soil types**. Its prolific seed production and the ability of its seed to germinate soon after dispersal from the plant lead to dense mats of the weed. It can be a problem in the garden, in soft fruit and in herbaceous and woody perennials, but is most commonly found after **soil cultivation** in vegetable plots and flower beds.

Life cycle. The seedling cotyledons are narrow, purple underneath, and the first true leaves have step-like teeth. The adult plant (Figure 16.9) has an upright habit and produces as many as 25 yellow, small-petalled flower heads. Flowering occurs in all seasons of the year. Each flower produces about 45 column-shaped seeds, 2 mm in length, densely packed in the fruit head. The seeds may germinate at any time of the year, with early May and September as peak periods. Since there may be more than **three generations** of groundsel per year (the autumn plants surviving the winter) and each generation may give rise to 1,000 seeds, it is clear why groundsel is one of the most successful colonizers of cultivated ground. Its role as a host of **white blister** fungus increases its significance.

Spread. The seeds are wind dispersed by means of parachute-like **pappus**. In dry weather they can parachute seeds along in air currents for many metres. In wet weather the seeds become sticky and may be carried on the feet of animals, including humans. The seeds survive digestion by birds and thus can be transported in this way.

Control. A combination of control methods may be necessary for successful control. **Physical** control is by hoeing on a dry day or hand weeding, particularly in spring and autumn to prevent developing seedlings from flowering. Care should be taken not to allow uprooted flowering groundsel plants to release viable seed. Mulching with a 5–10 cm deep layer of bulky organic matter in late winter will create a physical barrier to suppress its growth. The gardener can use the contact herbicides such as **acetic acid, fatty acids or pelargonic acid** in such situations as beds containing woody perennials, and in bush and cane fruit. Care is required to avoid spraying foliage of cultivated plants.

Figure 16.9 A mature groundsel plant with flowers and fruits (source: Shutterstock, Orest Iyzhechka)

Annual weeds

There are many annual weed species found in gardens. This chapter can cover only a few examples that illustrate the main points of life cycle, spread and control. Two species, annual meadow grass and speedwell, are described below.

Annual meadow grass (*Poa annua*)

This species is a member of the Poaceae family

Location. This is an annual/ephemeral weed. It has a compact **tuft-forming** habit and is found in herbaceous and woody perennial borders, on paths and vegetable plots and container-grown plants. It is commonly found on lawns (Figure 16.10) and golf courses where its flowering heads and its yellowish green foliage make it conspicuous. It does not thrive on acid soils or those low in

Figure 16.10 Annual meadow grass, an adult plant in a lawn (source: Shutterstock, Yeripza Cardenas)

phosphates. Despite its relatively compact habit, it often emerges in sufficient quantities to smother crop seedlings. Its seed may be present as an **impurity** in lawn grass seed and also a constituent of poor quality, commercially produced rolls of turf.

Life cycle. Flowers can occur at any time of year and are usually **self-pollinated**. About 2,000 seeds per plant are produced from April to September. Plants will flower and produce seed, even when mowed closely and regularly, because the weed is able to flower at a plant height of less than 1 cm. Seeds germinate from February to November, with the main peaks in early spring and in autumn. Some seed will germinate soon after their release; others can remain viable in soil for at least four years. It is **shallow rooted** and susceptible to drought making turf areas containing it likely to suffer in dry summers. It is also highly susceptible to fungal diseases in winter such as 'fusarium' patch (*Microdochium nivale*) and anthracnose (*Colletotrichum cereale*) making it a liability in managed turf.

Spread. This weed has no obvious dispersal mechanism. Most seeds fall around the parent plant and become incorporated into the soil, but heavy winds may spread the seeds a few metres. Seeds may be carried around on soles of shoes and wheels or rollers of mowing machinery. Worms may bring seeds to the soil surface in worm casts.

Control. This is achieved by a variety of methods. In flower beds and vegetable plots, the **physical** action of hand weeding or hoeing on a dry day normally controls the weed, especially when it is in the early stages. Deep digging-in of seedlings and young plants is also usually effective. Mulching to 5–10 cm deep in late winter with bulky organic matter is effective against germinating seeds in beds and borders.

In turf, the general advice is to avoid patches of bare turf, and to lift any weed flower heads flattened by mowing by **brushing** so that the mower can be more effective. Mowing in different directions helps to achieve the same result. Soil management practices on fine turf can be an effective **cultural control**, this includes maintaining a well-structured, free draining soil through regular aeration, this, along with regular scarification, minimizes accumulation of **thatch,** which favours establishment of annual meadow grass. Use of **low phosphate** fertilizers is also helpful in avoiding excessive phosphate levels in the soil. Not surprisingly, there is no herbicide available to control a grass weed in a lawn, or on sports surfaces. For control in bare ground, ornamental beds containing woody perennials and fruit bushes, the home gardener can use contact herbicides such as **acetic acid**, **fatty acids** and **pelargonic acid** but care is needed to avoid spraying the foliage of cultivated plants.

Speedwells: common field speedwell (*Veronica persica*) and slender speedwell (*Veronica filiformis*)

These species are members of the Plantaginaceae family.

Location. The first species, the common field speedwell (*V. persica*) is an annual weed in **vegetable plots**, crowding out young crop plants and reducing growth of more mature stages (Figure 16.11). The second species, slender speedwell (*V. filiformis*), is a slender, mat-forming, perennial weed, once considered a desirable rock garden plant has become a common **turf** problem (Figure 16.12).

Figure 16.11 Common field speedwell

Figure 16.12 *Veronica filiformis* – slender speedwell (source: Shutterstock, weha)

Life cycle. The seedling cotyledons are oval, while the true leaves are opposite, notched approximately triangular and hairy in both species. The leaves of *V. filiformis* are approximately half the size of *V. persica*.

Veronica persica produces blue flowers, 1 cm wide, throughout the year, but mainly between February and November. The mature plant can produce in excess of 2,000 seeds which are light-brown and boat-shaped (2 mm across). The seeds of this species germinate all year round, but most commonly from March to May, the winter period being necessary to break **dormancy**. Seeds may remain viable for more than two years.

Veronica filiformis produces small, self-sterile purple-blue flowers between March and May and spreads by means of prostrate stems which root at their nodes to invade fine and coarse turf, especially in damp areas. Segments of this weed which are cut by lawnmowers easily root and thus increase the species. Seeds are not important in its spread.

Spread. Seeds of *V. persica* falling to the ground may be dispersed by ants. Seed of this species can be spread as contaminants of crop seed. *V. filiformis* rarely produces seed. Its slow spread is mainly by means of lawnmower activity.

Control. Common field speedwell (*V. persica*) is controlled by a combination of methods. The **physical action** of hoeing, hand weeding or mechanical cultivation, particularly in spring, prevents developing seedlings from growing to mature plants and producing their many seeds. **Chemical control** involves the use of contact herbicides such as **acetic acid**, **fatty acids** and **pelargonic acid** for control in bare soil, and in such situations as ornamental beds containing woody perennials, as well as in cane fruit, but care is needed to avoid spraying foliage of non-target plants.

Slender speedwell (*V. filiformis*) represents a different problem for control. **Cultural controls** such as regular close mowing and spiking of turf remove the high humidity necessary for this species' establishment and development.

Grass clippings containing stems of slender speedwell must be composted thoroughly to ensure the stems have decomposed before using the compost in the garden. Chemical control includes the use of turf herbicides containing **fluroxypyr** which will have some effect, but grass clippings collected after application of any lawn herbicide must not be added to a compost heap and the product's label instructions must be followed.

Figure 16.13 Creeping buttercup spreads by means of runners which are modified stems (source: Shutterstock, Orest lyzhechka)

Perennial weeds

Seven species – creeping buttercup, ground elder, couch grass, Japanese knotweed, yarrow, dandelion and broadleaved dock – are described to demonstrate the different features of their biology, particularly the parts of the plant that allow the plant to survive from year to year and make them successful perennial weeds. The flowering period of these weeds is mainly between spring and autumn, but the main problem for gardeners is the plants' ability to survive and reproduce vegetatively. Horsetails, mosses and liverworts, which are not closely related to flowering plants, are covered in a separate section (see p. 253).

Creeping buttercup (*Ranunculus repens*)

This species is a member of the Ranunculaceae family.

Location. This is a perennial weed in **turf** (Figure 16.13) but can also establish in bare soil in borders especially when the soil is poorly drained and prone to waterlogging. It is most common in neutral pH conditions but may occur on slightly acid or alkaline soils, especially when bare patches have been left by machinery, sports games or animal damage.

Life cycle. Creeping buttercup may grow up to 60 cm in height at flowering time. Horizontally growing **runners** develop above ground in May and develop roots along their length, representing the main problem to turf grass. Each new rooting point gives rise to a new plant during the summer months, so that after a few years of uncontrolled growth, the buttercup colony occupies a large circular area of turf. The low growing point of this weed keeps it out of reach of most lawnmowers. Its long, fibrous roots reach deep down into the soil, giving it resistance to dry weather and presenting a problem for herbicide control.

Spread. The mature plant produces bright yellow flowers from May to August, and as many as 700 seeds can be produced by each plant. Seeds may fall to the ground and survive for up to seven years in the soil, or may be eaten by birds and thus spread further afield.

Control. Using **physical** controls in bare ground, borders and beds, creeping buttercup can be removed by hand weeding including the use of a hand fork or garden fork for established plants. **Cultural** controls include autumn aeration of lawns by spiking which helps to improve surface drainage of lawns and creates soil with a moisture content less conducive to the species and consequently less weed establishment. **Chemical** control involves the application of a selective herbicide containing a mixture of **2,4-D**, **dicamba** and **fluroxypyr** which is transported from the foliage down to the weed's roots without harming the turf.

Ground elder (*Aegopodium podagraria*)

This species is a member of the Apiaceae family.

Location. This is a perennial weed that crowds out plants in established ornamental beds and borders (Figure 16.1).

Life cycle. Ground elder is so named because it has a leaf shape superficially resembling that of the elderberry shrub. It is, however, not closely related to this species. Ground elder produces white underground **rhizomes** in the soil but close to the surface. It can rapidly form a dense matt of foliage and invade gardens by creeping under fences from neighbouring garden and wasteland. It can also be present in the compost of introduced container-grown plants that are then planted in the garden.

Control. **Physical controls** include carefully removing rhizomes with a garden fork, as they occur close to the surface of the soil. However, eradicating the plant completely needs meticulous attention as the small pieces of rhizome remaining in the soil can develop into a new plant.

One rather prolonged but effective control measure involves the temporary removal and potting of shrubs and herbaceous border plants from the weed-affected areas. This procedure is then followed by placing over the area a large sheet of thick **black polythene** that deprives the weed of light and moisture. It is recommended that this sheet be left in position for at least a year to achieve control. It may be necessary to repeat this procedure if the weed emerges above ground again. **Chemical controls** include application of **glyphosate**, a translocated herbicide that is effective against ground elder. The chemical is applied to foliage and moves down from the leaves to the underground roots and rhizomes, thus killing the whole plant. Glyphosate controls many types of broadleaved (eudicot) and monocotyledon weed species – any potentially susceptible garden plants need to be protected at the time of application; spot treatment of the weeds will be less likely to cause a problem. Mid-summer sprays achieve the best results as the weed has considerable leaf coverage and is in active growth at this time. In this way, the chemical is deposited over a large area of foliage and there will be effective absorption and transport of the active ingredient to the rhizomes and roots. However, a second application may be applied in late summer, if the weed has not been successfully controlled by the first application.

Couch grass (*Elymus repens*)

This species is a member of the Poaceae family.

Location. This grass is a perennial weed and is thought to have been introduced 2,000 years ago by the Romans as a medicinal plant and is a widely distributed species. It quite often becomes dominant in herbaceous plantings and regularly cultivated beds used for vegetable production. It cannot tolerate regular close mowing so tends not to be a problem in lawns.

Life cycle. The dull-green leaves can be confused, in the vegetative stage, with creeping bent (*Agrostis stolonifera*). However, the small 'ears' (ligules) at the base of each leaf are a distinguishing feature of couch. The plant may reach a metre in height and often grows in clumps. Seeds (9 mm long) are produced only after cross-pollination between different strains of the species, and the importance of the seed stage, therefore, varies from field to field. The seed may survive in the soil for up to 10 years.

Spread. Couch grass seeds may be carried in grass seed batches over long distances. From May to October, stimulated by the high light intensity, overwintered plants produce horizontal rhizomes (Figure 16.14) just under the soil; these white rhizomes may spread 15 cm per year in heavy soils and 30 cm in sandy soils. They bear scale leaves on the underground nodes that remain suppressed during the growing period and do not produce new stems. But, in autumn, rhizomes attached to the mother plant often grow above ground to produce new plants that survive the winter. If the rhizome is cut by cultivations such as digging or rotary cultivation, fragments containing a node and several centimetres of rhizome are able to grow into new plants. The rapid growth of couch grass creates severe competition for light, water and nutrients in any infested crop.

Control. This can be achieved by a combination of **physical** and **chemical** methods. Physical

16

Figure 16.14 Couch grass plant showing rhizomes which are modified stems

methods in bare soil include deep digging or ploughing (especially in heavy land) which exposes the rhizomes to drying. Further control by rotary cultivation when the weed reaches the one- or two-leaf stage disturbs the plant at its weakest point, and repeated rotary cultivation will eventually cut up the rhizomes into such small fragments which are unable to propagate. In planted areas, it may be necessary to lift cultivated plants and remove the rhizomes from around the roots of the cultivated plants and from the soil generally before replanting. A total translocated herbicide such as **glyphosate** can be used to control in such situations as bare soil or beds and borders containing woody perennials. Care is necessary to avoid contaminating the foliage of cultivated plants and spot treatment may be necessary.

Japanese knotweed (*Reynoutria japonica*)

This species is a member of the Polygonaceae family.

Location. This species is a perennial weed and was introduced to the Netherlands from Japan in 1849 by the German botanist Philipp Franz von Siebold. It was subsequently introduced to Kew Gardens in 1850 as an ornamental plant. Its highly invasive habit has resulted in it becoming a pernicious weed which is listed in Schedule 9 of the **Wildlife and Countryside Act 1981** and it is subject to

Figure 16.15 *Cirsium arvense* – creeping thistle (source: Shutterstock, LFRabanedo)

Section 14 of this Act. It is an offence to plant or cause it to grow in the wild. It was first recorded in the wild in Britain at Maesteg, South Wales, in 1886 and is now commonly found in locations such as gardens, roadsides, waste ground and railway embankments where it colonizes and crowds out other plants; it has a significant negative impact on built structures such a buildings, walls, roads, paths, and drainage infrastructure. It is now a common problem in many parts of Britain and Ireland.

Life cycle. Japanese knotweed is an **herbaceous perennial** species which can grow to 2 m in height and produces dense, bamboo-like clumps with large, heart-shaped leaves up to 14 cm in length (Figure 16.16a) The stems have purple flecking along their length and the nodes of the stems give rise to side branches; the stems die back to ground level in winter leaving behind clumps of cane-like growth. In the soil, it has extensive, deep growing and spreading **rhizomes**. It has white flowers which are produced in late summer and early autumn (Figure 16.16b).

Spread. It does not produce seed in Britain and Ireland as all plants are a male-sterile clone; however, it can hybridise with other closely related species. The plant spreads locally by direct rhizome growth which extends the existing plant. Vegetative propagation occurs when new plants grow from detached pieces of rhizome as small as 1 cm. The rhizomes can be distributed by human activity such as movement and disposal of contaminated soil or organic matter which contains the rhizomes; water can carry parts of rhizomes in rivers.

Control. It is not an offence for a property owner to have Japanese knotweed growing on their land and it is not necessary to report its presence. However, if Japanese knotweed is causing a nuisance there may be a civil liability. In the UK, the **Anti-social Behaviour, Crime and Policing Act 2014** does not explicitly refer to Japanese knotweed or other similar invasive non-native species, but **Community Protection Notices** can be issued by local authorities or the police to require someone to control or prevent the growth of Japanese knotweed or other plants which are capable of causing serious problems in a community. Legal action can be taken if the conduct of the property owner is having a detrimental effect of a persistent or continuing nature on the quality of life of those in the locality.

It is not advisable to treat knotweed without professional help unless the operator has the appropriate skills and expertise to undertake the task effectively. Companies which specialize in treating Japanese knotweed are **registered and accredited** by industry organizations such as **Invasive Non-native Specialists Association**, **British Association of Landscape Industries** and the **Property Care Association**. **Physical control** involves digging out all parts of the plant and burying it; the **Environment Agency** must be notified at least one month before it is buried. The dead brown canes of Japanese knotweed can be composted on site, as long as they are cut, but not pulled, a minimum of 10 cm above the crown. The plant material must be **buried** on the site it came from, including any ash from burning and soils containing potential rhizome **propagules** at a depth of at least **5 metres** if it has not been sealed with a geotextile membrane and at a depth of at least **2 metres** if it has been sealed within a geotextile membrane. It should be ensured that any **geotextile membranes** used for burial is undamaged and sealed securely and that it will remain intact for 50 years and is UV resistant. No parts of the plant can be legally disposed of in household waste or local authority green waste collection containers. Disposal of the plant can only take place at a licenced landfill site and only a registered waste carrier can be used to transport the waste, so it is not possible to undertake this type of disposal without professional help. Businesses who want to burn Japanese knotweed waste must inform the **Environment Agency** a minimum of **one week** before burning it, and the **Environmental Health Officer** of the local council must also be notified.

A **D7 exemption** from the Environment Agency which allows the burning of plant tissue in the open is also required as well as a requirement to follow local byelaws and not cause a nuisance to the community. **Chemical control** involves foliage application or treatment of the cut stems with a translocated herbicide such as **glyphosate**. As with any herbicide application, the product must be **approved** for the field of use. It usually takes at least three years to eradicate Japanese knotweed with herbicides as there is likely to be regrowth after the initial herbicide application. Rhizomes can remain dormant in the soil for years so repeat applications of herbicide will be necessary. Using a professional contractor who is qualified, accredited and has expertise in Japanese knotweed control and has access to the most effective herbicide is likely to be the best option for home gardeners.

Yarrow (*Achillea millefolium*)

This species is a member of the Asteraceae family.

Location. This attractive perennial species has strongly aromatic foliage. It is a common **wildflower** (Figure 16.17) often found on impoverished, dry soils. However, it can become a significant weed in managed **turf**.

Life cycle. The mature plant has dissected bipinnate leaves produced throughout the year on wiry, woolly stems, which commonly reach 45 cm in height, and which from May to September produce flat-topped white to pink flowers (see p. 252) Each plant may produce 3,000 small, flat seeds annually. The seeds germinate at the soil surface.

Spread. Seeds are dispersed short distances by wind. The plant produces underground rhizomes which can grow up to 20 cm long per year. In autumn, rooting from the nodes occurs and new stems appear.

Control. Control of this weed in lawns may prove difficult. **Cultural control** includes lawn management practices which develop a strong, healthy lawn including regular mowing at an appropriate height for the type of lawn, annual aeration and scarification practices, maintenance of adequate soil fertility and irrigation in dry periods to keep the grass alive. **Chemical control** involves the application of selective lawn herbicides containing **2,4-D**, **mecoprop-p** and **dicamba**.

Dandelion (*Taraxacum* sect. *Taraxacum*)

This genus is a member of the Asteraceae family.

Figure 16.16 (a) Dense growth of Japanese knotweed; (b) Japanese knotweed foliage and flowers (source: Shutterstock, simona pavan)

Figure 16.17 Yarrow, an attractive wild flower but can also be a pernicious weed in lawns (source: Shutterstock, Orest lyzhechka)

Figure 16.18 Dandelion growing in a grassed area (source: Shutterstock, Mariya Teneva)

Location. Dandelions are members of the taxonomically complex genus *Taraxacum*. Taxonomists now believe the plant commonly known as *Taraxacum officinale* is approximately 250 different microspecies in Britain and Ireland, and they are classified into eight different groups with most garden species likely to be members of *Taraxacum* section *Taraxacum*. They are rosette forming, evergreen herbaceous perennial plants with a deep **taproot** and are commonly seen in roadside verges, lawns (Figure 16.18), borders and paths. The perception of dandelion as a weed has, for many people, changed. Their significance for wildlife is being recognized as the flowers provide nectar and pollen from early spring onwards for insects such as bees and butterflies and subsequently the seeds provide food for birds such as finches.

Life cycle. Seedlings emerge mainly in spring. Flowers are produced from March to October and hundreds of seeds are produced by each plant. Most seeds survive for only a year in the soil. Established, mature plants can survive for many years.

Spread. Seeds are **wind dispersed** by means of tiny parachute-like **pappus** and can travel hundreds of metres. They are also able to spread in the **moving water** found of ditches and by **animals**

Figure 16.19 Broad-leaved dock (source: Shutterstock, Eileen Kumpf)

through their digestive systems. The plant may survive and regenerate from sections of roots, after they have been chopped up by garden tools and machinery.

Control. If control of dandelion is considered to be necessary, **physical control** involves removal of the whole plant including the taproot by use of a hand fork in borders, but this can leave bare gaps if undertaken in lawns and can lead to possible invasion by other weeds, so a localized light application of a bulky lawn top-dressing medium followed by overseeding with an appropriate grass seed mixture may be necessary. If **chemical control** is implemented in lawns, it involves the application of selective lawn herbicide products containing **2,4-D** and **dicamba** which are translocated to the taproot.

Broadleaved dock (*Rumex obtusifolius*)

This species is a member of the Polygonaceae family.

Location. This is a common perennial weed in vegetable plots and bare soil.

Life cycle. The mature plant is readily identified by its long (up to 25 cm) shiny green leaves (Figure 16.19). The plant may grow 1 m tall, producing a conspicuous **branched inflorescence** of small green flowers from June to October. The seed represents an important stage in this species' life cycle, surviving multiple years in the soil and most commonly **germinating** in spring. Like most *Rumex* spp., the seedling develops a stout, ultimately branched **taproot**, which may penetrate the soil down to 1 m in the mature plant, but most commonly reaches 25 cm. Segments of the taproot, **chopped** by garden tools and machinery, are capable of producing new plants.

Spread. The numerous plate-like fruits (3 mm long) may fall to the ground or be dispersed by seed-eating birds such as finches.

Control. High levels of seed production, a resilient taproot and a resistance to many herbicides present a problem in the control of this species.

In turf and meadows, **physical control** involves removal of the taproot by a garden fork, which is effective in vegetable plots and bare soil, but this leaves bare gaps in turf leading to possible invasion by other weeds. **Chemical control** involving the use of a product containing selective lawn herbicides such as **dicamba**, **MCPA** and **mecoprop-P** is effective, especially against young dock plants. **Triclopyr** is selective against broadleaved weeds in grass and can be used in situations such as **grassland**. In ornamental beds containing woody perennials, and in cane and bush fruit, it is not selective as many of the cultivated plants are likely to be broadleaved species so spot treatment with a product containing **glyphosate** applied to foliage of the target plants would be necessary; care is required to avoid contaminating the foliage of non-target plants.

Horsetails

Horsetails belong to the genus **Equisetum**, the only genus in the family *Equisetaceae*. The **field horsetail** (*Equisetum arvense*) can be a significant weed in established ornamental plantings (Figures 16.20 a,b), and especially in damp soil situations. It produces persistent **rhizomes**.

Spread. Underground rhizomes may be dispersed and can be a method of spread when the weed is dug up.

Control. **Physical control** involves regular cutting down of the foliage and **cultural practices** involve maintenance of good soil structure and appropriate levels of soil fertility and water and air content to allow cultivated plants to thrive and compete. Foliar application of a product containing translocated herbicides such as **glyphosate** can be successful on bare ground. Care is needed to avoid spraying the foliage of non-target plants.

Mosses

Mosses are primitive plants and include three common species which can be a problem in gardens. The small cushion-forming moss (*Bryum* spp.) grows on sand capillary benches in greenhouses and in turf in acidic soil conditions that has been closely mown. **Carpet moss** (*Hypnum* spp.) is common on less closely mown, unscarified turf, or on the surface of growing media in container-grown plants (Figure 16.21a). **Hair moss** (*Polytrichum* spp.), which looks quite different from the other two species, is erect and has a rosette of leaves (Figure 16.21c). It is found in dry acid conditions around golf greens. **Physical control** involves scarifying the turf and removing the moss

Figure 16.20 (a) Horsetail has an extensive system of rhizomes; (b) horsetail spore-bearing strobilus emerging from soil

Figure 16.21 (a) *Hypnum* moss on a lawn, (source: Shutterstock, Ahanov Michael); (b) *Pellia* liverwort (source: Shutterstock, simona pavan); (c) *Polytrichum* moss which may occur in turf

debris in September and October when the grass is still active and can recover before winter starts, followed by **cultural control** such as aeration to reduce soil compaction and improve drainage, application of fertilizer to maintain suitable nutrient status and reduction of shade where possible. **Chemical control** with a contact scorching moss-killer containing **sulphate of iron** can be applied in spring or autumn and is most effective if applied two weeks before scarification when the dead moss can be raked out. Moss on **sand benches** becomes less of a problem if the sand is regularly washed.

Liverworts

Liverworts are closely related to mosses and include several weed species in the *Marchantia* and *Pellia* genera. They are recognized by their fleshy, flat thalli growing on the surface of container-grown plants compost (Figure 16.21b). These plants thrive when the soil and compost surface is excessively wet and a low pH is present or when nutrients are so low as to limit plant growth. **Physical control**

can be carried out by hand removal of liverworts on the surface of the compost followed by top dressing with a layer of fresh growing media. **Cultural control** involves practices which create conditions that conspire against the establishment of liverworts, including use of a well-aerated, free-draining compost with adequate nutrients, appropriate pH and avoidance of over irrigation. **Chemical control** involves the application of a contact herbicide such as **pelargonic acid to the thalli** ensuring that the herbicide does not come into contact with any green plant parts, as they will be harmed.

Further reading

Adams, C., Brook, J. and Early, M. (2015) *Principles of Horticulture Level 3*. Routledge.

Barter, G. (2024) *RHS What's That Weed?* Dorling-Kindersley.

BCPC. (2002) *Weed Management Handbook*. Wiley-Blackwell.

Blamey, M., Fitter, R. and Fitter, A. (2013) *Wild Flowers of Britain and Ireland*. Bloomsbury Natural History.

British Crop Production Council. (2025) *UK Pesticide Guide*. BCPC.

Brown, L.V. (2008) *Applied Principles of Horticulture*. 3rd ed. Butterworth Heinemann.

Cliff, A. (2017) *The Value of Weeds*. The Crowood Press.

Hessayon, D.G. (2009) *Pest and Weed Expert*. Transworld Publishers.

Keble-Martin, W. (1991) *The New Concise British Flora*. Godfrey Cave Associates Ltd.

Mabey, R. (2012) *Weeds, The Story of Outlaw Plants* Profile Books.

Richards, G. (2021) *RHS Weeds*. Welbeck.

Stace, C. (2019) *New Flora of The British Isles*. 4th ed. G&M Floristics.

Thompson, K. (2009) *The Book of Weeds*. Dorling Kindersley.

Walker, J. (2016) *Weeds*. Earth-Friendly Books.

Wallington, J. (2019) *Wild About Weeds*. Laurence King Publishing.

Williams, J.B. and Morrison, J.R. (2003) *A Colour Atlas of Weed Seedlings*. CRC Press.

www.daera-ni.gov.uk/topics/plant-and-tree-health

www.gardenorganic.org.uk/expert-advice/garden-management/weeds

https://planthealthportal.defra.gov.uk/

www.rhs.org.uk/media/pdfs/advice/WeedkillersForGardeners. RHS Advisory Service

www.rhs.org.uk/prevention-protection/weeds-non-chemical-control

www.sasa.gov.uk/plant-health

www.teagasc.ie/crops/horticulture/plant-health/

The online material is accessible via the QR code and includes further information on many of the topics in the book, such as

- Hedge and field bindweed
- Common plantain

CHAPTER 17

Plant pests

Figure 17.1 A rabbit eating garden plants

This chapter includes the following topics:

The biology, damage and control methods of garden pests from the following groups:

- Mammals
- Birds
- Molluscs
- Insects
- Mites
- Other arthropods
- Nematodes

DOI: 10.4324/9781003581260-17

At some time, most gardeners or growers will have been concerned to see the appearance of their decorative plants or the quality of their crop adversely affected by the activities of pests. This chapter describes the main groups of plant pests and some of the most significant pest species seen in the gardens and protected structures of Britain and Ireland, ranging from relatively large mammals such as rabbits to microscopic nematodes. In addition to the descriptions of pest life cycles and the damage they cause, a range of prevention and control measures are described.

Mammal pests

There are relatively few mammal pest species in Britain and Ireland, but some species cause serious damage to plants in the garden and horticultural situations. The rabbit, grey squirrel and mole are described next.

A pest is an animal, usually a mammal, bird, insect, mite, mollusc or nematode, which causes unwanted damage to plants.

Rabbit (*Oryctolagus cuniculus*)

The rabbit (Figure 17.1) is common in most countries of central and southern Europe. It was thought to have been introduced to Britain and Ireland around the 12th century by the Normans and became an established pest in the 19th century, but recent evidence suggests rabbits were first introduced to Britain by the Romans 2,000 years ago. It is now common in most parts of Britain and Ireland.

Damage. A mature rabbit can consume 0.5 kg of plant food per day. They can quickly **graze** shoots on herbaceous plants and vegetables to ground level. Vegetables such as carrots, lettuce and ornamental annual bedding plants and herbaceous perennials are common targets for this pest (Figure 17.2). By standing on their hind legs, they can graze soft growth on woody plants to 50 cm high. Bark on the stems of young trees and shrubs of both ornamental and fruit species can be gnawed which can lead to ring-barking, where the vascular tissue just beneath the bark is destroyed all the way around the stem leading to an inability to transport water and nutrients up the plant; this can subsequently lead to death of the plant. Rabbits also scrape and **dig holes** in lawns and on golf courses, which impacts on playing quality, and also in beds and borders in gardens, which exposes plant roots to drying. Cereal crops can be badly affected, particularly when in the seedling stage, and quite large areas may be destroyed.

Figure 17.2 Rabbit grazing damage to tulip leaves (source: Shutterstock, photowind)

Life cycle. The rabbit's high reproductive ability enables them to maintain large numbers, even when continued control methods are being used. The female (doe), weighing about 1 kg, can reproduce within a year of its birth and may have three to five litters of three to six young ones in one year, commonly in the months of February to July. The young emerge from the underground nest after only a few weeks to find their own food. Large burrow systems (warrens) penetrating as deep as 3 m in sandy soils may contain as many as 100 rabbits. Escape or bolt holes running off from the main burrow system may allow the rabbit to escape from predators such as weasels.

Control. Rabbit control is **required by law** in England (except the City of London and Scilly Isles); the responsibility lies with the land owner. If this is not possible, the land owner must stop them causing damage to crops on adjoining land by constructing rabbit proof fencing to contain them. Preventive **physical controls** can be effective. Rabbit-proof fencing can be erected around the land containing the plants or area to be protected. It should be a 2.5 cm wire mesh and be at least 1.2 m high with the bottom buried 30 cm below ground level with this lower 15 cm bent facing outwards to stop rabbits tunnelling underneath. Repellent products such as **citronella extracts** and **aluminium ammonium sulphate** are constituents of commercially available repellent products and can be sprayed on to susceptible plants to deter rabbits. Shooting and humane trapping of rabbits is possible in commercial horticulture and agriculture but these methods should only be carried out by appropriately qualified professional operators and within **guidelines** published by Natural England. **Cultural control** involves selection of garden plants which show **resistance** to rabbits including *Acanthus* spp., *Agapanthus* spp., *Dahlia* spp., *Galanthus* spp., *Betula* spp., *Berberis* spp. and *Prunus* spp.

Grey squirrel (*Sciurus carolinensis*)

The grey squirrel was introduced into Britain and Ireland from North America in the late 19th century at a time when the native red squirrel population was being reduced by disease. The grey squirrel (Figure 17.3) has become dominant in many areas, but the red squirrel survives, particularly in central and southern Scotland, the Lake District, central and north Wales and the Isle of Wight.

Damage. The horticultural damage caused by grey squirrels varies with each season. In spring, bulbous plants coming into growth may be eaten, and the bark of many young tree species stripped off (**ring-barking**), which is a major problem in forestry and woodland situations leading to death of the plant. In summer, fruit and vegetables such as pears, plums, raspberries and peas may suffer. Autumn provides an alternative food source in the form of wild plant seeds, but apples and potatoes may be damaged at this time. Damage can also occur in lawns, beds and borders when squirrels dig holes to store and subsequently retrieve food which disrupts lawn and planted areas. Horticultural situations located close to wooded areas can be badly damaged by the grey squirrel.

Life cycle. Grey squirrels most commonly produce two litters of three offspring from March to June, in twig platforms (dreys) high up in trees. The female may become pregnant at an early age (six months). As the squirrels have few natural enemies and they live high above ground, control is difficult in most areas. In some northern parts of Britain and Ireland, however, pine martens (a tree-inhabiting relative of the weasel) are predators of grey and red squirrels.

Control. During the months of April to July is when most damage is seen. **Physical controls** such as the use of **cage traps** containing food such as maize seed can capture grey squirrels. The release of captured grey squirrels back into the wild is not permitted as it is illegal to release a non-native invasive species, so trapping presents a problem of how to humanely and legally treat a trapped animal. In practice, trapping is unlikely to have a significant effect on the grey squirrel population and the damage they cause. **Netting** of susceptible cultivated plants such as soft fruit and ornamental shrubs when they are of interest to grey squirrels can be effective. Grey squirrels can chew through plastic netting so the use of **wire netting** is the best option for permanent structures such as fruit cages. **Repellent** products such as citronella extracts and aluminium ammonium sulphate are present in commercially available preparations and can be used for short-term protection at the plant's most susceptible period.

Figure 17.3 An adult grey squirrel (source: Shutterstock, Nicola_K_photos)

Mole (*Talpa europaea*)

The mole is found in all parts of Britain but is not present in Ireland.

Damage. This dark grey, 15 cm long mammal (Figure 17.4a), weighs approximately 90 g when fully grown. It uses its spade-shaped feet to create an underground tunnel system 5–20 cm deep and up to 0.25 ha in extent. The most obvious sign of its presence are **molehills**, heaps of excavated soil

Figure 17.4 (a) An adult mole emerging from the ground (source: Wikimedia Commons); (b) molehills on a lawn (source: Shutterstock, Zigmunds Dizgalvis)

thrown up on the surface of lawns, borders and beds. The tunnel's contents are excavated and deposited on the surface of the soil to create the molehills (Figure 17.4b).

The resulting root **disturbance** to turf and crops can cause wilting and may result in damage to plants and loss of level surfaces, which can be a problem on **golf courses** and lawns. It should be noted that the plants themselves are not eaten by moles.

Life cycle. Throughout the year, in its dark environment, this solitary animal moves, actively searching for earthworms, slugs, millipedes and insects. Only in spring do males and females meet. In June, one litter of two to seven young are born in a grass-lined underground nest, often located underneath a dense thicket. Young moles often move above ground to find a new territory, reaching maturity at about four months. Moles can live for up to four years.

Control. Unlike the rabbit and grey squirrel, the mole is a native animal in Britain and an integral part of the **ecosystem**. Unless damage is severe or affecting the playing conditions in situations such as golf courses, it may well be the most appropriate option to accept a limited level of damage. Natural predators of the mole include tawny owls, weasels and foxes. **Physical control** such as trapping is possible but it can cause suffering, and a humane, instantaneous death is not certain. It should be undertaken by a skilled professional but is best avoided unless there is an extreme necessity to use this method. **Plastic mole netting** can be installed on the surface of the soil on new turf areas prior to laying the turf. This can prevent them from subsequently coming to the surface and reduces problems with molehills. Proprietary mole repellent powders can be applied into a tunnel between recently created molehills and also sprinkled on the surface of other areas as a preventative. The odour is harmless but unpleasant to the mole, but such products have a short-term effect and will probably require reapplication. **Ultrasonic deterrent** devices can be inserted into the soil and deliver a frequency of sound and vibration which is intolerable to the mole that causes the animal to move away from the disturbance.

Bird pests

Wood pigeon (*Columba palumbus*)

Damage. At 40 cm long when mature, the wood pigeon (Figure17.5) is known to horticulturists as a pest on numerous outdoor **edible crops**. In spring, seeds and seedlings of crops such as brassicas, beans and germinating grass seed can be systematically eaten. In summer, soft fruit receives its attention and tree fruits such as plums can be taken in large quantities as well as the weight of

Figure 17.5 An adult wood pigeon (source: Shutterstock, WildMedia)

the pigeons causing damage to the branches. In winter, brassicas are often attacked, particularly when snowfall prevents the consumption of other food. The wood pigeon is attracted to high-protein foods such as seeds when they are available.

Life cycle. Wood pigeons lay several clutches of two eggs per year from March to September. Eggs are laid on a nest of twigs situated deep inside the tree and hatch after about 18 days, the young remaining in the nest for 20–30 days. Predators such as jays and magpies eat many eggs, but the main population-control factor is the lack of food in winter. Numbers in Britain and Ireland are boosted a little by migrating Scandinavian pigeons in April, but the large majority of birds are **resident** in Britain and Ireland and do not migrate around the country.

Control. The wood pigeon spends much of its time feeding on **wild plants** and only a small proportion of its time is spent eating **crops**. Control of the whole population therefore seems ethically unsound and is both costly and impracticable. **Physical** control involves the protection of particular areas by means of disturbing or scaring devices. In gardens, plastic or stainless steel strips bearing pigeon spikes may be placed on vantage points on buildings and walls and reduce their damage to garden plants. **Ultrasonic deterrents** are effective over a small area. Five-centimetre mesh **netting** placed over an area prevents the pigeon's entry to such targets as young plants and soft fruit. Gel or liquid formulations of **aluminium ammonium sulphate** applied to plant areas or brushed on to pigeon vantage places such as roof ledges deters the bird. Life-size **model predators** such as owls or falcons may be purchased and placed in a suitably prominent place in the garden.

Bullfinch (*Pyrrhula pyrrhula*)

The bullfinch is another occasional bird pest. It can seriously affect soft and tree fruit yields by nipping out fruiting buds overwinter. Physical protection of susceptible plants should be implemented when required.

Molluscs (slugs and snails)

Slugs and snails belong to the phylum Mollusca.

Damage. Slugs, unlike snails, lack a **shell**, (Figure 17.6c) and this permits their movement under the soil in search of their food source: seedlings, roots, soft stems, leaves, flowers, fruits, tubers and bulbs of many species. They feed by means of a **radula**, a file-like tongue which cuts through the plant material held by the soft mouth. They can also scoop out cavities in affected plant tissue such as strawberry fruits which ruins the fruit and opens the tissue to potential fungal infections by grey mould (*Botrytis cinerea*). In moist, warm weather they often cause above ground damage to leaves and soft stems of many herbaceous perennials and annuals and vegetables, which disfigures the appearance and reduces photosynthetic capability of the plant (Figure 17.6a). Snails are also capable of causing considerable damage in gardens to young plants, particularly species such as lettuce and hosta and to crops being cultivated in greenhouses such as tomatoes and petunias.

Life cycle. Slugs and snails are **hermaphrodite** (bearing in their bodies both male and female reproductive organs). They mate in spring and summer and lay clusters of up to 50 round, white eggs in rotting vegetation, the warmth from which protects this sensitive stage during cold periods. Slugs range in size from the keeled slug (*Milax* spp.), 3 cm long, to the black garden slug (*Arion hortensis*) (Figure 17.6b), which reaches 10 cm in length. Slugs move slowly by means of an **undulating foot**, the slime trails from which may indicate their presence. The three species of mottled slug (*Testacella* spp.), occurring mainly in southern Britain, are quite common, but rarely seen, living predominantly underground. The mottled slugs are **carnivorous**, feeding on earthworms and sometimes on other slugs (Figure 17.6d).

Control. Slugs and snails play a significant role in the recycling of organic matter in the garden and are part of the garden's **ecosystem**. It should be recognized that not all slugs and snails are plant pests, and they provide an important **food source** for many other species of animals such as hedgehogs, frogs and toads, centipedes, ground beetles and birds such as song thrushes; any garden feature or practice which promotes the presence of these predators is likely to be beneficial in reducing slug and snail numbers to a tolerable level. **Cultural control** includes avoiding **excessive tidiness** in the garden and allowing dead stems and leaves of herbaceous perennials to remain over winter will provide hibernation sites for predators such as ground beetles. Hedgehogs which predate slugs can be encouraged by creating 30 cm

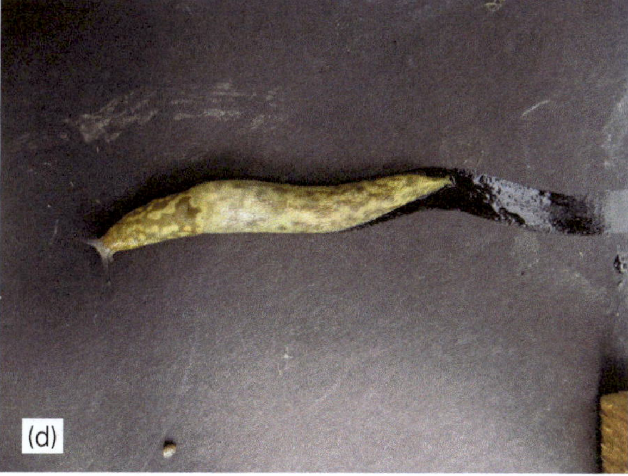

Figure 17.6 (a) Slug and snail feeding damage on the leaves of hostas (source: Shutterstock, Tony Baggett); (b) a black garden slug on a lawn (source: Shutterstock, Ion Mes); (c) a garden snail (source: Shutterstock, Baloost); (d) carnivorous *Testacella* slug which eats earthworms

square holes in the bases of garden boundaries which permit their access. Log and leaf piles and compost heaps provide potential hibernation places. Allowing wind-fallen fruit to remain on the ground will provide additional food. Shallow ponds with gently sloping sides and vegetation planted around it will provide a water source and cover which will encourage hedgehogs, frogs and toads. Song thrushes are one of the few birds that will predate snails – they are capable of smashing the snail on to a hard surface to break the shell; establishment of woody planted features such as trees, shrubs and hedges will help provide them with nesting sites and food in the form of fruits. The use of **resistant or less susceptible** species and cultivars is an effective way to limit the damage caused by slugs and snails. Potato tubers are particularly susceptible to damage from underground slugs although early cultivars are generally less prone to damage than maincrop ones, which should be lifted as soon as they are mature. Cultivars such as 'Charlotte', 'Pentland Dell' and 'Estima' show more resistance than many others. Many species of ferns and ornamental grasses suffer no damage at all. *Hosta* species, which are notoriously prone to slug and snail damage, may be grown with more confidence if blue-leaved cultivars of species such as *H. sieboldiana* are selected. Other herbaceous species which show some resistance include *Euphorbia* spp., *Alchemilla mollis, Bergenia* spp. and *Lamprocapnos spectabilis.* **Physical control** includes **handpicking** in the evening or early morning from spring to autumn when the weather is damp and the slugs and snails are most active. They can be placed in containers and relocated away from susceptible plants or humanely destroyed by securing in a polythene bag and freezing them. Barriers such as copper strips are sometimes placed around pots and materials such as horticultural grit, crushed eggshells and wool pellets placed round the base of susceptible plants with the aim of deterring the pests and reducing damage to plants. However, the evidence for the efficacy of these methods is largely anecdotal and there is very little scientific research to support the claims that they deter slugs and snails. A range of homemade and proprietary **traps** with baits are available and can be placed close to susceptible plants where slug and snail activity is expected; the trapped molluscs can then be destroyed humanely. This method of control can be effective on a small scale and is more likely to appeal to the home gardener. **Biological control** of slugs can be undertaken by the application of **parasitic nematode** such as *Phasmarhabditis californica* which is effective against most common species of slug in Britain and Ireland and can be purchased from a supplier. The parasitic nematodes are added to water and then drenched or sprayed onto the soil; they then move through the soil, enter the slug's body and release bacteria which quickly stops the slugs feeding, and they subsequently die below ground within 4–21 days. Like all biological control methods, the **environmental conditions** are an important factor in the success of the process. In this particular case, the nematode will be most effective when the soil is moist but not excessively wet and the temperature is between 5°C and 30°C, so the warmer months between spring and early autumn are the best time to apply them. This method of control has to be applied every six weeks to maintain ongoing effectiveness. **Chemical control** is limited to products containing the active ingredient **ferric phosphate**. When applied as a pellet and used in accordance with manufacturer's instructions it will present relatively few risks to other organisms. The use of other substances such as salt should not be undertaken. It is illegal for professional users to apply any substance which is not approved as a molluscicide. It is also an inhumane method of destroying the pest and can contaminate the soil.

Insects

Belonging to the large phylum of **Arthropoda** are invertebrate animals which includes woodlice, mites, millipedes and symphilids (Table 17.1). Insects are horticulturally the most significant arthropod group, both as **beneficial organisms** and, sometimes, as **pests**.

Insect structure and biology

The body of the adult insect is made up of segments and is divided into three main parts: the head, thorax and abdomen (Figure 17.7). The **head** bears the mouthparts of which there are two main methods of **feeding** (Figure 17.8). Caterpillars, sawfly larvae and beetles have **biting** mouthparts. In aphids and their relatives, the mandibles and maxillae are fused to form a delicate tubular **stylet** that sucks up liquids from soft plant tissues.

Insects remain aware of their environment by means of compound eyes which are sensitive to movement (of predators) and to colour (of flowers). The **thorax** bears three pairs of legs, and in most insects, two pairs of wings. The **abdomen** bears breathing holes (spiracles) along its length, which lead internally to a breathing system of tracheae. The blood is colourless, circulates digested food and has no breathing function. The digestive system, in addition to its food-absorbing role, removes waste cell products from the body by means of fine, hair-like growths located near the end of the gut.

Since the animal has an **external skeleton** made of tough chitin, it must shed and replace its 'skin' (**cuticle**) (Figure 17.7) periodically by a process called **ecdysis**, in order to increase in size.

Table 17.1 Arthropoda groups significant in horticulture

Group	Key features of class	Location	Damage
Woodlice (Malacostraca)	Grey, seven pairs of legs, up to 2 mm in length	Damp organic soils	Generally beneficial
Millipedes (Diplopoda)	Brown, many pairs of legs, slow moving	Most soils	Generally beneficial
Centipedes (Chilopoda)	Brown, many pairs of legs, very active with strong jaws	Most soils	Beneficial
Symphilids (Symphyla)	White, 12 pairs of legs, up to 8 mm in length	Glasshouse soils	Generally beneficial
Mites (Arachnida)	Variable colour, usually four pairs of legs (e.g. two-spotted spider mites)	Soils and plant tissues	Mottle or distort leaves, buds, flowers, tubers and bulbs; soil species are beneficial
Insects (Insecta) see below	Usually six pairs of legs, two pairs of wings		
Springtails (Collembola)	White to brown, 3–10 mm in length	Soils and decaying humus	Generally beneficial
Aphid group (Hemiptera)	Variable colour, sucking mouthparts, produce honeydew (e.g. greenfly)	All locations	Leaf distortion, transmit viruses
Moths and butterflies (Lepidoptera)	Large wings; larva with three pairs of legs, and four pairs of false legs and biting mouthparts (e.g. cabbage white butterfly)	Mainly leaves and flowers	Eat leaves, stems, roots and fruits
Flies (Diptera)	One pair of wings, larvae legless (e.g. leatherjacket)	All locations	Leaf mining, eat roots
Beetles (Coleoptera)	Horny front pair of wings which meet down centre; well-developed mouthparts in adult and larva (e.g. wireworm)	All locations	Eat leaves, roots and tubers and fruit
Sawflies (Hymenoptera)	Adult like a queen ant; larvae have three pairs of legs and more than four pairs of false legs (e.g. rose-leaf curling sawfly)	Mainly leaves and flowers	Eat leaves
Thrips (Thysanoptera)	Yellow and brown, very small, wriggle their bodies (e.g. onion thrips)	Leaves and flowers	Cause mottling of leaves and petals
Earwigs (Dermaptera)	Brown, with pincers at rear of body	Flowers and soil	Eat flowers

The **two main groups** of insect develop from egg to adult in different ways. In the first group, **exopterygota**, which includes the aphids, thrips and earwigs, the egg hatches to form a first stage (instar) called a **nymph**, which resembles the adult in all but size, wing development and possession of sexual organs. Successive nymph instars more closely resemble the adult. Two to seven instars (growth stages) occur before the adult emerges (Figure 17.9).

This development method is called **incomplete metamorphosis**. In contrast, in the second group of insects, **endopterygota**, which includes the moths, butterflies, flies, beetles and sawflies, the larvae (grubs/caterpillars) undergo a remarkable change (**complete metamorphosis**) within the pupae. The egg hatches to form a first **instar**, called a **larva**, which usually differs greatly in shape from the adult. For example, the larva (caterpillar) of the large cabbage white butterfly bears little resemblance to the adult butterfly (Figure 17.9).

Some damaging larval stages are shown in Figure 17.10 and these can be compared with the often more familiar adult stage.

The method of **overwintering** differs between insect groups. Outdoors, aphids survive winter mainly as eggs, while most moths, butterflies and flies survive as the pupa. The speed of increase of insects varies greatly between groups. Aphids can reach sexual maturity in days and complete their life cycle within weeks in summer, often resulting in vast numbers in the period May to September. However, wireworms, the larvae of click beetles, typically take three to four years to complete their life cycle. Insect groups are classified into their appropriate order (Table 17.1) according to their general appearance and life-cycle stages.

Insect pests

A selection of significant insect pests follows in which each species' life cycle features are given. While comments on specific control are mentioned here, the reader should also refer to Chapter 15 for details of general types of control (physical, cultural, biological, chemical, IPM etc.) and for explanations of terms used.

Aphids and their relatives (order Hemiptera)

This significant group of insects has the egg–nymph–adult life cycle (see Exopterygota, p. 263) and piercing and sucking mouthparts.

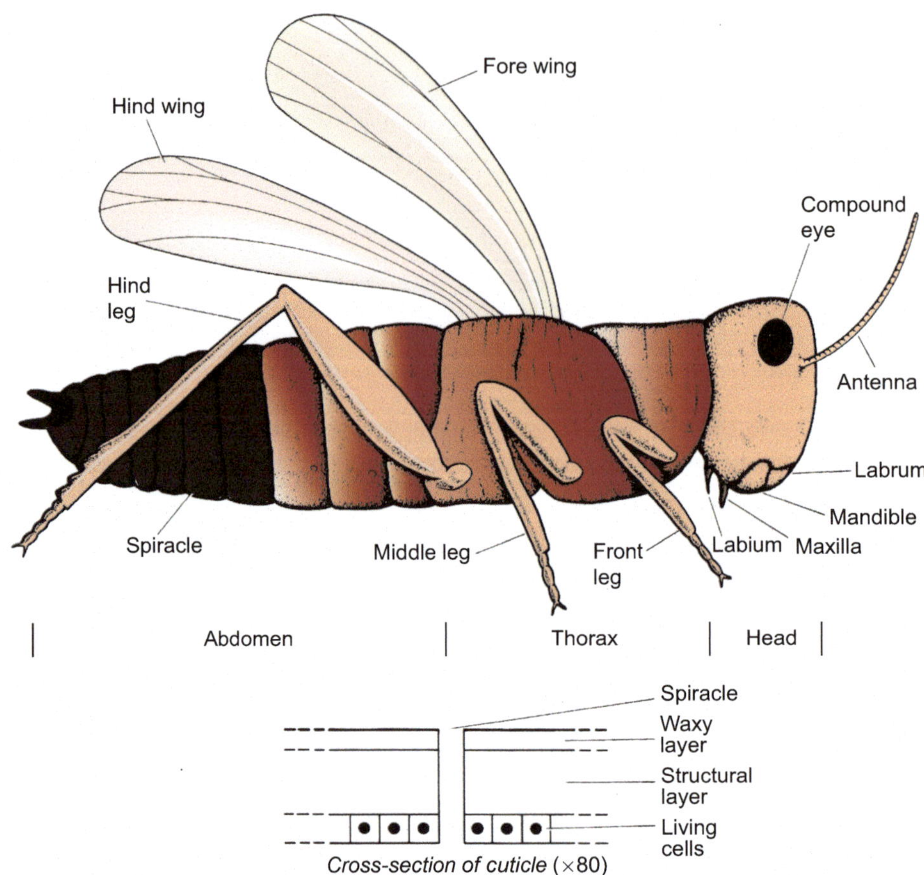

Figure 17.7 The external appearance of an insect. Note the mouthpart, spiracles and cuticle – the three main entry points for insecticides

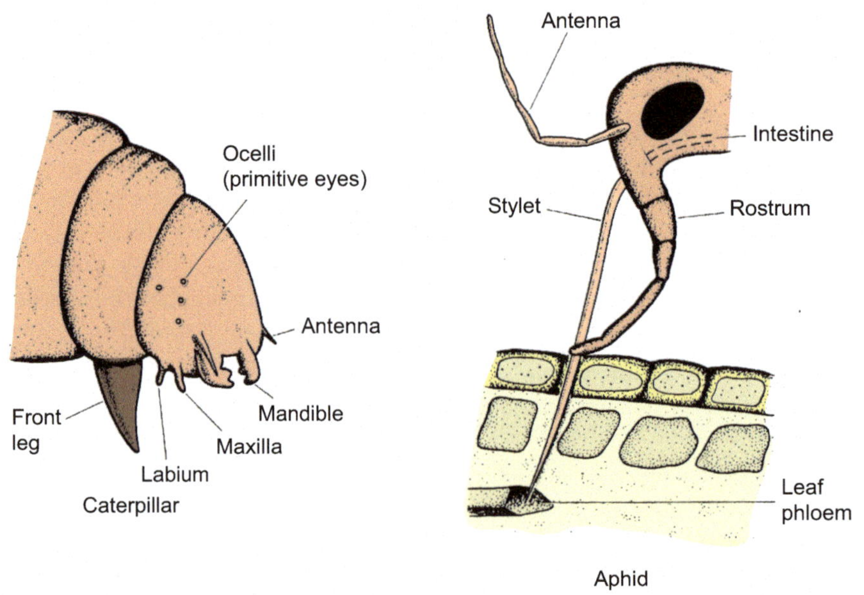

Figure 17.8 Mouthparts of the caterpillar and aphid. Note the different methods of obtaining nutrients. The aphid selectively sucks up dilute sugar solution from the phloem tissue

Peach-potato aphid (*Myzus persicae*)

Damage. This species commonly occurs in greenhouses all year round and outdoors in summer. The adult and nymph stages of this pest can cause various types of damage. Using the feeding **stylet** (Figure 17.8), the aphid sucks sap leading to a weakening of the plant (Figure 17.11a) and a reduction in vigour and crop yield; in the process it can inject a digestive juice into the plant tissues which leads to distortion of young, non-woody shoots. Having sucked up sugary

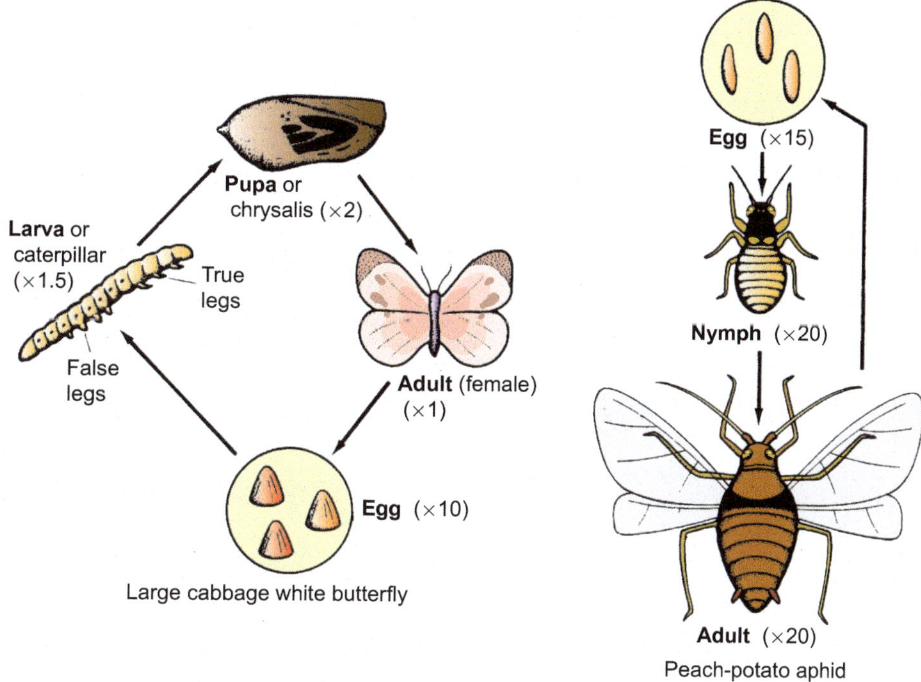

Figure 17.9 Life cycle stages of a butterfly and aphid pest. Note that the four stages of the butterfly life cycle are very different in appearance. The nymph and adult of the aphid are similar

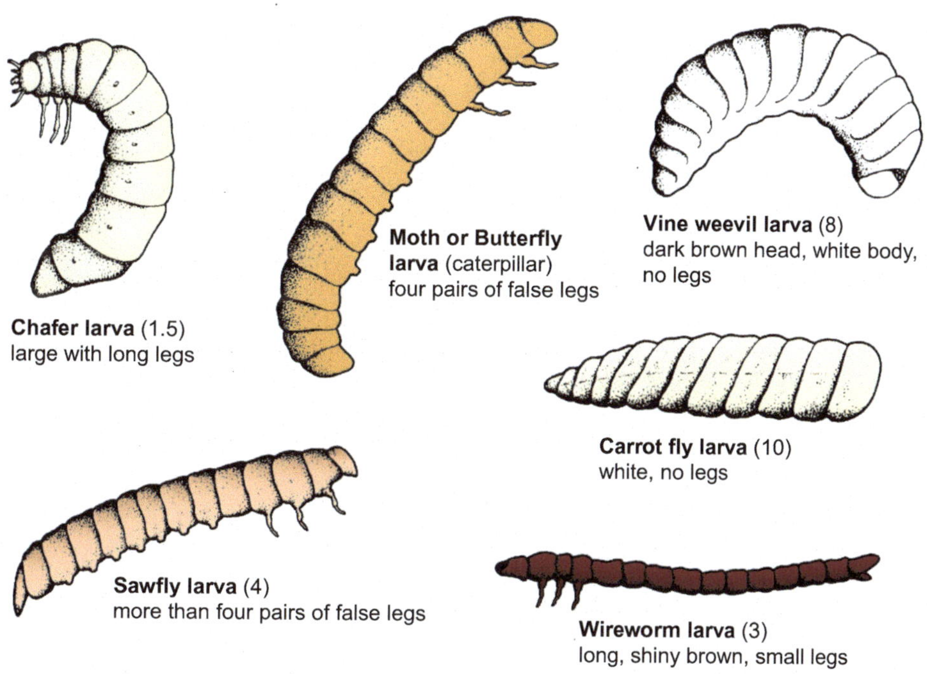

Figure 17.10 Insect larvae which can damage plants

sap, the aphid **secretes** a sticky substance called **honeydew** that can coat the foliage and block leaf stomata which reduces photosynthesis, particularly when the black-coloured **sooty mould** fungi (typically *Cladosporium* spp. and *Alternaria* spp.) grow on the honeydew; the sooty mould is also unsightly on ornamental plants. If the host plant is infected with a virus, the aphid can act as a **vector** and subsequently transmits it to other plants. It is known to transmit many types of viruses including **mosaic viruses** and **leaf roll virus** (see p. 305) on potato.

Life cycle (Figure 17.12) This aphid varies in colour from light green to pink, measures 3 mm in length (Figure 17.11a), and has an incomplete life cycle (Figure 17.9). In sheltered locations, it can **overwinter** as adults or nymphs on winter hosts such as brassicas, herbaceous species and weeds but also as **eggs** on peach trees or related species. The **summer hosts** include potato and a **wide range** of other species. From early spring, the emerging females give birth to nymphs directly without any egg stage (**vivipary**) and without fertilization by a male (**parthenogenesis**). Several generations of wingless aphids are produced before **winged aphids** appear from May onwards; the aphids then migrate to the **summer hosts**. Successive generations occur throughout the summer until late summer when, in response to decreasing daylength and outdoor temperatures, both male and female sexual forms are produced. These have wings and they fly to the **winter host**. Here, the female is fertilized and then lays thick-walled black eggs. In greenhouses, the aphid may survive the winter as the nymph and adult stage on plants such as *Begonia* and *Chrysanthemum*, or on weeds (Figure 17.12).

Spread. Occurs mainly in early summer by winged females but also on plants in transport.

Control. Aphids are an important food source for many beneficial organisms and some level of natural control inevitably takes place. **Cultural control** includes implementing garden management practices which encourage natural predators such as ladybird adults and larvae (Figure 15.8c), lacewing larvae (Figure 15.7b) and hoverfly larva and birds such as blue tits; all these organisms will naturally reduce aphid numbers, and the cultivation of attractant plants such as poached egg plant (*Limnanthes douglasii*) (Figure 17.11b) to attract hoverfly (Figure 17.11c) and pot marigold (*Calendula officinalis*) to attract ladybirds close to the affected plants is helpful. **Physical control** includes using fingers and thumb to squash isolated colonies or pinching out and destroying soft growing tips of badly infested plants. **Winter tree washes** containing plant and fish oils can be applied from November to February on to the stems and trunk of dormant woody plants such as fruit trees and bushes to destroy the overwintering aphid eggs by physical action. **Biological control** can be introduced into greenhouses; small parasitic wasps such as *Aphidius colemani* and *Aphelinus abdominalis* are available from suppliers, when released into the crop the wasps lay their eggs inside the bodies of aphids which subsequently die after being consumed by the developing wasp larvae. **Chemical control** can involve the application of insecticides on active nymphs and adults. Products containing active ingredients such as **pyrethrins** and synthetic pyrethroides such as **lambda-cyhalothrin** are found in products approved for this purpose, although the product label instructions must always be checked to ensure the product is approved for use on the plant or crop to be treated. The potential effects of insecticides on beneficial organisms and the environment should always be considered.

Figure 17.11 (a) Aphids feeding on leaves and soft stems (source: Shutterstock, Tomasz Klejdysz); (b) poached egg plant, pollen source for hoverflies; (c) hoverfly on *Nepeta* sp.

Black bean aphid (*Aphis fabae*)

This 2 mm long black aphid (Figure 17.13a) is commonly referred to as 'blackfly', although it is not a fly species. Most individuals are black, but some are dark olive green. The species found on beans should not be confused with the **cherry blackfly** (*Myzus cerasi*) which is often seen on edible and ornamental cherry trees.

Damage. The most commonly affected plant species is broad bean, but runner bean, French bean and beetroot can also be affected. The insect uses its stylet to suck the sugary sap from the soft young stems, leaves and flowers, leading to

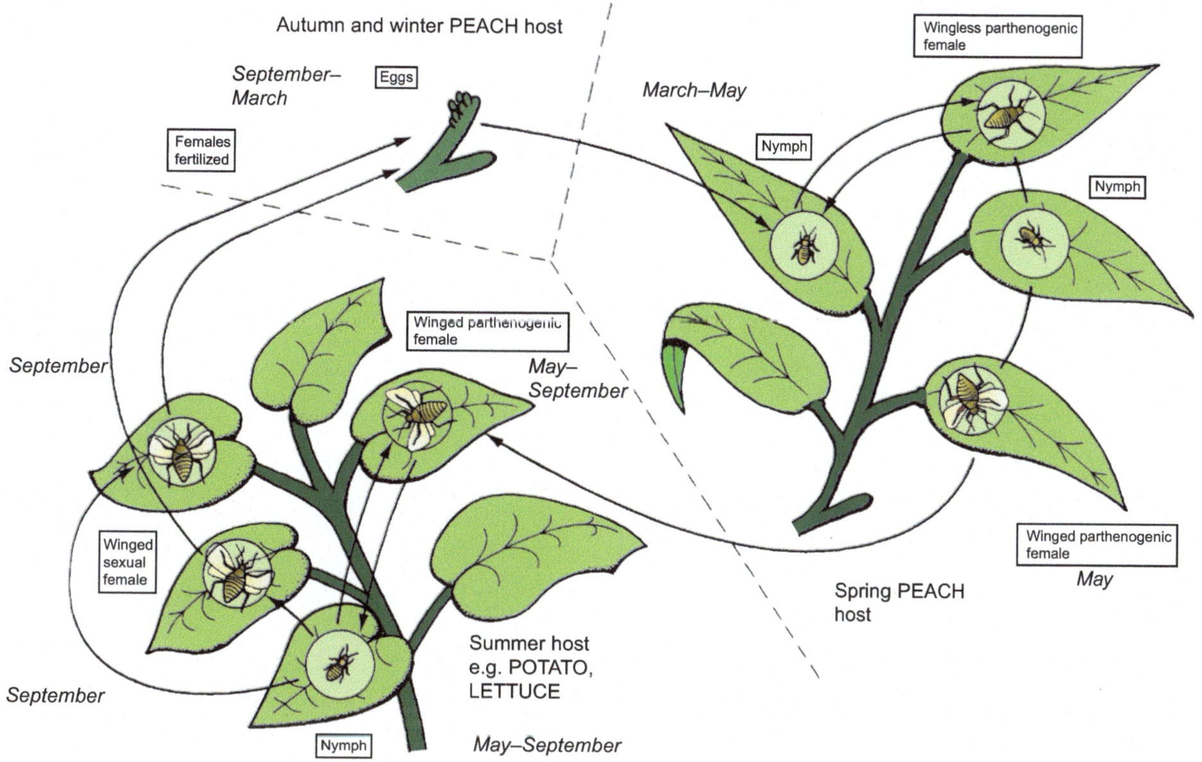

Figure 17.12 Peach-potato aphid life cycle. Female aphids produce nymphs on both the peach and the summer host. Males are produced only in late summer and autumn. Eggs survive the winter. In greenhouses, the pest may remain active throughout the year

secretion of **honeydew** and subsequent growth of **sooty mould**. The plant lacks **vigour** and becomes stunted, and the production of pods and beans is seriously affected with small, retarded pods being produced. **Ants** are often seen in the vicinity of the aphids collecting the honeydew. This aphid is also the vector of a virus which causes beet mosaic virus disease in sugar beet. As the nymphs develop, they moult, shedding their skins which leaves white casts on the surface of the plant; these are sometimes mistaken for whitefly.

Life cycle. This species overwinters mainly as the eggs on woody hosts such as *Euonymus europaeus* (Figure 17.13b), *Viburnum opulus* and *Philadelphus* spp. In early spring, wingless female nymphs emerge from the eggs and feed on the new, soft shoots, and they subsequently give birth to another generation of wingless females by **parthenogenesis**. By late spring, a third generation of winged females is produced; they eventually leave the winter host and migrate to the **summer host** plants such as broad bean, runner bean, French bean, beetroot and some weed species, such as fat hen, which act as alternative hosts. Feeding damage to the summer host occurs throughout the season. By late summer, winged female and males are produced and they fly back to the woody winter hosts where they mate and the females lay the dark-green, turning to black, overwintering eggs on the stems, fissures in bark and the axils of buds where they will remain until the following spring.

Spread. Natural spread is achieved in late spring by the migrating **winged females**. Movement of plants by gardeners can also contribute to spread of this insect.

Control. Control methods are similar to peach-potato aphid. **Cultural control** includes the same practices as those described for peach-potato aphid to encourage natural predators and parasites such as ladybirds, lacewings, and hoverflies. **Physical control** involves the fingers and thumb method to squash isolated colonies or pinching out and destroying soft growing tips on badly infested plants. Plant invigorators containing surfactants to act upon the insects physically and nutrients to stimulate plant growth can also be applied. **Chemical control** – as beans are edible plants, any insecticide product being used must have approval for use on edible plants and the instructions pertaining to application and dilution rates, maximum number of applications and the minimum interval between the last application and harvesting must be followed. Short persistence contact insecticides such as **pyrethrins** and more persistent synthetic pyrethroids such as **deltamethrin** are effective but they should never be applied to plants which are in flower as they can be harmful to pollinating insects such as bees.

Figure 17.13 (a) Black bean aphid; (b) spindle tree (*Euonymus europaeus*), winter host of black bean aphid

Figure 17.14 (a) Rose aphid; (b) apple woolly aphid; (c) aphids become swollen and golden-brown when parasitized by an *Aphidius* wasp

Other aphids

There are many other horticulturally significant aphid species. The rose aphid (*Macrosiphum rosae*) attacks young shoots of rose (Figure 17.14a). The **cabbage aphid** (*Brevicoryne brassicae*) affects cabbages and other brasssicas (Figure 17.14c). The **apple woolly aphid** (*Eriosoma lanigerum*) causes galls on apple stems (Figure 17.14b).

Glasshouse whitefly (*Trialeurodes vaporariorum*)

This insect (Figure 17.15a) is not a true fly but is a relative of other **sap-sucking** pests such as aphids, scale insects and mealybugs. The adult looks like a tiny white moth with wings held horizontally when at rest. It was originally introduced from tropical and sub-tropical America but now causes serious problems on a wide range of greenhouse edible and ornamental plants including pot plants, cut flowers and bedding plant crops. It should not be confused with the very similar, but slightly larger, **cabbage whitefly** on brassicas (Figure 17.15b). (Another glasshouse pest, **tobacco whitefly**, is mentioned briefly after glasshouse whitefly.)

Damage. Plants which are seriously affected include cucumber and tomato, fuchsias, chrysanthemums, pelargoniums and poinsettia. Chickweed and other **weeds** in glasshouses may harbour the pest over winter in all stages of the pest's life cycle. All stages of the insect's life cycle, except the egg, have stylets which are used

Figure 17.14 (Continued)

to suck the sap from soft tissues. If the number of insects is high, feeding on sap can result in distorted growth of shoots and leaves can wilt and eventually drop. The damage can result in a loss of vigour and reduction in crop yield. The excess sugar in the form of **honeydew** can be secreted by the whitefly on to the surface of the leaves and other parts of the plant. Sooty mould subsequently develops, which reduces photosynthesis and further affects vigour and crop yield. The plant's appearance and produce can be spoiled which can lead to both ornamental and edible crops becoming unsaleable. There is also the potential for the spread of viruses.

Life cycle. The adult glasshouse whitefly is about 1 mm long, with white wings and body and is able to fly from plant to plant. They are usually found towards the top of the plant and on the underside of leaves. The life cycle goes through six stages, **egg**, **first**, **second**, **third** and a **fourth larval** stage which is sometimes referred to as '**pupa**', and finally an **adult**. The adult female lays approximately 200 tiny oval shaped eggs in a circular pattern on the under-surface of the leaf; they are initially yellowish but turn black within three days. The eggs hatch in approximately nine days at 21°C to produce a first instar larva, referred to as a 'crawler', which is mobile and moves away to locate a suitable place to feed, usually on the underside of a leaf. It then becomes immobile and inserts its stylet into the leaf to start feeding on sap. It subsequently develops into the second, third and fourth instar larval stages; the fourth instar (Figure 17.15a) is a non-feeding stage from which the adult emerges. Within days of emerging, the female starts to lay eggs. At a temperature of 21°C, the development from egg to adult takes approximately 27 days, and adults have a life span of three to six weeks. The skins of the egg and 'pupa' stages are relatively thick and are more difficult to control with insecticides.

Spread. This is mainly introduced by **infected plants** being brought into the glasshouse or by chance arrivals of adults through doors or vents. Within a greenhouse, the flying adults are highly mobile.

Control. **Cultural control** can involve the removal of weeds in greenhouses, such as chickweed or sow thistle which harbour the pest. Thorough inspection is required for all new plant material for all stages of the pest when it first arrives at the nursery. Scouting with regular inspections of the **lower** leaf surfaces of plants is important in identifying the presence of the pest. **Physical control** can be partially achieved by yellow sticky traps being placed above susceptible plants; these traps are used principally to **monitor** the presence of the adults but some control does take place. These sticky traps do have disadvantages – the flying adults of biological control agents can also be attracted to and trapped by them so they should be removed if flying biological control agents are to be introduced. Plant invigorators containing surfactants to act upon the insects physically and nutrients to stimulate plant growth can also be applied, and they are effective against the adults and early larval stages but the egg and older larval stages are more resistant. **Biological control** – several organisms are available for control of glasshouse whitefly. A minute parasitic chalcid wasp, *Encarsia formosa*, is available for purchase. It can be introduced on small cards with parasitized whitefly larvae attached. The cards are hung up in the crop and adult wasps emerge from the whitefly larvae on the card; they then fly out into the crop with females looking for whitefly larvae to parasitize. The female adult wasp lays an egg inside the third and fourth larval stages of the whitefly. The developing whitefly is eaten away by the wasp larva, the whitefly larva turns black and subsequently dies and the young adult wasp eventually emerges as the next generation. Control is most effective if *Encarsia formosa* is applied when whitefly numbers are low and the temperature is between 20°C and 25°C. Some chemical controls used against other pests and diseases can harm the parasitic wasp if not carefully selected. Other biological control agents which are effective against glasshouse whitefly include *Amblyseius andersoni*, a predatory mite which feeds on the eggs and nymphs of glasshouse whitefly and can also be introduced to glasshouse crops to control whitefly. **Chemical control** can involve the application of short persistence contact insecticides such as **pyrethrins** but repeat application would be necessary for long-term effectiveness. **Synthetic pyrethroids** such as deltamethrin will have longer effectiveness. These insecticides are not compatible with biological control agents and can harm the beneficial control organisms. A **biological insecticide** containing *Lecanicillium muscarium* **Ve6**, an entomopathogenic fungus, can be applied on glasshouse crops being affected by whitefly. When applied, the spores of the fungus attach to the whitefly and germinate; fungal growth then enters the insect's body and kills it. The fungus develops on the dead whitefly before producing more spores which can infect other whiteflies. This type of product is also effective on aphids and thrips. In the appropriate environmental conditions, the pest dies within seven to ten days. This type of product is classed as a pesticide and should be only be used by appropriately trained and certified professional operators.

Tobacco whitefly (*Bemisia tabaci*)

This species is closely related to the glasshouse whitefly. However, Britain and Ireland are designated as tobacco whitefly–free areas, but it could establish in protected environments where it has the potential to be a major pest, particularly of glasshouse salad crops such as tomato and cucumber. It is similar in size (about 1 mm in length) but may be recognized by the more vertical ('tent-like') way it holds its wings, revealing its light yellow-coloured body. The female, unlike

Figure 17.15 (a) Glasshouse whitefly adults and larvae on a leaf (source: Shutterstock, Tomasz Klejdysz); (b) cabbage whitefly; (c) bud blast; (d) capsid damage on an apple fruit

glasshouse whitefly, lays eggs in a random fashion. The main problem with tobacco whitefly is its ability to vector some significant and destructive viruses such as **tomato yellow leaf-curl virus**. Tobacco whitefly is a **quarantine** (notifiable) pest in Britain and Ireland, and suspected outbreaks should be reported to the relevant plant health authority. In England and Wales, an **APHA** Plant Health and Seeds Inspector should be contacted. In Scotland, contact the Scottish government's **Horticulture and Marketing Unit**, and in Northern Ireland, contact **DAERA Plant Health Inspection Branch** and in the Republic of Ireland the **Department of Food, Agriculture and the Marine**.

Leaf hoppers

These squat light brown or green insects are approximately 8 mm long when adults and feed on a wide variety of plant species. The adults, as their name suggests, can avoid threats by hopping off the leaf. They are also able to fly from plant to plant. They live and feed on the **under-surface** of leaves but very little obvious feeding damage is apparent. The nymphs and adults of the rhododendron leaf hopper (*Graphocephala fennahi*) feed on rhododendrons, and they are active and visible from spring to autumn when the feeding activities cause insignificant direct damage to the leaves and the plant more generally. However, in autumn the females create small incisions in the flower buds developing for next year where they lay their eggs. The **bud blast fungus** *Seifertia azaleae* is thought to infect the buds through these wounds. This fungus causes the damaging and unsightly bud blast disease which kills the affected flower buds (Figure 17.15c). However, the validity of a link between the activities of the leaf hopper and this disease is now being questioned.

Control of rhododendron leaf hopper may not be necessary or desirable in most cases. **Cultural control** can involve the encouragement of natural predators of the leaf hopper such as birds and ladybirds by establishing planting and habitats which attract the predators. **Physical control** of bud blast can be undertaken by **picking off** flower buds which are infected by bud blast, and the buds should be taken away and destroyed to prevent the spores being dispersed to other healthy buds. **Chemical control** of the leaf hopper should be avoided if possible, but it can involve the application of short persistence contact insecticides such as **pyrethrins** or more persistent **synthetic pyrethroides** such as deltamethrin; both are effective when applied directly on to the insects when they are active but should never be sprayed on flowering plants as they are harmful to pollinating insects.

Common green capsid (*Lygocoris pabulinus*)

This light green insect resembles a large (5 mm) aphid. It occurs in small numbers on a range of woody and herbaceous species. Although it is rarely seen, it can cause considerable damage to buds, leaves and fruit by injecting toxic juices when it feeds, causing damage to plant growth (Figure 17.15d). Physical, cultural and chemical controls are generally the same as those described for peach-potato aphid.

Thrips (order Thysanoptera)

Owing to their increased activity during warm humid summer weather, thrips are sometimes referred to as 'thunder-flies'. Damage is caused when they puncture the soft tissues of a cultivated plant with their mouthparts and suck the contents; this subsequently causes the surrounding tissue to die and leads to mottling of leaves and a reduction in photosynthetic ability. The vigour and yield of the plant is subsequently affected, and viruses such as **tomato spotted wilt virus** can be transmitted by onion thrips. They can be significant pests in glasshouses.

Onion thrips (*Thrips tabaci*)

Puncturing and sucking feeding damage on onions and leeks results in a silver mottling on the leaves. A range of other edible and ornamental species such cucumber, tomato, dahlia, and chrysanthemum are also affected. In heavy infestations, fruits such as cucumber can be damaged directly. **Physical control** can be implemented in glasshouses by positioning **blue sticky traps** above susceptible plants to trap adult thrips and monitor the presence of the insect. Application of **plant invigorators** containing surfactants to act upon the thrips physically and nutrients to stimulate plant growth are suitable for thrips control. **Biological controls** include *Orius laevigatus*, a predatory bug whose adults and nymphs feed on thrips larvae, and *Amblyseius andersoni*, a predatory mite which also feeds on the larvae; both can be effective in glasshouses. **Chemical control** involves the application of a bio-insecticide containing *Lecanicillium muscarium* **Ve6**, an entomopathogenic fungus which can be applied on glasshouse crops affected by onion thrips by appropriately trained and certified professional operators. Short persistence contact insecticides such as pyrethrins can also be effective, but repeat application may be necessary. The insect is small and elusive and can be difficult to target with chemical controls.

Western flower thrips (*Frankliniella occidentalis*)

This pest is found in outdoor and glasshouse flower and vegetable crops. It is the most significant vector of tomato spotted wilt virus and causes damage by its puncturing and sucking feeding activities. Plant vigour and yield is reduced and the aesthetic appearance of ornamental plants such as chrysanthemum and roses are compromised. Control measures are generally the same as those described for onion thrips.

Moths and butterflies

This insect group characteristically contains adults with four large wings and curled feeding tubes. The larva (**caterpillar**), with six small legs and eight false legs, is modified for a leaf-eating habit (Figure 17.8). Some species are specialized for feeding inside fruit such as **codling moth** larva in apple fruits, underground such as **cutworms** (Figure 17.18), inside leaves such as **horse chestnut leaf miner** (Figure 17.20) and inside tree branches such as **leopard moth**. The horticulturist may find large webbed caterpillar colonies of the **lackey moth** (*Malacosoma neustria*) on fruit trees and hawthorns or the **juniper webber** (*Dichomeris marginella*) causing webs and defoliation of junipers.

Large cabbage white butterfly (*Pieris brassicae*)

Damage. This is a common pest on the leaves of numerous **brassica** plants including edible plants such as cabbage, cauliflower, Brussels sprouts, broccoli and ornamental hosts such as, sea kale (*Crambe maritima*) and weeds such as shepherd's purse. The leaves are progressively eaten away by the larvae (caterpillars). The defoliating damage of the **larvae** can result in skeletonized leaves, with only the main veins showing which significantly impacts the ability of the plant to photosynthesize. Large amounts of **frass** (excrement) is deposited on the plant.

Life cycle. The **adults** (butterfly) emerge from the overwintering **pupae** (chrysalis) in April and May (Figure 17.16a) and, after mating, the females lay batches of 20–100 yellow, conical shaped eggs mainly on the lower surfaces of leaves. Within 14 days, the **larvae** emerge and start feeding (Figure 17.16b) but remain in colonies and subsequently moult to produce the later instars, which reach up to 25 mm in length and are yellow and black with visible hairs. In June, the fully fed larvae then crawl away to find a sheltered location such as crevices and bark fissures of trees where they can **pupate** (Figure 17.16c). A **second generation** of adults emerge in July and August, giving rise to a more damaging infestation of caterpillars than the first. The second pupal stage of the season overwinters away from the food plants on tree trunks, fences and eaves of buildings. There can be a third generation if the season is favourable.

Spread. The species is spread by the **winged adult** butterfly.

Control. **Cultural control** – if possible, small populations of this insect should be tolerated because of its importance as a food source for other animal species and the beneficial pollination activities of the adult butterfly. Any horticultural practices which encourage natural predators of the caterpillars, including birds such as **blue tits and starlings**, into the garden should be considered; this includes the cultivation of dense stands of woody plants such as hedges. A small **parasitic wasp** (*Cotesia glomerata*) lays its eggs inside the caterpillar, and the yellow cocoons emerge (Figure 17.16d). **Physical control** includes regular **scouting** of the crop checking the underside of leaves during the periods when eggs are laid and the caterpillars are emerging; they can then be picked off by hand. Susceptible crops can be grown under **small mesh netting** or horticultural fleece which prevents the butterflies from laying eggs on the host plants; both these methods can be effective in a relatively small-scale production but the mesh netting must in place before the adults start mating and it must be kept clear of the leaves as the female butterfly can lay eggs on to the leaves through the mesh. Proprietary calcium-based **repellent sprays** can be applied to the foliage of susceptible plants to deter the adults from laying their eggs. **Biological control** can include the application of parasitic nematodes such as *Steinernema feltiae* to the affected plants when the caterpillars are feeding; these products are most effective in cool, damp conditions. Naturally occurring parasites such as the wasp *Cotesia glomerata* may parasitize the larvae of the large cabbage white (Figure 17.16d). **Chemical control** can include the application of **contact insecticides** which are effective on the caterpillar stage of the insect but may need several applications to be effective. All products used must have approval for use on edible plants if intended for use on vegetables. Products containing **pyrethrins** and **synthetic pyrethroids** such as deltamethrin are approved for such use. A bio-insecticide containing the bacterium *Bacillus thuringiensis* **subsp.** *kurstaki* strain is available for use only by appropriately trained and certified professional horticulturists. After application to the foliage of the affected plant, the bacterium is ingested by the caterpillars when feeding; they rapidly stop feeding and die within three days.

Winter moth (*Operophthera brumata*)

Damage. This insect can be a significant problem on top fruit, such as apples, but also affects

Figure 17.16 (a) A female large cabbage white butterfly (source: Shutterstock, petrovichlili); (b) large cabbage white caterpillar (source: Shutterstock, Ewa-Saks); (c) a large cabbage white pupa (source: Shutterstock, Florian Teodor); (d) *Cotesia glomerata* wasp parasitizes and kills cabbage white caterpillars

other woody plants such as currants, roses and beech. The **caterpillars** (Figure 17.17a) feed on **young leaves** in spring and early summer and they often form other leaves into loose webs, reducing the plant's ability to photosynthesize. They occasionally feed on the blossom and young developing apple fruits.

Life cycle. The adults emerge from soil-borne pupae from October to December, most emerging in late November; hence the species' common name. The male is a greyish-brown moth, 2.5 cm across its wings, while the female is **wingless**, looking at first sight rather like a spider (Figure 17.17b). The females crawl up the tree and lay the light green eggs around the buds of smaller branches and bark fissures. Each female lays between 100 and 200 eggs. The eggs hatch in spring at **bud burst** to produce small green larvae with black heads which then start feeding, as they develop white stripes on the back and sides. These larvae move in a characteristic **looping** fashion and when fully grown, by late May, they descend on silk threads to the ground before pupating in the soil until October.

Spread. This is slow because the females do not have wings. The small caterpillars can be blown from tree to tree by wind.

Control. On ornamental plants, the larvae do not cause any significant damage to the plant, so tolerate a limited amount of damage if possible and accept that the caterpillars are a natural food source for other animals including birds such as blue tits. **Cultural control** can involve implementing garden practices which encourage natural predators such as birds, ground beetles and hedgehogs. **Physical control** can be implemented by the application of a **vegetable oil-based** insect glue band (Figure 15.2a) around the trunk of each tree and any tree stakes to trap the wingless females as they crawl up the tree to lay their eggs on the branches of the tree. The glue bands should be applied annually in October at a height of approximately 45 cm above ground level and go all the way round the trunk. **Chemical control** involves the application of insecticide sprays with products containing **pyrethrins** or **synthetic pyrethroids** such as deltamethrin in spring when the caterpillars are active but before the flowers open, as these active ingredients can be harmful to pollinating insects.

Figure 17.17 (a) Winter moth caterpillar moves in a looping fashion; (b) wingless winter moth female looks similar to a spider

Cutworm (e.g. *Noctua pronuba*)

Cutworms are caterpillars (larvae) of certain moths in the Noctuidae family which feed on plants at ground level.

Damage. The caterpillars of the **large yellow underwing moth** (Figure 17.18), unlike most other moth larvae, live in the soil and **emerge at night**, nipping off the stems of young plants and eating holes in leaves of non-woody plants, such as bedding plants, potatoes, celery, turnips and conifer seedlings. This damage reduces the plant's photosynthetic ability and disfigures ornamental plants.

Life cycle. The adult moth, 2 cm across, with brown forewings and yellow to orange hind wings, emerges from the shiny, soil-borne, chestnut brown pupa from mid-June to August and lays about 1,000 eggs on the stems of a wide variety of weeds. The young caterpillars, having fed on weeds, descend to the soil and live there from late summer until the following May when they pupate and in the later stages cause the damage described earlier, eventually reaching about 3.5 cm in length. They are grey to grey-brown in colour, with black spots along the sides. Several other cutworm species, such as **heart and dart moth** (*Agrotis exclamationis*) and **turnip moth** (*Agrotis segetum*), may cause damage similar to that of the yellow underwing moth. In all three species, their typical caterpillar-shaped larvae should not be confused with the legless **leatherjacket**, which is also a common underground larva.

Spread. The larvae are able to crawl from plant to plant, but most spread is by the mobile, flying adults. *Control*. **Cultural control** – the adult females tend to lay their eggs in areas of dense vegetation, therefore rigorous **weed control** is beneficial. **Heavy irrigation** of the soil prior to sowing or planting can destroy young cutworm larvae. Larger, more established **transplants** are more tolerant of cutworm damage than younger transplants or seedlings. **Physical control** includes removal of cutworm larvae by **hand** or natural predators such as birds like blackbirds, starlings and robins when they are exposed during soil cultivation prior to sowing or planting. On a small scale, problems can be reduced by growing susceptible plants under **small mesh netting** to exclude the adults and reduce egg laying in the crop. **Chemical control** – there are, currently, no insecticides available to home gardeners. Professional horticulturists who are appropriately trained and certified can apply a bio-insecticide containing *Bacillus thuringiensis* **subsp.** *kurstaki* to the foliage of susceptible plants.

Figure 17.18 A male yellow underwing moth (source: Shutterstock, Tomasz Klejdysz)

Box tree caterpillar (*Cydalima perspectalis*)

Damage. This Asiatic species of moth was first recorded in the UK in Kent in 2007. It was found in gardens for the first time in 2011. It has spread across the south-east of England and further north and west with reported sightings in Wales, Scotland and Ireland. In Britain and Ireland, it has only been recorded feeding on *Buxus* spp. Typical disfiguring damage is caused by the larva feeding on the leaves (Figure 17.19), often only leaf skeletons remain. Also webbing of the branches with frass and residues of moulting being deposited on the plant is also present. The loss of a significant amount of foliage also reduces the plant's photosynthetic ability.

Life cycle. There are at least **two generations per year**. The caterpillars emerge from the **eggs** in autumn, and they are initially a greenish yellow colour with black heads. As they develop the head stays black and their green body develops dark brown stripes. They overwinter as small inactive caterpillars on the leaves. In early spring, the caterpillars become active and feed throughout spring before pupation occurs in **late spring** under white cocoons on the leaves and stems. The pupae are between 1.5 and 2.0 cm long. They are initially green but become brown as they develop. By **June**, adult moths with white wings with a brown border and a wingspan of 4 cm emerge from the pupae and mate, the females then lay clusters of 5 to 20 eggs on the underside of the leaves. Initially the eggs are pale yellow and difficult to see; as they mature the eggs develop a black spot. When eggs hatch, the next generation will continue to develop through larval and pupal stages to become adults by **late summer**. The fertilized females lay eggs and the caterpillars hatch and **overwinter** on the plant. Generations overlap and the adult moths and caterpillars can be seen on box plants throughout spring and summer.

Spread. The adult moth is winged and highly mobile locally. Plant transportation is responsible for the insect being spread long distances.

Figure 17.19 Box tree caterpillar eating foliage of a box tree (source: Shutterstock, Gertjan Hooijer)

Control. **Cultural control** can involve the use of non-susceptible species as replacements for box plants and is becoming common. Similar species which will not be affected by box caterpillar include *Ilex crenata*, *Berberis darwinii* 'Compacta' and cultivars of *Euonymus fortunei*. **Physical control** involves regular **scouting** for caterpillars and removal by handpicking. **Box moth traps** containing a box moth–specific pheromone dispenser can be placed at 1.5 m over the box plants; they release a pheromone which lures the males into the trap. Once in the trap they cannot escape and can be identified and counted to establish the level of infestation. Trapping the males also prevents them mating, which reduces the number of eggs laid by females. **Biological control** can be introduced in the form of *Steinernema carpocapsae*, an entomopathogenic nematode. A product containing the nematodes is diluted and sprayed on to the foliage of the affected plants when the caterpillars are feeding. The nematodes enter the caterpillars' bodies and release bacteria which stops them feeding and ultimately leads to their death. **Chemical control** can be undertaken by the application of a **mating disruption** product which contains the active ingredient **hexadecenal**, vegetable oil and natural wax. It is applied directly to the stems of box plants to saturate the area with female pheromones; as a result male moths are unable to find the females, consequently mating and egg laying are disrupted leading to a decrease in the number of caterpillars. Each application lasts for three months. This product is approved in the UK for use by appropriately trained and certified professional operators. Contact insecticides containing **pyrethrins** can be applied to target the caterpillars directly. Due to the short persistence of this active ingredient, repeat applications may be necessary.

Other moth pests

Other moth pests which are now causing problems in Britain and Ireland include oak processionary moth (*Thaumetopoea processionea*), leek moth (*Acrolepiopsis assectella*), holm oak leaf-mining moth (*Phyllonorycter messaniella* and *Ectodemia heringella*) and horse chestnut leaf miner (*Cameraria ohridella*) (Figure 17.20).

Flies

This group of insects typically have only a single pair of functioning wings. The hind wings are modified into little stubs which act as balancing organs. The **larvae** are legless and elongated, and

Figure 17.20 Horse chestnut leaf miner (damage is caused by small moth caterpillars) (source: Shutterstock, IanRedding)

their mouthparts, where present, are simple hooks. The larvae are the only stage of the life cycle causing plant damage to plants.

Chrysanthemum leaf miner (*Chromatomyia syngenesiae*)

Damage. This leaf miner is the larva of a small fly species which can cause serious damage to susceptible crops by mining in the leaf (Figure 17.21a). This leads to a reduction in photosynthesis and a weakening and a **lack of vigour** in the affected plant. In extreme cases the leaves become desiccated which affects the appearance and saleability of ornamental plants. It is found on many members of the **Asteraceae** plant family, including *Chrysanthemum* × *morifolium*, florist's 'cineraria' (*Pericallis* × *hybrida*), *Jacobaea maritima* and *Gazania* species.

Life cycle. The adult flies emerge at any time of the year in greenhouses, but normally only between July and October outdoors. The adults, which measure about 2 mm in length and are grey-black with yellow underparts, fly around with short hopping movements. The female lays about 75 minute eggs singly inside the leaves, causing small white spot symptoms to appear on the upper leaf surface. The larval stage is greenish white in colour and tunnels into the **pallisade mesophyll** of the leaf, leaving behind the characteristic mines (Figure 17.21a). On reaching its final stage, the 3.5 mm long larva develops into a brown pupa within the mine. The adult soon emerges from the pupa. The total life cycle takes approximately three weeks during the summer months.

Spread. This insect is spread locally by the winged, mobile **adult stage** and over long distances by **transport** of plants for horticultural purposes.

Control. Leaf miners can be food for other animals in the garden including blue tits which feed on the larvae. Consequently, it may be possible to tolerate low levels of infection, particularly in non-crop plants which are not being cultivated commercially. Regular scouting of commercial crop plants to help identify problems will allow prompt action to be taken. **Cultural control** – alternative host weeds such as groundsel and sow thistle should be rigorously controlled in the greenhouse and surrounding areas. **Physical control** – affected leaves can be removed by hand and crushed. **Yellow sticky traps** (Figure 15.2b) can be paced 30 cm above crop plants in glasshouses to monitor and trap the flying adults. **Biological control** – a parasitic wasp, *Diglyphus isaea*, can be introduced into greenhouses to control the leaf miner. **Chemical control** – leaf miners are difficult to control with insecticides. The larvae and pupae are protected within the tissues of the leaves making the use of insecticide control methods inappropriate in most cases.

Other fly pests

The **American serpentine leaf miner** (*Liriomyza trifolii*) can cause damage to a range of **greenhouse plants** including chrysanthemum and gerbera crops and has created problems for commercial growers in recent years.

Carrot fly (*Psila rosae*) is a widespread and significant pest on **Apiaceae** family crops such as carrot, celery, parsnip and parsley, where tunnelling by the larvae makes the roots useless (Figure 17.21b). The **cabbage root fly** (*Delia radicum*) causes similar damage in **Brassicaceae** family crops such as cabbage, cauliflower, Brussels sprouts, swede, turnip and others.

Leatherjacket (*Tipula* spp.) is an underground pest, and the larvae of the **crane fly** (daddy-longlegs) are seen in late summer. It is a natural inhabitant of **grassland** and causes most problems on lawns and sports turf, but also damages roots of young strawberry plants and brassicas.

Sciarid fly (fungus gnats) (*Bradysia* spp.) larvae are small (3 mm) and translucent with a black head. They feed on fine roots of **greenhouse pot plants** such as *Cyclamen persicum*, orchid and *Freesia* spp., especially when the plants are overwatered and the growing medium is excessively wet. *Lycoriella* spp. are pests in mushroom crops. **Biological control** by application of *Steinernema feltiae*, an entomopathogenic **nematode** into the growing medium, and introduction of *Stratiolaelaps scimitus*, a predatory mite, is available to the grower. The flying adults are also caught by **yellow sticky traps**.

Agapanthus gall midge (*Enigmadiplosis agapanthi*) is a recently introduced pest; the larvae

Figure 17.21 (a) Chrysanthemum leaf mines; (b) carrot fly damage

feed inside the flower buds of *Agapanthus* spp. which results in deformed flowers that can turn brown and fail to open fully.

Beetles

This group of insects has adults with hard, horny front wings (elytra) which, when folded, cover the delicate hind wings used for flight. The meeting point of these hard wing cases produces the characteristic straight line down the beetle's back over its abdomen. The thick skin (cuticle) of beetles enables many of them to live successfully **underground**. Many beetle species are **beneficial**, helping in the pollination of flowers such as *Magnolia* species, the breakdown and **recycling** of organic matter, and as **predators** of plant pest species (see p. 226) as well as being a **food source** for other species of wild animals. Some beetles such as vine weevil, click beetle (wireworm), chafers and, in recent years, lily beetle, asparagus beetle and rosemary beetle can cause significant feeding damage to cultivated plants. Some beetles can be **vectors** of plant pathogens, a notable example being the transmission by **elm bark beetles** of the devastating *Ophiostoma novo-ulmi*, the **Dutch elm disease**–causing fungal pathogen.

Vine weevil (*Otiorhyncus sulcatus*)

This species belongs to the beetle group, but as with all weevils, possesses a **longer snout** (rostrum) than other beetles. It is native to Britain and Ireland and found widely in gardens and plant nurseries.

Damage. The feeding activity of the larvae between autumn and spring causes the most significant damage. Roots and tubers of **container-grown** pot plants in greenhouses such as *Cyclamen persicum* and *Primula* spp. can be seriously damaged. Outdoors, container-grown **herbaceous perennials** such as *Heuchera* spp., *Hylotelephium* spp., and also many **trees and shrubs**, such as *Hydrangea* spp. and *Taxus baccata*, are commonly affected. The larval damage to roots causes yellowing of foliage and wilting similar to symptoms of root diseases such as *Verticillium* wilt. In serious cases, the plant can **collapse and die**. Digging around the plant's root zone will, however, usually reveal the distinct creamy-white larvae (Figure 17.22a). The adults are **nocturnal** and therefore only active at night. They are rarely seen during the day when they hide in vegetation and under pots and benches in glasshouses. The adults feed on **foliage** causing characteristic irregular notching around the edges of leaves (Figure 17.22c) on many species, such as evergreen *Euonymus*, *Rhododendron* spp. and herbaceous perennials such as *Bergenia* spp., which disfigures the plant and leads to a reduction in photosynthesis.

Life cycle. The adult is approximately 9 mm long, black in colour, with a rough textured cuticle (Figure 17.22b). The forewings are fused together making it incapable of flight. There is one generation per year; all vine weevils are **female** and they reproduce **parthenogenetically** by laying unfertilized eggs in the soil or compost next to the roots of a preferred plant species in late summer; the female insect can lay hundreds of eggs in this way. The larvae hatch within 10–15 days, and they are white and legless with a characteristic chestnut-brown head and C-shaped body when disturbed. There are six **larval instar stages** and overwintering occurs in the larval stage. As the temperature rises in

17

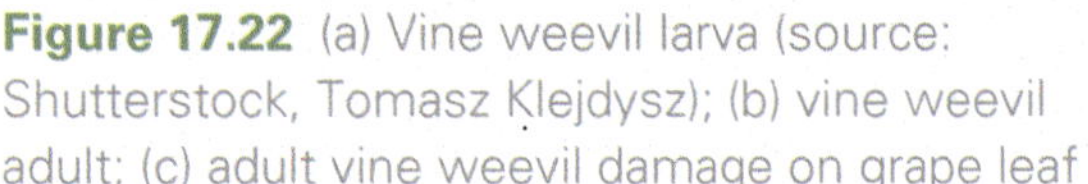
Figure 17.22 (a) Vine weevil larva (source: Shutterstock, Tomasz Klejdysz); (b) vine weevil adult; (c) adult vine weevil damage on grape leaf

late winter, the larvae become more active and pupation occurs in the soil in spring. The adults emerge outdoors in May and June in Britain and Ireland but often earlier in greenhouses where the life cycle becomes more irregular. The young adults are initially white but subsequently become darker in colour.

Spread. This is achieved locally by the adults **crawling** around at night, but the spread over longer distances occurs by movement and transport of plants in containers with the larvae in the growing medium.

Control. Thoroughly inspecting newly arrived plants and regular **scouting** of crop plants for signs of the pest and its damage is an important way to permit early action in the event of vine weevil being present. **Cultural control** can involve creating garden features or planting which will encourage **natural predators** of the vine weevil into the garden. Frogs and toads, hedgehogs, ground beetles and birds such as song thrush are all natural predators. **Physical control** includes measures such as **handpicking** the adult weevils when they are active on summer evenings. In greenhouses, they can be found under benches or pots. The larvae can be discovered and removed by hand when potting container-grown plants and then fed to birds. It is possible to **trap** the adults – traps with rolls of corrugated paper can be placed near affected plants in glasshouses, and the adult weevils enter the rolls as a hiding place and can be removed and destroyed. **Biological control** can be introduced in the form of parasitic nematodes such as *Steinernema kraussei, Steinemena carpocapsae* and *Heterorhabditis bacteriophora*. Various products containing these nematodes are available from suppliers; they are diluted in water and applied as a drench or spray onto the compost or soil to control the **larvae**. This form of control is most effective if applied in late summer/early autumn when the larvae are still small and the soil temperature and moisture content is favourable for effective action by the nematodes. **Chemical control** – the insecticide acetamiprid can be watered on to the growing medium of container-grown **ornamental plants** to give protection against the larvae for up to four months.

Wireworms

Wireworms are the soil-inhabiting larvae of click beetles. There are approximately 70 species of click beetle in Britain and Ireland but most damage appears to be caused by **three species**: *Agriotes lineatus*, *A. obscurus* and *A. sputator*. They are commonly found in grassland, but the wireworm (Figure 17.10) will bore through underground organs of plants such as potatoes and carrots. Crop rotation is an important cultural practice to prevent them establishing. The parasitic nematode *Heterorhabditis bacteriophora* is available for soil application as a **biological control** of the larvae.

Garden chafer (*Phyllopertha horticola*) and Welsh chafer (*Hoplia philanthus*)

Chafers are common beetles which, when **adults**, feed on **leaves** of trees such as oak and hazel, but the larvae are a problem on turf such as lawns and golf courses. The **soil-inhabiting** larvae (Figure 17.10) eat the grass roots leading to bleached-yellow patches appearing in the lawn, notably in **summer** when the larvae are becoming fully grown and active. The quality of the turf can reduce dramatically and in periods of low rainfall, the grass plants may die leaving bare patches. **Secondary damage** can be caused by animals such as badgers, foxes and birds such as crows when they **dig up** the turf to feed on the chafer larvae. While healthy, well-fertilized and irrigated turf often shows minimal damage, it may be necessary to introduce a **biological control**, involving a soil drench of a product containing the **parasitic nematode** *Heterorhabditis bacteriophora*.

Lily beetle (*Lilioceris lilii*)

This introduced beetle has steadily increased in numbers in Britain and Ireland since the mid-20th century and is now widespread in many areas. The **adult** is bright red; its head and legs are black; and it is 8 mm in length (Figure 17.23b). The **larvae**, which often have a dark slimy appearance as they cover themselves in their own excrement, are 8–10 mm in length, orange in colour, with black heads. Both **adults and larvae** feed on the foliage of a wide range of lilies and fritillaries during spring and summer which is disfiguring and reduces the plant's ability to photosynthesize so the plant becomes progressively weaker. Regular checks and handpicking of adults, larvae and eggs, which are laid on the underside of the leaves, can be effective on a relatively small scale. Non-insecticide repellents containing a calcium chloride solution can be sprayed on to the foliage of susceptible plants to deter the adults and larvae from feeding. **Contact insecticides** such as pyrethrins can be sprayed on to the adults and larvae but repeated application may be necessary as pyrethrins have a sort persistence. They are also potentially harmful to a wide range of other non-pest insects so should be used with caution and not applied to flowers.

Other beetle pests

Spring time damage by **flea beetles** (*Phyllotreta* and *Psylliodes* species) which are up to 3 mm long, black in colour (Figure 17.23a). The adults feed on the leaves of young members of the **Brassicaceae** family including **vegetables** such as cabbage, turnip, radish and **ornamentals** such as wallflower and stock. The small holes created by the beetle's feeding activity are unsightly and reduce photosynthetic ability, which weakens the plant and creates openings that fungal pathogens may infect. Fine horticultural mesh draped over young brassica plants helps to prevent damage.

In **recent years** in Britain and Ireland, other beetles which have become common garden pests include **viburnum beetle** (*Pyrrhalta viburni*) which is 5 mm long and light brown in colour; adults and larvae are found feeding on foliage of *Viburnum opulus*, *V. tinus* and *V. lantana*. The adult of **rosemary leaf beetle** (*Chrysolina americana*) is 8 mm long, metallic green with purple stripes and, along with the larvae, cause feeding damage on lavender (*Lavandula* spp.) rosemary (*Salvia rosmarinus*) and thyme (*Thymus* spp.). **Asparagus beetle** (*Crioceris asparagi*) is 7 mm long, has a metallic black body with four white spots and is common in asparagus beds; adults and larvae cause feeding damage. All of these leaf-eating beetles weaken and disfigure the plants. **Regular checks** scouting in spring and summer and handpicking of these insects can be effective on a small scale. Insecticide sprays of products containing **pyrethrins** can be used on these beetle pests but should not be applied to flowering plants and must have label approval for use on edible plants.

Sawflies

Sawflies, together with bees, wasps and ants, are members of the **Hymenoptera** order which has adults with two pairs of translucent wings, with forewings and hind wings locked together by fine hooks. The slender waist-like first segments of the abdomen give these adults a characteristic appearance. Adult sawflies resemble flying ants. The larvae of some species resemble caterpillars (Figure 17.10), but some such as **pear and cherry**

slugworm (*Caliroa cerasi*) are black, have no visible legs and look rather 'slug like'.

Gooseberry sawflies. Three species of sawfly can be pests on gooseberry plants, redcurrants and white currants, but not blackcurrants. The **common gooseberry sawfly** (*Nematus ribesii*) is the most significant with the larvae causing extensive damage to foliage, often leaving only the main leaf veins of the leaves uneaten (Figure 17.24). The reduction in the plant's photosynthetic ability will lead to weakening of the plant and a reduction in crop yield and quality. The pale-green larvae with black spots are easily recognized. Often three life cycles per year occur. The other two species are the **pale spotted gooseberry sawfly** (*Nematus leucotrochus*) and **small gooseberry sawfly** (*Pristiphora appendiculata*). Regular **scouting** and **handpicking** larvae from spring onwards is recommended. Sprays of insecticidal products containing pyrethins are effective on the larvae. **Biological control** can be introduced as sprays of diluted products containing the **parasitic nematode** *Steinernema carpocapsae* when the larvae are feeding.

Large rose sawfly. The larvae of two species, *Arge pagana* and *A. ochropus*, skeletonize rose leaves and can be controlled from spring onwards by **handpicking** eggs and larvae. Sprays of products containing deltamethrin or pyrethrins are also effective against the larvae but should not be applied to flowers.

Berberis sawfly, (*Arge berberidis*), was first recorded in the south-east of England in approximately 2002 and has now spread to many parts of England and into Wales. The larvae **defoliate** the foliage of *Berberis thunbergii* and species of *Berberis* formerly known as '*Mahonia*'. Regular checks and **handpicking** of the feeding larvae should take place in late spring and early summer.

Ants. Two species, the **black ant** (*Lasius niger*) and the **red ant** (*Myrmica rubra*), are commonly found in British and Irish gardens. They cause little direct damage to plants but encourage aphids that produce honeydew. **Ant heaps** located in lawn areas can spoil the appearance of the turf and create ideal germination sites for weed seeds. Physical removal of the ant heaps in lawns by brushing on dry day avoids the use of potentially hazardous chemical controls.

Springtails (order Collembola) are wingless insects, 2 mm in length, that are very common in many types of soils. Their feeding activity on organic matter helps in the breakdown of soil **organic matter**. They can occasionally cause damage to plants but are considered to be generally beneficial in gardens.

Figure 17.23 (a) Flea beetle adult and leaf damage (source: Shutterstock, Tomasz Klejdysz)

Figure 17.23 (b) Lily beetle adult and leaf damage (source: Shutterstock, Amelia Martin)

Figure 17.24 Gooseberry sawfly larvae and leaf damage (source: Shutterstock, Heiti Paves)

Mite pests

Mites are in the order **Acarina** and are grouped with ticks, spiders and scorpions in the class **Arachnida**. Although similar to insects in some respects, they are distinguished from them by the possession of **four pairs of legs**, a **fused body** structure (no clear abdomen or thorax) and the **absence of wings** (Figure 17.25). Many species are **beneficial** including some soil-inhabiting mites which are involved in **breaking down** organic matter; some other species are **predators** on plant pests and are commonly introduced as **biological control** agents. Several above ground species can be significant pests on plants. The **life cycle** is usually composed of egg, larva, nymph and adult stages.

Two-spotted spider mite (red spider mite) (*Tetranychus urticae*)

Damage. The **larvae**, **nymphs** and **adults** feed by piercing leaf tissue and sucking out the cell contents in a wide range of greenhouse plants including cucumber, tomato, peach, nectarine, poinsettia, *Fuchsia* and *Pelargonium* species. The damage leads to an unsightly **mottling** of the upper leaf surface (17.26a) which **reduces photosynthesis** and affects plant vigour and crop yield. In large numbers, the mites can destroy **leaves** and eventually the whole plant dies. The nymphs and adults also produce **fine, silken webbing** which covers leaves and stems and spoils the appearance of ornamental plants (Figure 17.26b). Most of the damage occurs in **greenhouses** in spring and summer; in hot, dry summers the mites can also become a problem on outdoor plants such as strawberries.

Life cycle. This mite thrives in **warm, dry** greenhouses. It lives mainly on the under-surface of leaves. The adult is 1 mm long and yellowish in colour, with two black spots on its back (Figure 17.25) but becomes red in autumn. The female lays about 100 tiny spherical eggs on the underside of the leaf which hatch as larvae followed by two nymph stages and then finally develops into an adult. The life-cycle development varies according to temperature with the average length of the life cycle from egg to adult being 55 days at 10°C but 12 days at 21°C. In autumn, the female (with eggs inside her) **hibernates** (diapause). She often emerges in March or April. A second, closely related species, the **carmine spider mite** (*Tetranychus cinnabarinus*), is commonly found on greenhouse tomato and chrysanthemum crops; it is orange-red in colour without any black spots on its back.

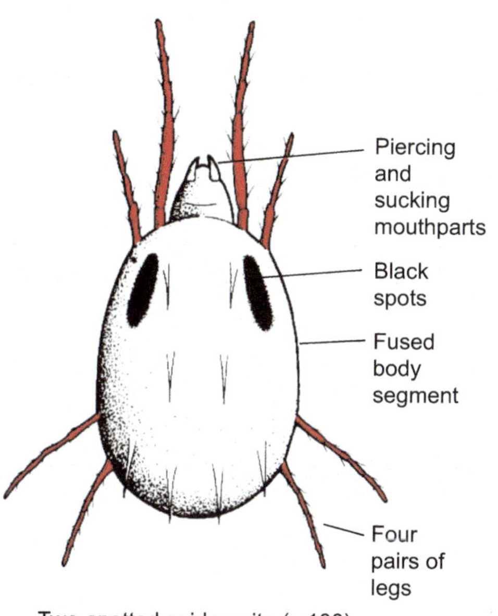

Figure 17.25 A two-spotted spider mite

Figure 17.26 (a) Fine mottling symptoms caused by two-spotted spider mite on a palm leaf; (b) fine strands ('webbing') produced by the mites

Spread. This occurs when adults and nymphs crawl over short distances from plant to plant. It is moved over long distances by the transportation of infected plants and often goes unnoticed because of its small size.

Control. Growers should carefully check incoming plants for the presence of the mite, using a **10× hand lens** if necessary. **Cultural control** can involve creating a **high humidity** environment in glasshouses, which inhibits the presence of the mite, by regular wetting of paths and misting of plants with water; reducing ventilation also increases humidity but can result in other undesirable environmental effects such as excessively high temperatures and fungal diseases such as grey mould. Control of weeds such as chickweed in and around the greenhouse can remove alternative hosts for the mite. **Physical control** can involve the removal and destruction of infected plants or parts of plants from the glasshouse. Thorough cleaning of empty glasshouses in winter with a horticultural disinfectant such as peroxyacetic acid can destroy overwintering adults. Application of plant invigorators containing surfactants to act upon the mites physically and nutrients to stimulate plant growth are suitable for the two spotted spider mite control. **Biological control** can be introduced into glasshouses and includes **natural predators** of the two-spotted spider mite such as the predatory mites *Phytoseiulus persimilis* (Figure 15.10) and *Amblyseius andersoni* and the predatory midge *Feltiella acarisuga*. **Chemical control** products containing the acaricides **pyrethrins** and **acetamiprid** are available for control of two-spotted spider mite, but most products containing acetamiprid which are only approved for use on **ornamental plants** must not be applied to edible plants unless the label instructions permit that use.

Blackcurrant gall mite (*Cecidophyopsis ribis*)

This microscopic mite lives most of its life inside the **dormant buds** of blackcurrant bushes where it feeds on sap. It causes swollen, spherical buds in winter (Figure 17.27) leaves, and flowers are distorted when they open in spring, consequently there will be reduced flowers from which fruits can form and a reduction in plant vigour because of damage to leaves. The mites emerge between April and June and can transmit blackcurrant **reversion virus** disease as they are dispersed by wind and rain and on the bodies of aphids. The mites enter the developing buds again later in the summer where they continue to live and breed until the following spring. The selection of new plants which have been propagated under the requirements of the **Fruit Propagation Certification Scheme** helps to ensure pest and disease-free planting stock. Blackcurrant 'Ben Hope' is a cultivar with resistance to the blackcurrant gall mite. Cutting out and destroying stems with infected buds in winter is necessary.

Figure 17.27 Blackcurrant gall mite causes swollen, spherical buds (source: Shutterstock, salarko)

Cyclamen mite (*Phytonemus pallidus*)

This microscopic mite feeds by piercing and sucking out cell contents, and in the process it injects harmful **saliva** into pot-grown *Cyclamen persicum* and other species such as *Fuchsia*, *Gerbera* and *Chrysanthemum*. In greenhouses it causes a range of symptoms, including stunting, flecking of petals and distortion of leaves. Care should be taken to prevent the introduction of infested plants and propagation material into glasshouses. The **predatory mite** *Amblyseius californicus* can be introduced as biological control for cyclamen mite in greenhouse crops. Effective **chemical control** is difficult to achieve as the mites inhabit inaccessible places.

Other mite pests

Other horticulturally significant mites which require a mention include the **fruit tree red spider mite** (*Panonychus ulmi*) which can cause serious leaf mottling of ornamental apple, pear, plum and cherry. **Bryobia mite** (*Bryobia rubrioculus*) is found in fruit trees including apple, pear and cherry and also causes damage to greenhouse crops such as cucumbers, if blown in from neighbouring trees. In recent years there has been a number of novel mite pest species arriving in Britain and Ireland. Hazel shrubs are affected by **hazel big bud mite** (*Phytoptus avellanae*) which leads to rounded, swollen buds that fail to develop normally and **fuchsia gall mite** (*Aculops fuchsiae*) which was first observed in Britain in 2007 and causes young *Fuchsia* leaves to become red and distorted and deforms the development of flowers. Checking plants regularly and promptly and cutting out and destroying infected stems and buds should be undertaken.

Other garden arthropods

In addition to insects and mites, the phylum Arthropoda contains four other horticulturally relevant groups, the animal sub-phylum **Crustacea** (including woodlice) and the classes **Symphyla** (symphilids), **Diplopoda** (millipedes) and **Chilopoda** (centipede), which are generally beneficial but some can very occasionally cause damage to plants (Table 17.1).

Woodlice are common in gardens, and there are approximately 35 species in Britain and Ireland. They are generally beneficial in the breakdown of organic matter by their feeding activities. Infrequent damage is confined mainly to greenhouse plants where occasionally stems and lower leaves of seedlings and fruits such as strawberry can be damaged. This crustacean (Figure 17.28a) should be considered an essential part of the garden's ecosystem and not be destroyed.

Garden symphilid (*Scutigerella immaculata*) resembles tiny white millipedes and can occasionally cause damage by feeding on the root hairs of young plants. They are relatively insignificant as pests in most garden situations.

Millipedes are slow-moving creatures with many legs (two pairs to each body segment). They are useful in breaking down soil organic matter, but the flat millipede (*Brachydesmus superus*) sometimes feeds on soft tissue such as seedlings. It is, however, relatively insignificant as a garden pest.

Centipedes superficially resemble millipedes. They help to control soil pests by feeding on insects, mites and nematodes in the soil (Figure 17.28b), they should be considered as benefical organisms.

Figure 17.28 (a) Woodlice are generally beneficial and are involved in breaking down organic matter (source: Shutterstock, Hwall); (b) *Geophilus* centipede, a useful predator on soil pests (source: Shutterstock, Dan Olsen)

17

Nematode pests

This group of organisms are members of the animal phylum **Nematoda** and are commonly referred to as **eelworms.** They are found inhabiting almost every environment in the world. They range in size from relatively large animal parasites, such as *Ascaris* (about 20 cm long) in the guts of livestock, to the tiny soil-inhabiting species (about 0.5 mm long). Non-parasitic species in soil are usually **beneficial**, feeding on plant remains and soil bacteria and helping in the formation of **humus** or by being parasitic on plant pests – they are commonly used in products for **biological control** of plant pests. The general structure of the nematode body is shown in Figure 17.30.
A feature of the plant parasitic species is the **spear** in the mouth region, which is thrust into plant cells. Salivary enzymes are then injected into the plant and the plant juices sucked in by the nematode.
In suitable conditions, nematodes are very active, moving in a wriggling fashion in soil moisture films, most actively when the soil is at **field capacity** and more slowly as the soil either waterlogs or dries out. Symptoms on plants caused by these minute animals are sometimes confused with those caused by fungi or bacteria.

Two horticulturally damaging species are described next. Four others are mentioned.

Potato cyst nematodes (*Globodera rostochiensis* and *Globodera pallida*)

Damage. These significant species of nematodes are found in most soils that have grown **potatoes** in Britain and Ireland. Leaves of infected plants become yellow, photosynthesis is reduced and the plants become stunted (Figure 17.29) and occasionally die, and the crop yield is consequently reduced. The distribution of damage in the plot is characteristically in patches. **Tomatoes** grown in greenhouse soil and outdoors may be similarly affected. The pests may be diagnosed in the field

Figure 17.29 Potato plants stunted by potato cyst nematode

by the tiny, mature white or yellow females (**cysts**) that are attached to the potato roots. A **10x hand lens** is necessary to see them clearly.

Life cycle. A proportion of the eggs in the soil hatch in **spring**, stimulated by chemicals which are released by potato roots. After hatching, the juvenile nematodes enter and feed inside the potato roots affecting the plant's ability to take in water and nutrients. If they fail to locate a root to enter, they die. When the female nematodes are fully developed, they move to the outside of the root and the now swollen female leaves only her head inserted in the plant tissues (Figure 17.30). After fertilization by the males, which have left the roots and are free living in the soil, the white female swells and becomes almost spherical, about 0.5 mm in size, and contains 200–600 eggs. As the potato crop reaches harvest, the female changes colour. In *G. rostochiensis* (golden nematode), the change is from white to yellow and then to dark brown, while in the other species, *G. pallida* (white cyst nematode), no yellow phase is seen. The significance of the species difference is addressed later in the text in the information on selection of resistant cultivars. Eventually, the female changes to a dark brown colour and falls from the root into the soil when the crop is lifted. This stage, which looks like a minute brown onion, is called the **cyst**, and the many eggs inside this protective shell can survive in the soil for up to **ten years**.

Spread. This nematode spreads as cysts, with the movement of infested soil, on boots and vehicle tyres. The cysts also be transported in water if the land floods or in wind-blown soil.

Control. **Cultural control** – use of crop rotation as an effective way of helping to prevent problems. Although the cysts can survive for up to ten years in the soil, growing potatoes on the same site for no less than one year in four will help reduce the incidence of this pest. Cultivation of the annual species *Solanum sisymbriifolium* as a **trap plant** can possibly reduce the rotation length on the plot. This close relative of potato and tomato releases chemicals which stimulate egg hatching but, because they do not host the nematode, the juvenile nematodes die without completing their life cycle and so the numbers in the soil are depleted. Some potato cultivars, such as 'Pentland Javelin', Blue Danube and 'Maris Piper', show **resistance** to golden nematode strains found in Britain and Ireland, but not to white cyst nematode. However, the cultivars 'Nadine' and 'Sante' have some dual resistance to both nematode species, but resistance to white cyst nematode is not complete. Small-scale cultivation of potatoes and tomatoes in **containers** with a soil-less or sterile growing medium will avoid any problems with this pest. **Physical control** involves the removal and destruction of all material from infested plants. All footwear and tools which have been in contact with soil containing infested plants should be sterilized with a horticultural disinfectant such as peroxyacetic acid. For **chemical control**, fluopyram is an active ingredient in a **nematicide** product which is approved for control of potato cyst nematodes, but its use is only permitted by appropriately trained and certified professional growers; no chemical control is available for home gardeners.

Other nematode pests

Stem and bulb nematode (*Ditylenchus dipsaci*) inhabits plant tissues and damages bulbs of **ornamentals** such as *Narcissus* spp., *Tulipa* spp. and *Hyacinthus* spp. and **vegetables** such as onions, garlic, carrots and French and runner beans. Bulbous species develop swollen bulbs and stunted growth with distorted leaves. *Narcissus* bulbs show brown rings when cut across and their emerging leaves have slightly raised yellow streaks. The damage can lead to further bacterial and fungal infections. Affected carrots have swollen and rotting leaf bases. Stems of beans are swollen and distorted. **Control methods** are predominantly **cultural**, including weed control of the chickweed alternative host. Crop rotation with **resistant crops** such as lettuce and brassicas helps to reduce pest numbers. Planting material from infested soil should be avoided. Purchasing bulbs from reputable suppliers who provide healthy plants is essential.

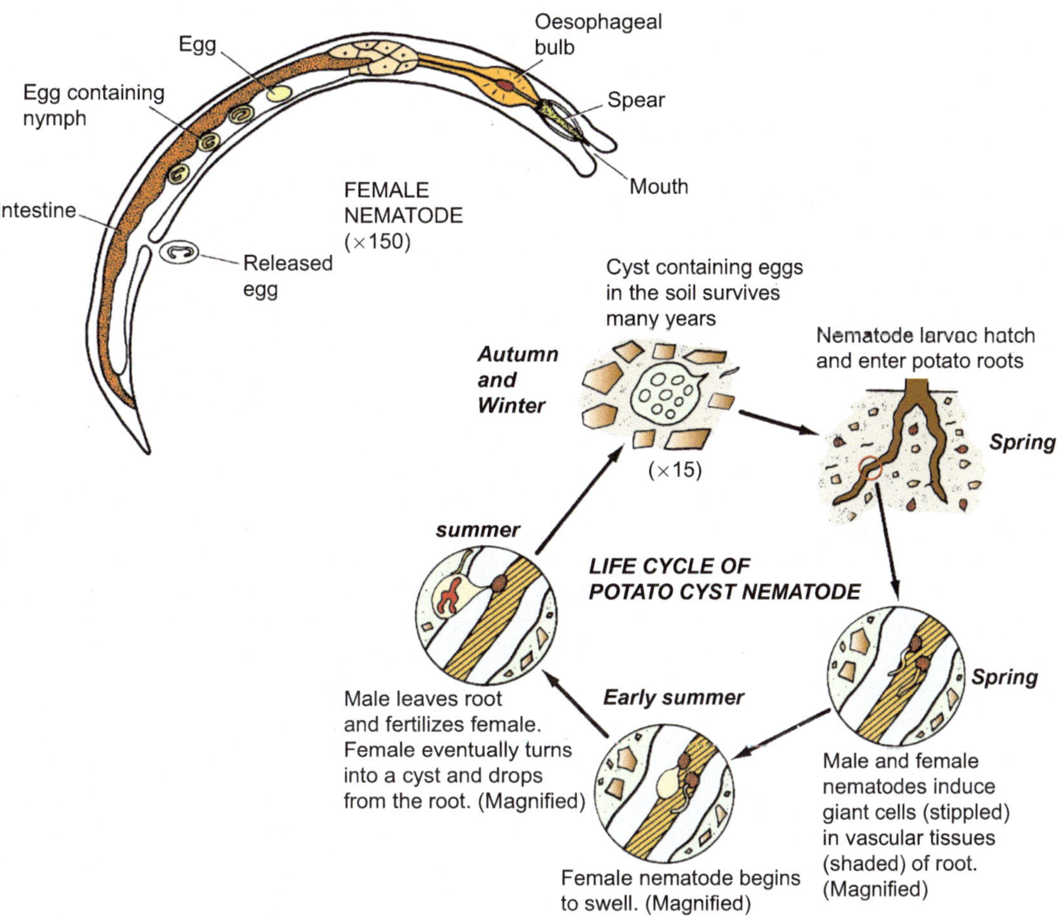

Figure 17.30 Generalized structure of a nematode, and life cycle of a potato cyst nematode

Chrysanthemum eelworm (*Aphelenchoides ritzemabosi*) is a foliar nematode which affects leaves and buds of chrysanthemums but can be present in the soil, too. It causes a blotching and purpling of the leaves that spreads to become a dead brown, V-shaped area between the veins of the leaves; the lower leaves are worst affected and growth is stunted. **Cultural control** methods include propagation from **healthy stock-plants**, cultivation in a **sterile growing medium** and avoidance of overhead irrigation to reduce the risk of spreading the nematodes. **Mulching** around the base of plants can provide a barrier to prevent the nematodes from moving up from the soil. **Physical control** includes removal and destruction of all affected plant debris. **Warm-water treatment** of dormant chrysanthemum stools at 46°C for five minutes is effective for outdoor-grown plants.

Root knot eelworm (*Meloidogyne* **spp.**) is a worldwide important pest in tropical countries. It used to be a serious pest in greenhouse crop production in Britain and Ireland, but the widespread cultivation of crops such as tomato and cucumber in soil-less hydroponic systems has reduced its significance. The pest causes large root **galls** up to 4 cm in size, on the roots of plants such as chrysanthemum, cucumber and tomato, with resulting wilting and poor plant growth and crop yield. **Partial-soil sterilization** helps reduce the problem in glasshouse soil-grown crops. For growers of tomato in soil, the rootstock '**KVNF**' with other cultivars grafted on to it can resist the nematodes.

Migratory plant nematodes

Unlike the nematodes described earlier, the migratory species feed only from the **outside** of the root. The dagger nematodes (e.g. *Xiphinema diversicaudatum*) and needle nematodes (e.g. *Longidorus sylphus*), which reach lengths of 0.4 and 1.0 cm, respectively, attack the young roots of crops such as rose, raspberry and strawberry and cause stunted growth. In addition, these species vector important viruses – **arabis mosaic** on strawberry and **tomato black ring** on ornamental cherries. The nematodes also can survive on the roots of a wide variety of weeds. The use of crop rotation, clean planting stock and efficient weed control helps keep these problems at bay.

Further reading

Adams, C., Brook, J. and Early, M. (2015) *Principles of Horticulture Level 3*. Routledge.

Alford, D.V. (2002) *Pests of Ornamental Trees, Shrubs and Flowers*. Wolfe Publishing.

British Crop Production Council. (2025) *UK Pesticide Guide*. BCPC.

Brock, P.D. (2021) *Britain's Insects. A Field Guide to the Insects of Great Britain and Ireland.* Princeton University Press.

Brown, L.V. (2008) *Applied Principles of Horticulture.* 3rd ed. Butterworth Heinemann.

Buczacki, S. and Harris, K. (2010) *Pests, Diseases and Disorders of Garden Plants.* HarperCollins.

Chinery, M. (2009) *Collins Complete Guide to British Insects.* Collins.

Cloyd, R. (2016) *Greenhouse Pest Management.* CRC Press.

French, J. (2007) *Natural Control of Garden Pests.* Aird Books.

Greenwood, P. and Halstead, A. (2018) *RHS Pests and Diseases.* Dorling Kindersley.

Helyer, N. (2003) *A Colour Handbook of Biological Control in Plant Protection.* Manson Publishing.

Hessayon, D.G. (2009) *Pest and Weed Expert.* Transworld Publishers.

Ives, J. (2022) *A Gardener's Guide to Biological Pest Control.* The Crowood Press Ltd.

Watson, G. (2013) *Tree Pests and Diseases: An Arborists Field Guide.* Arboricultural Association.

www.daera-ni.gov.uk/topics/plant-and-tree-health

https://forestrycommission.blog.gov.uk/category/tree-health/

https://planthealthportal.defra.gov.uk/

RHS. (2024a) *Pesticides Available for Home Gardeners*. Online: www.rhs.org.uk/advice/pdfs/pesticides-for-home-gardeners.pdf RHS Advisory Service

RHS. (2024b) *Rabbit Resistant Plants.* Online: www.rhs.org.uk/prevention-protection/rabbit-resistant-plants apps.rhs

www.sasa.gov.uk/plant-health

www.teagasc.ie/crops/horticulture/plant-health/

SUPPORT MATERIAL

The online material is accessible via the QR code and includes further information on many of the topics in the book, such as

- Mealybugs
- Codling moth

CHAPTER 18

Plant diseases and disorders

Figure 18.1 Honey fungus toadstools growing on an infected tree base

This chapter includes the following topics:

- Structure and biology of fungi, bacteria and viruses
- Fungal diseases
- Bacterial diseases
- Viral diseases
- Plant disorders

DOI: 10.4324/9781003581260-18

The climate of Britain and Ireland, along with the wide range of wild species, garden plants and cultivated crops grown, creates opportunities for a diverse range of disease-causing pathogenic fungi, bacteria and viruses to thrive and spread. These pathogenic organisms always affect plant growth adversely, sometimes causing minor problems but in more serious cases resulting in severe damage or death of the plant. This chapter describes the whole range of disease-causing groups which the horticulturist may encounter. It explains their biology and the symptoms and signs which accompany them and the prevention and control measures which can be used. Plant disorders are sometimes overlooked, or confused with diseases, so their various causes and how these might be dealt with are also discussed in this chapter.

Structure and biology of fungi, bacteria and viruses

A fungus is composed, in most species, of **hyphae** (microscopic strands), which may occur together in a loose structure, **mycelium**, form **sclerotia** (Figure 18.3), dense resting bodies, or produce **rhizomorphs** (Figure 18.18), complex underground root-like strands. The group of organisms causing diseases such as damping off (e.g. *Pythium* spp.), potato blight (*Phytophthora infestans*) and downy mildew of brassica (*Hyaloperonospora brassicae*) have been reclassified as a separate fungus-like group, **'Oomycetes'**, and are distinct from true fungi although they tend to be treated in the same way as fungal diseases.

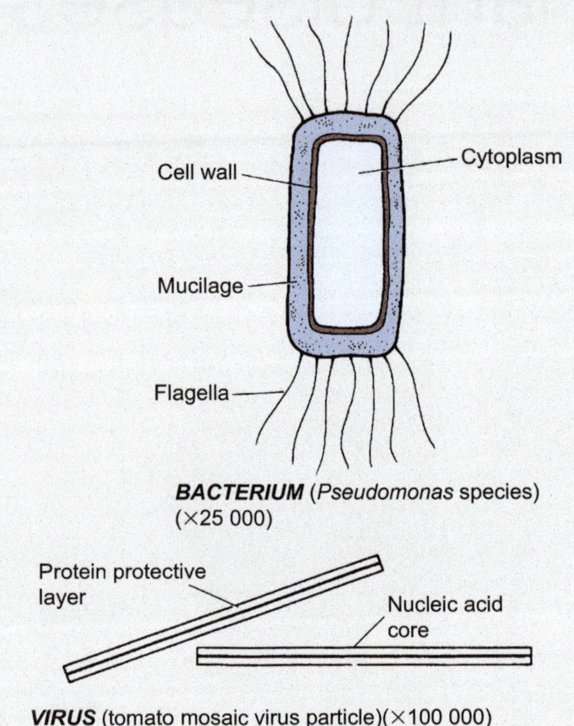

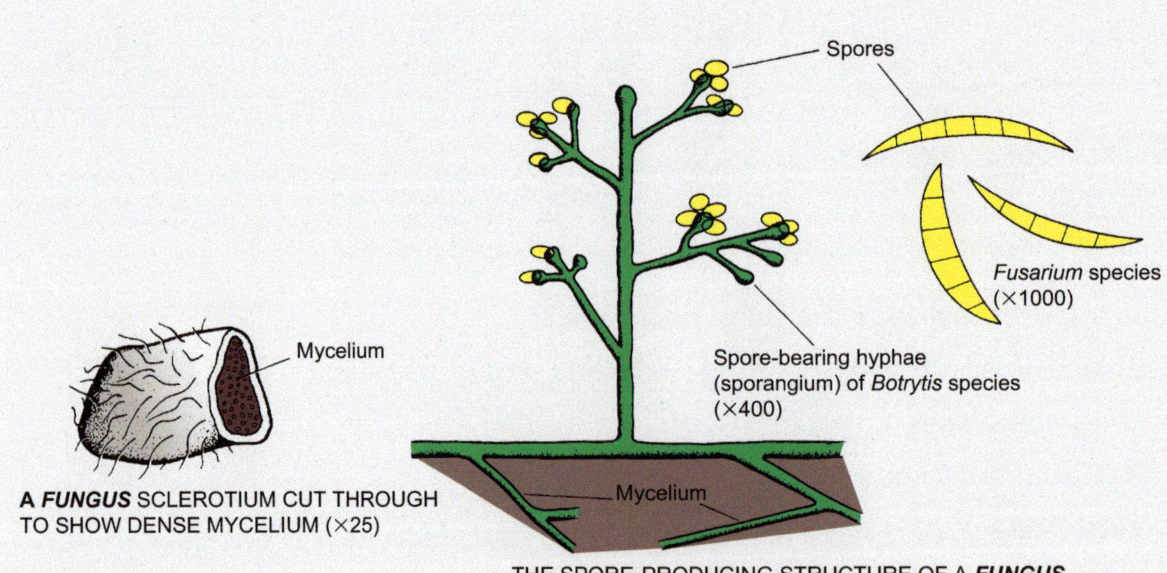

Figure 18.2 Microscopic structure of fungi, bacteria and viruses

Also, the organism causing brassica club root (*Plasmodiophora brassicae*) is different from fungi; it produces a jelly-like structure (plasmodium) inside the cells of the host plant.

The hyphae in most fungal species are able, in the appropriate conditions, to produce **spores**. Two soil-borne fungi – *Rhizoctonia solani* that causes a **black scurf** disease on potatoes and *Stromatinia cepivora* that causes **white rot on onion** (Figure 18.3) – are notable exceptions, where the hyphae only very rarely produce spores. Wind-borne spores are generally very small (about 0.01 mm), not sticky and often borne by special hyphae protruding above the leaf surface such as those of **grey mould** (*Botrytis cinerea*), so that they catch wind currents. Water- or rain-borne spores can be sticky, such as those of damping off. Minute asexual spores produced without fusion of two hyphae commonly occur in seasons favourable for disease increase – for example, humid weather for **downy mildews** and dry, hot weather for **powdery mildews**. Sexual spores, produced after hyphal fusion, commonly develop just before unfavourable conditions occur (e.g. a cold, damp autumn). They are produced as single spores in the oomycetes. In the powdery mildews and other species in the Ascomycota division, spores are produced in groups within a protective hyphal **spore case** (like a tiny black spherical flask less than 1 mm in size) often observable to the naked eye. Different genera and species are identified by microscopic measurement of the shape and size of the spores or the spore-bearing spore cases, or by the appearance of the hyphae.

Horticulturists without microscopes must use the **symptoms** in the host plant or **signs**, visible manifestations of the fungal pathogen, to identify the particular problem.

Figure 18.3 White rot on onion; note the black sclerotia which enable this pathogen to survive for many years in the soil

A **plant disease** is a harmful disturbance in a plant's normal functions caused by a pathogen such as a fungus, bacterium or virus.

While disease-causing, or **pathogenic**, fungi are the main concern of this chapter, it should be noted that in the garden and ecosystem generally, most fungi are beneficial. **Saprophytic** fungi break down organic material such as dead plant tissue like tree stumps, dead animals and microorganisms. In addition, many species of **symbiotic** mycorrhizal fungi exist in the soil and form mutually beneficial relationships with the roots of many plant species. Other species are formulated and used as **bio-pesticides** to protect cultivated plants from pathogens or pests. Also, the toadstools of edible species of fungi are the basis of commercial crop production throughout the world, and some important medications to treat human diseases such as bacterial infections are derived from fungal organisms such as *Penicillium* species.

Infection

The spore of a leaf-infecting fungal pathogen, after landing on the leaf in damp conditions, produces a germination tube, which being delicate and easily dried out must enter through the cuticle, the stomata or a wound within a few hours before dry, unfavourable conditions occur. Within the leaf, the hyphae grow, absorbing food until, within a period of a few weeks, they produce a further crop of spores (Figure 18.5). Foliage diseases such as potato blight often increase very rapidly when conditions are favourable.

Roots may be infected by **spores** (e.g. in damping off), by **hyphae** (e.g. in wilt diseases), by **sclerotia** (e.g. in *Sclerotinia* rot) or by **rhizomorphs** (e.g. in honey fungus). Root diseases are generally less affected by short periods of unfavourable conditions and can sometimes increase at a slower, more constant rate; although in hydroponic systems (see p. 213), increase can be much more rapid.

Phyllosphere

On the surface of leaves and stems lives a population of microorganisms (bacteria and fungi) which occupy a microhabitat commonly called the **phyllosphere.** These bacteria may be 'casual' or 'resident'. Casual organisms such as *Bacillus* spp. mainly arrive from soil, roots and water and are more common on leaves closer to the ground. These species are capable of rapid increase under favourable conditions, but then may decline. Resident organisms such as *Pseudomonas* spp. may be weakly parasitic on plants, but more commonly persist (often for considerable periods) without causing damage, and on a wide variety of plants. Phyllosphere bacteria can reduce the infections of **fungal diseases** affecting leaves including powdery mildew, downy mildew and grey mould; a **bio-fungicide** product containing a strain of the bacterium *Bacillus amyloliquefaciens* has been developed and formulated to help control these diseases. There remains the general principle that a healthy, well-nourished plant will be more likely to have organisms on the leaf surface available to reduce fungal infections.

There are also phyllosphere fungi such as *Alternaria* and *Cladosporium* which are found on leaf surfaces, and these may be useful in combating fungal disease infection in a similar way to the bacterium described previously.

Fungal diseases

Practical relevance of fungal classification

The classification of fungi has some practical implications in understanding fungal disease **life cycles and control.** Species within a fungal division can sometimes have similar methods of spread and survival. Their similarity of spore structure, hypha structure and biochemistry also means that they are often susceptible to control by the same or similar fungicide active ingredients. Because of the horticulturally practical aspects of fungal classification, the appropriate fungal division is given for each disease as it appears in the text (for example, the fungal-like group of **oomycetes** is mentioned against potato blight, damping off and downy mildew).

Potato blight (*Phytophthora infestans*)

This is the most significant disease of **potato** in Britain and Ireland. The oomycete *Phytophthora infestans* pathogen affects foliage, stems and tubers and has a major economic impact in **commercial production**.

Damage. This disease is a constant threat to potato production in **warm, humid summers** and was a contributing factor to the Great Famine in Ireland in the 19th century. The first symptoms are seen as **yellowing** of the foliage which then produces a **white bloom** on the under-surface of the leaf in damp weather. The leaves and stems may then go brown and shrivel, leading to their death (Figure 18.4a). The tubers may have dark surface spots that, internally, appear as a dry, deep red-brown rot. This pathogen can also infect **tomato** plants, the most notable symptom being the dark brown blisters on the fruit (Figure 18.4b).

Life cycle. The pathogen survives the winter as mycelium and sexual spores in the potato **tubers** or debris from the previous summer's crop. The spring emergence of infected shoots results in the production of asexual spores (Figure 18.5).

Spread. The spores, which are spread by **wind** from infected crops, land on potato leaves or stems and can, after infection, result in a further crop of

Figure 18.4 Potato blight: (a) leaf symptoms on potato; (b) fruit symptoms on tomato (source: Shutterstock, Radovan1)

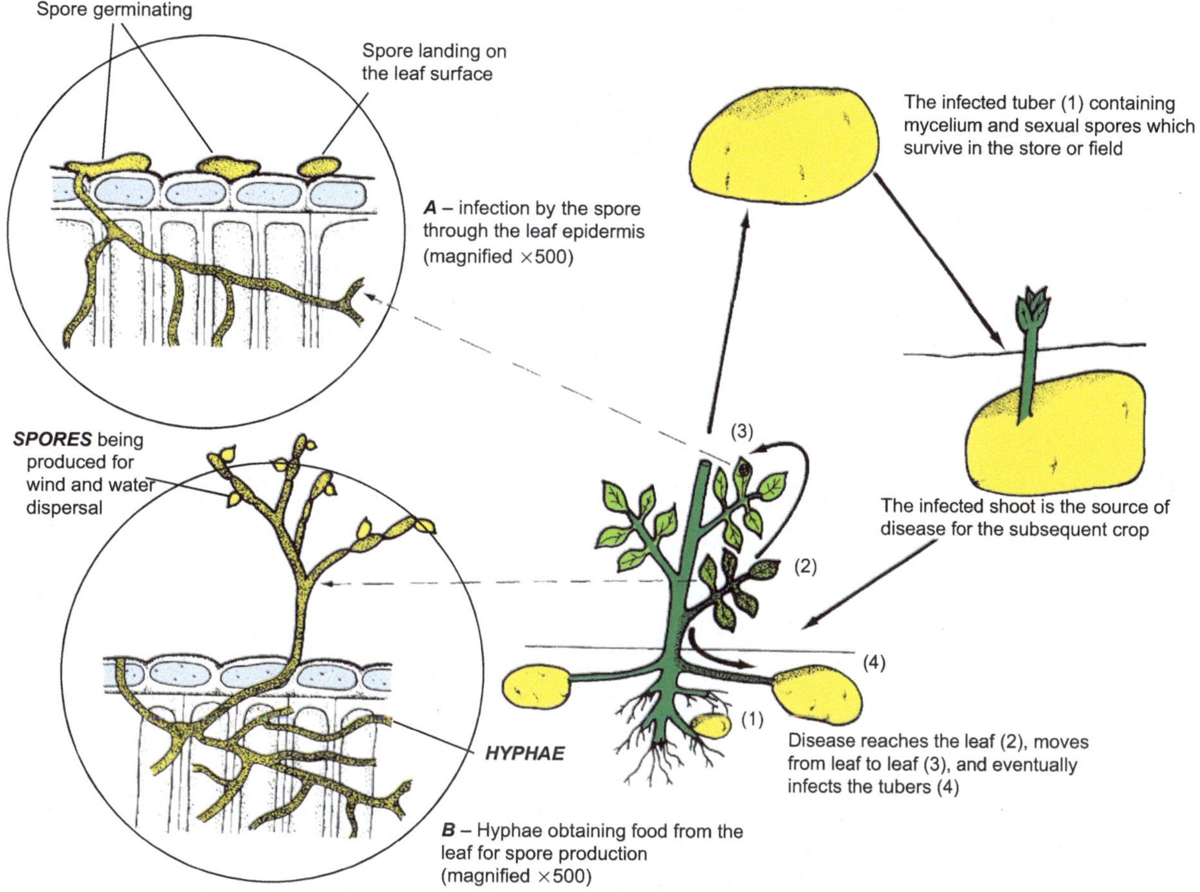

Figure 18.5 Potato blight life cycle

spores within a few days under **warm and humid** weather conditions.

The disease can spread very quickly. Later in the crop, badly infected plants may have tuber infection as water from irrigation or rain washes spores down into the soil.

Control. **Cultural controls** dominate the approach to managing this disease including the cultivation of potatoes and tomatoes on a minimum **four-year rotation**. Selection of cultivars which show some **resistance** such as those in the 'Sarpo' range, for example 'Sarpo Axona' and 'Sarpo Mira', can help reduce the risk of severe infection. **First-early cultivars**, although susceptible to blight, are less likely to suffer as the environmental conditions which facilitate the disease's development are less likely to occur until after they have been harvested. Seed potatoes produced under the **Seed Potato Classification Scheme** should be bought fresh each year from a reputable supplier. Tubers harvested from infected plants should be used quickly and not stored for long periods. Thorough 'earthing-up' of potatoes can help prevent spores of the pathogen from being washed down to the tubers in the soil. **Physical controls** involve removing from site and destroying all infected plant tissue with no composting taking pace on site. **Chemical control** – there are no chemicals approved for home gardeners to use. Appropriately trained and certified professional growers can apply a protectant and systemic fungicide product containing the active ingredients **fluopicolide** and **propamocarb hydrochloride**.

Forecasting blight outbreaks

National warnings to **forecast the risk** of potato blight are given to commercial growers by government advisory services when temperature and humidity measurements meet the **'Hutton Criteria'**. That is, when two consecutive days each has a minimum temperature of 10°C and each day has at least six hours with a relative humidity greater than 90%. Protectant sprays of fungicides can be applied before infection becomes established.

Damping off (*Pythium, Phytophthora, Fusarium* and *Rhizoctonia* species)

Pythium and *Phytophthora* species are fungal-like oomycetes. *Fusarium* and *Rhizoctonia* species are true fungi.

Damage. These genera are all soil-borne organisms and can cause considerable losses to the seedling stage of the plant. The infection may occur at or below the soil surface, but most commonly the emerging seedling plumule is infected at the soil surface, causing it to topple and quickly die (Figure 18.6). Occasionally the roots of mature plants such as cucumbers are infected and turn brown and soggy, and the plants die.

Life cycle and spread. Pythium, Phytophthora, Fusarium and *Rhizoctonia* occur naturally in soils as **saprophytes**, but under damp conditions they produce the asexual spores that cause infection. These spores are spread by **water**. Sexual spores (oospores) are produced in infected roots and can survive several months of dry or cold soil conditions.

Control. **Cultural control** – prevention is best achieved by effectively managing the growing environment. Providing a pathogen-free, **free-draining** growing medium is important, and the use of **clean** pots and containers and fresh compost, or the partial sterilization of loam in soil-based composts with heat, is essential. Sowing seeds **sparingly** is important as over-sowing leads to increased risk of an infection. Excessive irrigation of seedlings leading to a **saturated** growing medium should be avoided as the pathogens increase rapidly under these conditions. Reducing **humidity** by effective ventilation also helps reduce the risk of infection. Water tanks with open tops, harbouring rotting plant debris, are a common source of **infected water** and they should be covered and cleaned out regularly. Sand and capillary matting on benches in greenhouses should be regularly replaced or cleaned by washing with a horticultural **disinfectant** containing an ingredient such as peroxyacetic acid. **Physical control** involves prompt removal and destruction of containers of infected seedlings. **Chemical control** – a **bio-fungicide** containing the fungus *Trichoderma harzianum* **T-22** can be applied as either a drench or granules to growing media of ornamental and edible plants grown under protection by appropriately trained and certified professional horticulturists. The fungus in this product can help protect plants against infection by soil-borne damping off pathogens and is also parasitic on some of the pathogens. No **chemical control** products are available for home gardeners.

Figure 18.6 Tomato seedlings affected by damping off (source: Shutterstock, AmBNPHOTO)

Downy mildew of brassica (*Hyaloperonospora brassicae*)

This pathogen is a member of the oomycete group of fungus-like organisms. Downy mildews are not closely related to powdery mildews, which are true fungi.

Damage. This disease causes a **white bloom**, mainly on the **lower leaf surface** (Figure 18.7) where a humid environment favours infection and spore production. The upper surfaces develop irregular, **yellow patches** which reduces photosynthesis and leads to a weakening of the plant resulting in a reduction in quality and quantity of produce; ornamental plants are disfigured. Members of the Brassicaceae family including **ornamentals** such as Brompton stock (*Matthiola incana*) and wallflower (*Erysimum × cheiri*) and **vegetables** such as cauliflower, broccoli and

Figure 18.7 Downy mildew of brassica symptoms on Brompton stock (*Matthiola incana*) leaves; note that the infection is on the lower leaf surface

occasionally **weeds** such as shepherd's purse are infected by this pathogen. This disease is most damaging at a time when seedlings have just **germinated**, particularly in spring when the young tissues of the host plant are susceptible and favourable damp conditions may combine to severely damage developing plants.

Life cycle and spread. Asexual **spores** (zoospores) are produced by hyphae present on the lower leaf surface, mainly in spring and summer, and are spread by wind currents. Thick-walled sexual spores (oospores) produced within the leaf tissues fall to the ground with the death of the leaf and **survive the winter**. Spring infections occur when rain splash carries the spores up from the soil to the lower leaf surface of seedlings and young plants.

Control. **Cultural control** includes crop rotation with a minimum **four-year rotation** for brassicas being implemented. Routine control of **brassica weeds** which can host the disease such as shepherd's purse and hairy bitter cress is important. Sowing and planting distances between individual plants should be generous to allow good air movement and the avoidance of **high humidity** which favours development of the disease. **Physical control** involves removal and **destruction** of any infected plant tissues; it should not be composted on site. **Chemical control** – there are no products available for home gardeners to apply for this disease. Appropriately trained and certified professional horticulturists can apply fungicide products containing **azoxystrobin** as the active ingredient.

Other crops such as lettuce and onions are infected by different species of downy mildews (*Bremia lactucae* and *Perenospora destructor*, respectively). These pathogens are **host specific**, and no cross-infection is seen between crops such as brassicas, lettuce and onions, which belong to different plant families.

Clubroot (*Plasmodiophora brassicae*)

The clubroot pathogen is a soil-borne organism which is classified separately from fungi – it is a close relative of the slime moulds.

Damage. It causes serious damage to most members of the **Brassicaceae** family, which includes **edible** crops such as cabbage, cauliflowers and Brussels sprouts and **ornamentals** such as Brompton stock (*Matthiola incana*) and wallflower (*Erysimum × cheiri*). Weed species such as shepherd's purse are also susceptible. Infected plants show signs of **yellowing** of leaves and **wilting** in dry, warm conditions, and the whole plant often becomes

Figure 18.8 Clubroot symptoms on cabbage roots

stunted. On examination, the roots appear **swollen and stubby** with a lack of fibrous roots (Figure 18.8) and may show a wet rot.

Life cycle and spread. The clubroot pathogen can survive in the soil for up to 20 years as minute **spores** which can germinate to infect the root hairs of susceptible plants. The organism forms plasmodium, a jelly-like mass, within the plant's root tissues. The plasmodium stimulates root cell division and causes cell enlargement, which produces the **swollen roots**. The flow of food and nutrients in phloem and xylem is disturbed, with consequent poor growth of the plant. At plant maturity, the spores produced by the plasmodium within the root are released as the root decomposes. The disease is favoured by **high soil moisture**, high soil temperatures and soils with a **pH below 7**.

Although this organism does not spread far in undisturbed soils, it can be carried longer distances on infected plants and on tools, footwear and wheels of garden machinery such as rotary cultivators.

Control. There are no chemical or biological controls for this disease. Control measures rely on **cultural controls** such as a minimum four-year crop rotation, ideally as long as possible, while simultaneously preventing Brassicaceae weeds from establishing on the plot. Improvement of **drainage** by practices to develop good soil structure or, on a small scale, growing brassicas in a free-draining soil in raised beds will help in maintaining appropriate soil water levels. Managing soil pH by the **liming** of soil with a material such as calcium carbonate to achieve a pH no less than 7 in advance of seed sowing and planting will inhibit pathogen activity. Selection of cultivars which show some **resistance** to clubroot such as summer cabbage 'Kilaxy'F1, cauliflower 'Clapton'F1, swede 'Invitation' and Brussels sprouts 'Crispus' F1 can help reduce the incidence of the disease. Transplants grown in a clubroot-free **seedbed** or plants propagated and grown in **containers** before planting out are less likely to be affected when planted into the vegetable plot. **Physical control** – infected plants should be immediately removed and destroyed, and composting of infected plants should be avoided. Sterilization of potentially contaminated tools or machinery with a horticultural **disinfectant** containing peroxyacetic acid should take place. There are no chemical controls available for clubroot.

Powdery mildew on strawberry (*Podosphaera aphanis*)

This belongs to the Ascomycota division of fungi. Powdery mildews should not be confused with downy mildews (see p. 292), which are an unrelated group of fungus-like organisms. Powdery mildews generally prefer **hot, dry** conditions whilst downy mildews occur most commonly in **cool, damp** conditions.

Damage. The first symptoms on strawberry are often **purple spots** seen on the upper leaf surface and, sometimes, on the lower leaf surface. A slightly fluffy dry, white infection (Figure 18.9) is then seen gradually covering areas of the upper leaf surface. Darkened areas on the upper leaf may indicate infection on the lower leaf surface. Leaves eventually turn brown and leaf edges curl. The loss of photosynthesis in the leaves leads to **reduced** vegetative growth and a **decrease** in fruit quality and quantity. The mildew can infect strawberry

Figure 18.9 Strawberry powdery mildew on leaves (source: Shutterstock, Kittisak Chysree)

flowers, causing a deep pink colouration. Affected fruit may appear distorted and often look rather dull in colour with the seeds protruding out from the fruit surface. Infection of young fruit may cause small, shrivelled strawberries.

Life cycle and spread. The fungus produces chains of asexual spores (conidia) from infected **leaf surfaces**. These spores are carried by **wind** to uninfected leaves. Above about 15°C, these spores are able to infect leaf tissue (mainly the upper surface) without there being a surface layer of water, although **humid conditions** and dampness do favour the disease. Consequently, powdery mildews are well adapted for infection in **summer conditions**. Strawberry powdery mildew becomes more serious towards the end of summer as nighttime humidity increases. It is also more serious in **polytunnel-grown** strawberries, where temperatures and humidity tend to be higher than outdoors.

The fungus overwinters in two ways. During autumn, tiny (1 mm) black spherical spore cases, which contain many spores, develop on the infected leaves and survive on dead leaf material. In mild winters, the fungus may survive within the green shoots of the dormant plant. Spread is by means of summer spores. **Transportation** of plants may introduce the disease to new locations

Control. **Cultural control** includes the selection of cultivars such as 'Pegasus and 'Florence' which show some **resistance** to powdery mildew. Organic matter added to the soil helps to **maintain soil moisture** levels in summer, thus reducing the risk of infection. The soil should not be allowed to dry out as the weakened condition of the plants favours infection. When irrigating, apply water **directly** to the soil and not the leaves. Avoid **overcrowding** of plants as it speeds up disease spread. Avoid over-feeding the plants with **nitrogen** fertilizers as this leads to softer leaf growth and increased susceptibility. **Physical control** includes **removal** of any infected shoots when the disease is seen early in summer; avoid shaking this plant material because it may release spores. Consider removal of weak and badly infected plants in autumn as they will be the most likely ones to harbour the disease over winter. All infected material should be **destroyed** and not composted on site. **Plant invigorators** can be applied, and they include a range of surfactants, plant nutrients and seaweed extracts which have a physical action on the powdery mildew fungus and act as a bio-stimulant to the plant to help resist the damaging effects of the fungus.

Chemical control – fungicide products containing the active ingredient penconazole, a **systemic fungicide**, are available for appropriately trained and certified professional horticulturists to apply for protection and treatment of strawberry powdery mildew. Home gardeners can use products containing water-dispersible **sulphur granules**, a protectant fungicide, for use on a range of **edible** and **ornamental** plants including strawberry plants.

Other powdery mildews

Apple powdery mildew (*Podosphaera leucotricha*) survives the winter as **mycelium** within the buds, which often appear small and shrivelled on twigs that have a dried, silvery appearance. Fruit may develop a russeted surface. This fungus may affect other **related species** such as pears, quinces, medlars and ornamental *Malus*. Pruning of silvered twigs and sprays with a **plant invigorator** can be used when mildew begins to spread through the tree.

Rose powdery mildew *Podosphaera pannosa*, (Figure 18.12), **gooseberry powdery mildew** (*Podosphaera mors-uvae*) occurring mainly on the gooseberry fruit (Figure 18.11) and **cucurbit powdery mildew** (*Podosphaera fuliginea*) on cucumber and courgette (Figure 18.10) are three other common powdery mildews.

Although hot dry summers may simultaneously lead to outbreaks of powdery mildew species in all the aforementioned plant species, **cross-infection** of the disease does not occur between strawberries, apple, rose, gooseberry and cucumber plants, as each distinct fungal pathogen is host specific.

Rose black spot (*Diplocarpon rosae*)

This belongs to the Ascomycota division of fungi.

Damage. This is a common disease of many species and cultivars of garden and greenhouse **roses** including bush roses (large-flowered and cluster-flowered types), climbers and patio roses. It is first seen in **early summer** as dark brown leaf spots, initially on older leaves, which may be followed by general leaf yellowing and then premature leaf drop (Figure 18.13). The infection doesn't kill the plant, but it has a disfiguring impact on the ornamental plant and reduces its photosynthetic ability which gradually **weakens** the plant.

Life cycle and spread. Spores are produced within spore cases embedded in the leaf spots on and

Figure 18.10 Cucurbit powdery mildew on a cucurbit leaf (source: Shutterstock, AJCespedes)

Figure 18.12 Rose powdery mildew on foliage (source: Shutterstock, aRTI01)

Figure 18.11 Gooseberry powdery mildew affecting the fruit (source: Shutterstock, Omfotovideocontent)

Figure 18.13 Rose black spot with leaf-yellowing symptom that accompanies the black spots (source: Shutterstock, Alex M1)

are released in wet and mainly warm weather conditions, and are then spread short distances by rain drops or irrigation water before beginning the cycle of infection again. The fungus **overwinters** in an inactive state on fallen leaves and on dormant stems and buds. In spring, spores are produced which then establish infections on the new leaves.

Control. **Cultural control** – resistance to this disease is not common but the selection of cultivars with less susceptibility is helpful. Some shrub roses such as the Japanese rose (*Rosa rugosa*) show good **resistance**. **Physical control** involves removal and destruction of **fallen leaves** and any **pruning material** which may contain the overwintering fungus. **Chemical control** involves application of a fungicide product with active ingredients such as **triticonazole or tebuconazole**, which are available to home gardeners and should be applied at first signs of the disease; repeat applications may be needed in each summer season.

Grey mould (*Botrytis cinerea*)

This belongs to the Ascomycota division of fungi and affects many plant species.

Damage. Grey mould is commonly recognized by a **grey-brown mould** of soft plant tissue and the fluffy, light grey fungal mass which follows its infection. It can occur on all soft tissue above ground, including stems, leaves, flowers and fruit, and on bulbs and tubers. In lettuce, the whole plant rots off at the base, and the plant subsequently goes yellow and dies. In tomatoes, infection in damaged side shoots and light-yellow spots (ghost spots) on the unripe and ripe fruit are found. Infected strawberry fruit may be covered by the fungus (Figure 18.14). In flower crops such as chrysanthemums, petals are often infected. Even if the plant survives a localized infection, it is likely to be weakened and disfigured.

Life cycle and spread. The fungus usually requires **wounded tissue** for infection, which explains its significance in crops which are **de-leafed**, such as tomatoes, or **disbudded** such as chrysanthemums. **Damp conditions** are essential for its infection and spore production and although it can occur at any time of year, it is particularly common in **autumn and winter**, especially in poorly ventilated and cool humid conditions sometimes found in greenhouses and **polytunnels**. The millions of spores are spread by wind to the next wounded surface. Black sclerotia (see p. 289), about 2 mm across and produced in badly infected plants, often act as the **overwintering** stage of the disease after falling to the ground and are particularly infective in unsterilized soils on young seedlings and soft, delicate plants such as lettuce.

Figure 18.14 *Botrytis cinerea* infecting a strawberry fruit

Control. **Cultural control** – grey mould can develop very quickly when the environmental conditions are favourable. Regular **scouting** of susceptible plants in periods of cool, humid weather is important in identifying early stages of infection and action should be taken to prevent development of the disease. Attention to **greenhouse humidity** control such as correct ventilation reduces the dew formation that is so important in the disease's development. Appropriate irrigation practices help to reduce the risks, and prolonged periods of **wetness** on foliage, fruits and flowers should be avoided as this will increase the risk of infection. **Generous crop spacing** to facilitate airflow in pot plants such as *Cyclamen persicum* is routinely practiced. **Good hygiene** in greenhouses and polytunnels is essential, including maintenance of a clean growing environment and removal of weeds and dead plant material which can host the fungus. Use of pathogen-free growing media, clean pots, containers and tools is essential. **Physical control** involves the removal of badly infected plants or localized infected tissue and destroying it well away from other susceptible plants. **Chemical control** – there are no fungicides approved for home gardeners to use. A **bio-fungicide** product containing the bacterium *Bacillus amyloliquefaciens* is approved for appropriately trained and certified professional horticulturists to apply to some horticultural crops. The fungal organism in this product produces anti-microbial metabolites which are antagonistic to the grey mould fungus, and it competes on leaf surfaces against the pathogen and offers enhanced

Figure 18.15 (a) Apple canker – with swollen branch and exposed wood; (b) coral spot – with swollen orange pustules

plant resistance to the disease. This product also has approval for control of downy and powdery mildews.

Apple canker (*Neonectria ditissima*)

This belongs to the Ascomycota group of fungi.

Damage. In fruiting and ornamental *Malus* (Figure 18.15a), *Pyrus* and *Sorbus* species, this fungus causes **cankers** in stems which appear as sunken, dark brown areas in the bark surrounded by swollen wound tissue and flaking bark. On younger stems, infections can develop to cause **girdling** of the whole stem leading to death of all tissue beyond the canker. Small red (perithecia) or white (conidia) fruiting bodies may be present on infected tissue. Wood may **fracture** in high winds, and fruits of apple and pear can become infected and develop rot prior to harvest and in storage.

Life cycle and spread. The fungus enters through **leaf scars** in autumn or through **pruning wounds** during winter. Care is therefore necessary to prevent infection, particularly in **susceptible** apple cultivars, by avoiding pruning in mild, wet conditions as spores produced in the spore cases found embedded in the canker tissue are spread by rain splash.

Control. **Cultural control** involves selection and planting of less susceptible cultivars. Although no cultivar of apple is completely resistant, some **dessert** cultivars such as 'Merton Russett' and some **culinary** cultivars such as 'Lord Derby' show some resistance. The popular dessert cultivar '**Cox's Orange Pippin**' is susceptible. Excessively wet and acidic soils are considered to increase the incidence of the disease so management of soil water and pH to provide optimum conditions is recommended. **Physical control** involves complete **removal** of smaller branches and spurs with cankers present to prevent further infection; in cankers of larger structural branches, cutting brown infected tissue back to healthy tissue may allow retention of the branch. Removed wood should be taken away from the garden and destroyed by burning if possible. Tissue exposed by removal of infected parts should be immediately protected by the application of a latex-based **pruning sealant** to the wound, especially in areas with high rainfall, which favours spread of the disease. Pruning tools and knives should be sterilized after use on infected tissue with a **horticultural disinfectant** containing peroxyacetic acid. **Chemical control** – there are no fungicide products approved for treatment of apple canker available for home gardeners. Products containing the active ingredients **cyprodinil** and **fludioxonil** can give some control of canker and are approved for appropriately trained and certified professional horticulturists.

Coral spot (*Nectria cinnabarina*)

This belongs to the Ascomycota group of fungi.

Damage. This disease is found on **dead branches** of trees and shrubs. Closer inspection shows a mass of **pink/orange pustules** (about 1 mm across) protruding from the dead wood (Figure 18.15b).

The disease is commonly seen on **woody broadleaf** species such as maples (*Acer* spp.), horse chestnut (*Aesculus hippocastanum*), beech (*Fagus sylvatica*), hornbeam (*Carpinus betulus*), lime (*Tilia* spp.), walnut (*Juglans* spp.), *Magnolia* spp. and *Elaeagnus* spp. Bush fruits such as blackcurrant and gooseberry are also susceptible. It is often seen in hedges where the regular hedge cutting process leaves many small branch stubs open to the risk of infection. Conifers are not usually affected.

Life cycle and spread. This fungus is a **weak pathogen**, often living as a saprophyte on dead wood. The aforementioned trees and shrubs are infected by spores landing on dead wood or through lenticels (see p. 58) of stems. The fungus is then able to progress further into living wood. This infection is often made worse when the host plant has been subjected to stress such as drought, prolonged waterlogging or root disease. The pink/orange pustules develop when the infected area of wood has died. The spores released from the pustules are spread by both **wind and water** splash. An established, mature plant is likely to withstand infections by coral spot but it can act as a source of infection which can spread to other plants.

Control. **Cultural control** involves pruning trees and shrubs in **dry weather** if possible, particularly when pruning in winter, to reduce the risk of infection. **Physical control** – infected stems should be cut out completely to leave healthy tissue and to minimize further spore production. Infected material should be taken away from the area and burnt. Pruning wound sealants are not usually recommended as they might interfere with the plant's natural healing process, but an application of a latex-based product may be of value when pruning in **wet weather** or when rain is forecast following pruning to reduce the risk of spores infecting the plant by water splash. Pruning cuts should be made close to junction points with the main branch or stem and just above a bud on narrower stems, as wound healing is most effective in these locations; branch stubs are liable to develop dieback and subsequent infection. There are no **chemical controls** available for coral spot.

Chalara ash dieback (*Hymenoscyphus fraxineus*)

This belongs to the Ascomycota group of fungi.

Damage. Having first appeared in Europe in the 1990s, it was first recorded in Britain and Ireland in 2012 and it is now **widespread**. It is a disease principally of common ash (*Fraxinus excelsior*) but also affects other mainland European ash species such as *F. angustifolia* and *F. ornus*. Some infections of Asiatic ash species have been reported but they appear to show **tolerance** of the disease and develop only mild symptoms. Infections have also been reported in Britain on other non-ash genera of the Oleaceae family such as *Chionanthus* and *Phyllyrea*, but these genera are not widely cultivated in Britain and Ireland. Leaves and shoots **wilt, blacken and die** (Figure 18.17) in mid- to late summer. **Premature** dropping of leaves occurs but the infection may have already progressed into the branches and, subsequently, the main trunk. Diamond shaped, light brown **lesions** or cankers can then form in the bark and these can girdle the stem leading to death of branches and eventually the whole tree. **Secondary pathogens** such as honey fungus (*Armillaria* spp.) can attack infected trees and cause root rot and plant death.

Life cycle and spread. Local spread is by **windblown** spores which are produced in large numbers from **mid-summer to autumn** on fallen, infected leaves of the previous year. The spores can be blown up to 30 km, landing on leaf surfaces which initially causes **leaf infections** that then progress into the branches. Spread over longer distances can take place by the transport of infected trees and logs, but spores can also be distributed in soil on tyres of vehicles and shoes of humans

Control. Chalara ash dieback is now so widely distributed in Britain and Ireland that it should be reported online only to the Forestry Commission, or the Forestry Division of the Department of Agriculture, Food and the Marine in the Republic of Ireland, if it is suspected in an area in which it has not been **previously recorded**. Maps showing previous recorded infections are posted online. Imports of *Fraxinus* species to Britain and Ireland from countries outside the EU and the UK are prohibited. Land owners or occupiers are not legally required to undertake any action on infected trees unless a government plant health authority serves a Statutory Plant Health Notice requiring action. **Physical control** – horticulturists can help to slow the local spread of the disease by collecting and burning (where permitted), **burying or deep composting** fallen ash leaves which can reduce the fungus's ability to spread. The Forestry Commission recommends that, when composting, the leaves should be covered with a **10 cm layer** of soil or a 15–30 cm layer of other plant material, and the heap should remain undisturbed for one year to prevent spore dispersal. **Biosecurity guidance** on the measures which are required pertaining to Chalara ash dieback is issued by the Forestry Commission for managers of woodlands and local authorities and other public agencies which manage trees. There are no chemical controls for this disease.

Rust fungi

Rusts belong to the Basidiomycota group of fungi.

They produce characteristic orange, brown or black raised leaf spots caused by the fungal tissue breaking through the leaf epidermis of the host (Figure 18.16b).

Rusts are a distinctive group of fungi often with complex life cycles involving up to five different spore forms within the same species.

Hollyhock rust (*Puccinia malvacearum*)

This is a common disease of hollyhock (*Alcea rosea*) in Britain and Ireland. It is especially damaging in wet summers. Hollyhock's related

Figure 18.16 (a) Hollyhock rust; (b) leek rust (source: Shutterstock, Graham Corney)

garden species in the Malvaceae family, including the genera *Abutilon*, *Hibiscus*, *Lavatera* and *Sidalcea*, may be affected by this disease.

Damage. **Leaves** are usually the worst affected plant parts. Numerous raised **grey spots** (see

Figure 18.17 Chalara ash dieback; new shoots wilt, blacken and die in mid-summer (source: Shutterstock, IanRedding)

Figure 18.16a) initially appear on the **lower leaf** surface. Later in the summer, **raised reddish-orange spots** may be produced on the **lower leaf surface**. **Flat yellow spots** may show on the **upper leaf** surface indicating where the infection has taken place directly below. After severe levels of infection, the leaves often **shrivel and die** and the plant is seriously disfigured. Stems may show a **cankerous appearance** after infection. The **calyces** on the outer part of the hollyhock flower can be affected in a similar way to the leaf. Continued infection in wet summers often leads to weakened plants which become stunted.

Life cycle and spread. Two types of spores are involved in infection. Teliospores are relatively large and are able to survive the winter, and then, in damp conditions, infect young hollyhock shoots in late spring to begin the disease cycle. In mild winters, the fungus may live as **mycelium** (p. 288) within the leaf and shoot tissues. In warmer months, the infection cycle takes about 14 days from infection to mature leaf spot production; at this time, the mass of hyphae and spores that have developed push through the lower epidermis to create the typical **raised spot** symptom. A new cycle of infection begins if the weather is damp. In the early summer period, the raised leaf spots are **grey** in colour from the numerous tiny **basidiospores** that are produced from the teliospores.

The smaller size of these spores facilitates their spread by **wind**. Towards the end of the summer, spore production changes as teliospore production becomes more common, giving the leaf spots a richer reddish-orange colour. These much larger spores are carried more easily by **rain droplets** than by wind currents.

Control. **Cultural control** – although hollyhocks are short-lived perennials, it is advisable that they be treated as **biennials** and **removed** after their first flowering period. Avoid **dense stands** of hollyhock where the resulting higher humidity can favour infection. Avoid cultivation of other susceptible malvaceous plants such as *Hibiscus* spp., *Lavatera* spp., *Abutilon* spp. and *Sidalcea* spp. near to hollyhocks. **Physical control** –regular checks for infection should take place and infected material should be **removed and burnt**. If infection is serious, the whole plant should be removed. Growers **should not** use 'self-saved' seed as this can be infected by the fungus. **Chemical control** – several fungicides such as tebuconazole with trifloxystrobin and triticonazole are approved for control of rust diseases on ornamental plants. Regular applications may be required to cope with newly arriving spores and developing infections, but the number of fungicide applications must not exceed the number stated on the label. Application of the fungicide should not commence until symptoms of the disease are apparent.

Honey fungus (*Armillaria* species)

This genus belongs to the Basidiomycota group of fungi. Seven species of *Armillaria* occur in Britain with *A. mellea* and *A. ostoyae* being responsible for most honey fungus infections. It is a widespread and common disease in gardens and parks with established **woody plant** collections.

Damage. Several pathogenic species of *Armillaria* cause root rot of primarily trees and shrubs including susceptible broadleaved plants such as *Betula* spp., *Forsythia* spp. and *Rhododendron* spp. and conifers such as ×*Hesperotropis leylandii* (Leyland cypress) and *Thuja* spp.; very occasionally, non-woody species such as strawberry may be affected. In summer periods of hot, dry weather the foliage initially wilts, turns yellow and then dies. A gummy exudate weeps from the bark of the lower part of the stem. Bark removed from close to the base of an infected plant can reveal sheets of **white mycelium** between the bark and wood. In autumn, honey-brown **toadstools** (Figure 18.1) can appear singly or in clumps around infected plants or tree stumps and also directly above infected roots. Below ground, strands of black rhizomorphs can be found growing through the soil around infected plants and under the bark of dead trees. Death of the plant will ultimately occur and can take from a few months to years depending on the species, age and health status of the plant, but even if the plant continues to withstand infection, the damage to the roots will destabilize the plant which may result in it being a structural hazard to humans.

Figure 18.18 Honey fungus 'bootlaces' (rhizomorphs) that enable the fungus to spread underground from a dead stump to infect other plants

Life cycle and spread. The infection process occurs underground as the **rhizomorphs** ('bootlaces') spread out through the soil from infected plants or stumps (Figure 18.18) for a distance of up to 30 m at up to 45 cm deep. An infected stump may remain a source of infection for many years. The rhizomorphs are the main means of spread locally and occur when they make direct contact with roots of susceptible plants. The nutrients they conduct provide the considerable energy required for the infection of woody plant roots. Mycelium then moves up the stem beneath the bark to a height of several metres and is visible when the bark is pulled away. The microscopic **spores** are produced by the **toadstools** and spread by wind, and they are capable of spreading the pathogen over longer distances.

Honey fungus often establishes itself in newly planted trees and shrubs which have been planted too **deeply**. The plants undergo stress due to the roots being too deep and the lower part of the stem being beneath soil level; consequently, they are more vulnerable to infection.

Control. **Cultural control** involves selecting and planting **resistant** species (Table 18.1) which are appropriate to the prevailing soil and site conditions. Species such as *Hydrangea* spp., *Salvia rosmarinus*, *Ginkgo biloba* and *Taxus baccata* all show resistance. Planting at the correct depth to prevent plants becoming stressed is vitally important, and for trees and shrubs the **root collar** should be at soil level. Planting in well-structured soil with an appropriate pH for the species, adequate levels of water, air and nutrients will ensure the plant is as healthy as possible and consequently less stressed. All bark- or wood-based mulching materials should be purchased from **reputable** commercial suppliers to ensure that infected mulch materials are not

Table 18.1 Levels of resistance to honey fungus in selected shrubs and trees

Common name	Scientific name	Resistance level
Alder	*Alnus* spp.	Susceptible
Honeysuckle	*Lonicera* spp.	Resistant
Birch	*Betula* spp.	Susceptible
Rosemary	*Salvia rosmarinus*	Resistant
Cedar	*Cedrus* spp.	Susceptible
Willow	*Salix* spp.	Susceptible
Leyland cypress	×*Hesperotropis leylandii*	Susceptible
Maidenhair tree	*Ginkgo biloba*	Resistant
Holly	*Ilex aquifolium*	Susceptible
Privet	*Ligustrum* spp.	Susceptible
Hydrangea	*Hydrangea* spp.	Resistant
Witch hazel	*Hamamelis* spp.	Susceptible
Camellia	*Camellia* spp.	Resistant
Rowan	*Sorbus aucuparia*	Susceptible
Apple	*Malus* spp.	Susceptible
Bay laurel	*Laurus nobilis*	Resistant
Rhododendron	*Rhododendron* spp.	Susceptible
Strawberry tree	*Arbutus unedo*	Resistant
Lilac	*Syringa vulgaris*	Susceptible
Tamarisk	*Tamarix* spp.	Resistant
Yew	*Taxus baccata*	Resistant

being imported into the garden. Research has indicated there is a small but possible risk of the fungus being introduced in this way. **Physical control** includes prompt removal of infected plants including the excavation of as much of the root system as possible. All infected material should be removed from site and **burnt** or disposed of at a **landfill** site. Underground physical barriers can be installed to prevent the rhizomorphs spreading into unaffected areas. A 45 cm minimum deep vertical strip of a flexible, impermeable fabric sheet such as butyl rubber or thick polythene can be buried in the soil around a designated area to prevent the rhizomorphs entering; it should extend to 3 cm above the soil surface. There are **no chemical controls** available to treat honey fungus.

Bacterial diseases

These microscopic organisms (Figure 18.2) measure approximately 0.001 mm and occur as single cells that divide rapidly in favourable environmental conditions. Many bacteria are **beneficial**, and some species are important in the breakdown of organic matter (see p. 185) leading to the production of humus and release of essential plant nutrients. Some bacteria such as *Rhizobium* species live **symbiotically** in the roots of some legumes and fix gaseous nitrogen in the air converting it to ammonia. Some parasitic species are **pathogenic** and cause disease which leads to serious damage or loss of cultivated plants.

Fireblight (*Erwinia amylovora*)

Damage. This disease was first recorded on pear trees in Kent in 1957, and it subsequently spread throughout Britain and Ireland. It can cause serious damage to species of the **pome-fruit** group of the **Rosaceae** family including the genera *Malus*, *Pyrus*, *Sorbus*, *Amelanchier*, *Crataegus*, *Cotoneaster*, *Pyracantha* and *Photinia*. Obvious symptoms are usually present between **May and September**; they include wilting and death of **blossom** followed by rapid dieback of shoots with dead leaves hanging on, the leaves rapidly turn a chestnut brown colour with shoot tips **curling over**. Infected plants can produce a white **gummy bacterial slime** which oozes from branches in warm, humid weather. If the outer bark of an infected stem is sliced off, a **reddish-brown** stain will often be seen in the wood immediately beneath the bark. **Cankers** may be present on infected branches and appear as areas of dead, sunken bark. When the disease reaches the main trunk, it can spread to other branches and may cause death of the tree within months of the initial infection. The general appearance resembles that of a **burnt plant** (Figure 18.19).

Figure 18.19 Fireblight on cockspur hawthorn (*Crataegus crus-galli*)

Life cycle and spread. Flowers are a main point of entry. The bacteria are spread by bees when they visit the flowers but infection can also occur through natural openings such as lenticels on woody stems and through wounds caused by pruning cuts. The bacteria spread rapidly in **summer** and destroy the phloem and cambium tissue. The **slime** which oozes from infected branches contains bacteria, and it can be dispersed to nearby plants by rain splash and wind or by plants touching. Humid conditions and temperatures in excess of 18°C, which occur from May to September, favour the spread. The bacteria **overwinter** in the cankers on the stems. Fireblight was once notifiable nationally but must now be reported only in 'fireblight pest-free areas' such as the Isle of Man. Prior to fireblight host plants being issued with valid plant passports for entry into a fireblight pest-free area, they must have been produced or maintained for at least two complete growing seasons in an area which has been inspected and verified as being free from fireblight.

Control. **Cultural control** involves selection and planting of less-susceptible or resistant species, especially in fruit growing areas. Avoidance of planting susceptible hawthorn (*Crataegus monogyna*) hedges in a commercial fruit growing area is essential. The '**Saphyr**' range of *Pyracantha* cultivars have been selected for their resistance to fireblight. **Physical control** involves **scouting** for early infection and should take place regularly during late spring and summer followed by prompt removal of any infected tissue. Cutting back 60 cm beyond any signs of infection on larger branches and 30 cm on smaller branches is necessary. All infected tissue should be removed and burnt. Pruning tools should be sterilized with a horticultural disinfectant such as peroxyacetic acid between individual plants and on completion of the work. The whole plant should be removed and destroyed if symptoms of infection appear in the main stem. There are no chemical controls for fireblight.

Bacterial canker of *Prunus* (*Pseudomonas syringae* pv. *morsprunorum* and *P. syringae* pv. *syringae*)

Damage. Two pathogenic bacteria cause this disease of the genus *Prunus*, which includes ornamental and fruiting species such as plum, cherry, peach and apricot. Symptoms include **cankers** which typically appear on the stem as **sunken areas** of bark which exude a light brown gum (Figure 18.20). The angle between branches is the most common site for the disease. In severe infections, **girdling** of the stems cause death of tissues above the infection and the resulting brown foliage. In May and June, leaves may become infected; dark brown leaf spots 2 mm across develop and the infected area may be blown out by heavy winds to give a '**shot-hole**' effect.

Figure 18.20 Bacterial canker, a canker on the trunk of a cherry tree

Life cycle and spread. Bacteria are present in the cankers and mainly carried by wind-blown **rain droplets**, infecting leaf scars and pruning wounds in autumn and young developing leaves in summer.

Control. **Cultural control** involves the selection of resistant cultivars – the edible cherry '**Merton Glory**' and plum '**Queen's Crown**' show some resistance. If possible, prune susceptible trees in **July and August** when there is less risk of infection occurring. **Physical control** involves cutting back to healthy wood any tissue which shows infection as soon as it is identified; this should be followed by the application of a latex-based pruning sealant to the wound. All infected tissue should be removed and **burnt**. Pruning tools should be sterilized with a horticultural **disinfectant** such as peroxyacetic acid, between plants and on completion of the work. There are no chemical controls for bacterial canker.

Soft rot (*Pectobacterium carotovorum*)

This bacterium affects stored potatoes, carrots, bulbs and corms, where the bacterium's ability to dissolve the cell walls of the plant results in a mushy soft rot. **High temperatures** and **humidity** caused by poor ventilation promote infection through lenticels. A related strain of this bacterium causes blackened stems (black leg) on potato plants. Spread is mainly caused by infected planting material. During storage of susceptible vegetative material, the bacteria may be spread by insects or by liquid oozing from infected plants. Preventive control measures are important. Plant material should be damaged as little as possible when harvesting, and diseased or damaged specimens should be removed before storage. Hot, humid conditions should be avoided in store. No curative measures are available.

Xylella fastidiosa

This bacterial pathogen infects **xylem vessels** of the plant's vascular system which becomes blocked leading to a range of **symptoms** such as wilts, diebacks, stunts and leaf scorch. The pathogen was first detected in Europe in Italy in 2013 and is spread by insects, such as the meadow **froghopper** (*Philaenus spumarius*), a common species in Europe. This disease has not yet been recorded in Britain and Ireland but is considered to be one of the most serious worldwide threats to plant health – hundreds of host plants have been found to be infected with the various strains of *Xylella fastidiosa*, and **woody perennial** plants such as *Prunus* spp., *Salvia rosmarinus*, *Lavandula* spp., *Nerium oleander and Olea europaea,* many other broadleaved trees and shrubs are susceptible to infection. Herbaceous plants may become infected as symptomless disease reservoirs. It is difficult to predict which plant hosts could be vulnerable to infection should it occur in Britain and Ireland where it is classed as a **quarantine** (notifiable) disease. Consequently, any suspected outbreaks must be reported immediately to the relevant UK and Irish plant health authorities.

Structure and biology of viruses

Viruses are extremely small; much smaller even than bacteria (Figure 18.2). The light microscope is unable to focus in on them, but they can be seen as rods or spheres when seen under an electron microscope. The virus particle is composed of a DNA or RNA core surrounded by a protective protein coat. On entering a plant cell, the virus takes over the organization of the **cell nucleus** to produce many more virus particles which are then spread around the host plant. Since the virus itself lacks any cytoplasm cell contents, it is often considered to be a non-living unit.

The virus's close association with the plant cell nucleus presents difficulties in the production of a curative virus control chemical that does not also kill the plant. No established commercial 'viricide' has yet been produced to treat plant viruses, consequently there are no **chemical** controls available. Crops which are propagated vegetatively (such as potatoes) may continue indefinitely with the virus inside them if there is no strategy to remove the virus.

Virus particles have been isolated as the cause of many diseases such as cucumber mosaic and potato leaf roll viruses. Other agents of disease which cause virus-like infections include **viroids** such as the pathogen which causes **chrysanthemum stunt disease.** They are the smallest known pathogen of plants and consist of a single-stranded RNA molecule but, unlike viruses, they lack a protein shell. **Phytoplasmas** also cause virus-like diseases such as aster yellows; they are not related to viruses, but are a group of small bacteria that induces symptoms similar to those produced by viruses.

Spread. A number of living organisms (**vectors**) spread viruses from plant to plant, sometimes over long distances. Peach-potato aphid is known to transmit many types of viruses, including potato leaf roll virus, to different plant species. The aphid stylet (Figure 17.8) injects salivary juices containing **virus particles** into the parenchyma and phloem tissues, enabling the virus to travel to other parts of the plant. '**Persistent virus transmission**' occurs when a vector acquires a virus and continues to remains **infective** for life. It is seen in some vector/virus combinations such as peach-potato aphid with potato leaf roll virus and *Xiphinema* dagger nematode which transmits arabis mosaic. However, in some vector/virus combinations such as plum pox, the virus survives only briefly as a contaminant on the insect's stylet, with the result that the vector does not spread the virus very far amongst a population of plants. Other examples of vector/virus combinations include bean weevils/ broad bean stain virus and *Olpidium* soil fungus/big vein virus of lettuce. Other significant methods of spread include **vegetative propagation** material (e.g. chrysanthemum stunt viroid and plum pox), **infected seed** (e.g. bean common mosaic virus and tomato mosaic virus) and **mechanical transmission** by hands of growers and on tools such as knives (e.g. tomato mosaic virus).

Symptoms. The presence of a damaging virus in a plant is recognizable to horticulturists only by

Figure 18.21 Cucumber mosaic virus symptoms on a cucumber leaf

means of its **symptoms** as the virus does not produce any physical manifestations which are visible to humans. Commercial growers may need to consult a plant **pathologist**, whose identification techniques will take place in a laboratory to accurately diagnose the specific virus.

Many viruses cause general **lack of vigour** in the infected plant accompanied by a reduction in crop quality and quantity, for example reversion virus in blackcurrant. **Leaf mosaic**, a yellow mottling, is a common symptom (cucumber mosaic virus; Figure 18.21). Other symptoms include leaf distortion into feathery shapes (cucumber mosaic virus), flower colour streaks (tulip break virus), fruit blemishing (tomato mosaic and plum pox), internal discoloration of tubers (tobacco rattle virus, causing 'spraing' in potatoes) and stunting of plants (chrysanthemum stunt viroid). Symptoms similar to those previously described may be caused by contamination due to inaccurate herbicide application, genetic 'sports' and environmental conditions such as poor soil fertility and unsuitable pH for the plant species, causing nutrient deficiencies which can look similar to some viral infections and mite damage (on cucumber leaves).

In the following descriptions of seven horticulturally significant viruses, scientific names of genus and species are not included as the binomial system is not used to name viruses.

Cucumber mosaic virus (CMV)

Damage. In addition to cucumber, the following may also be affected: spinach, celery, tomato, *Pelargonium* spp. and *Petunia* spp. On cucumbers, a mottling of young leaves occurs (see Figure 18.21) followed by a twisting and curling of the whole foliage, and fruit may show yellow sunken areas. On the shrub *Daphne odora*, a yellowing and slight mottle is commonly seen on infected foliage, while *Euonymus* leaves produce bright yellow leaf spots. Infected tomato leaves are reduced in size (fern-leaf symptom).

Life cycle. The virus may be spread by **contaminated hands** of growers, but more commonly an **aphid** (e.g. peach-potato aphid) is involved. Many crops such as lettuce, maize, *Pelargonium* and privet, and weeds such as fat hen (*Chenopodium album*), may act as a **reservoir** for CMV.

Control. **Cultural control** – examples of cultivars which show some **resistance** to CMV which are available to the grower are courgette 'Defender'F1, 'aubergine 'Bonica'F1 and cucumber 'Crispy Salad'F1. Selection of clean, **healthy stock** is vital in vegetatively propagated plants such as *Pelargonium*. Control of **aphid vectors** is important where **susceptible crops** such as lettuce and cucumbers are grown in succession, and when crops are next to other susceptible species. Removal of **susceptible weeds** such as chickweed, particularly from greenhouses, may help prevent widespread infection. Potentially infected tools should sterilized with a **horticultural disinfectant** such as peroxyacetic acid after use. **Hands of growers** should be washed thoroughly before and after working on susceptible plants. **Physical control** involves the immediate **removal and burning** of infected plants, as they will not recover and can act as source of further infection.

Tomato mosaic virus (ToMV)

Damage. This disease may cause serious losses in the **Solanaceae** family of plants which includes tomatoes and peppers. Infected seedlings have a stunted, spiky appearance. On more mature plants leaves have a pale green mottled or sometimes a bright yellow appearance. The stem may show brown streaks in summer when growing conditions are poor, a condition often resulting in death of the plant. Overall growth is stunted and fruit yield and quality may be reduced, the green fruit appearing yellow/bronze and the ripe fruit hard, making the crop inedible.

Life cycle and spread. ToMV can be spread within **infected** seed and also transmitted **mechanically** and spread in the sap from an infected plant on tools such as knives, the hands of growers and plant-to-plant contact. **Wounds** such as broken leaf hairs or the broken bases of de-leafed side shoots are common methods of entry. The period from plant infection to symptom expression is about 15 days. Particles of the virus can remain viable in infected crop debris while the debris remains alive.

Control. **Cultural control** includes purchasing **fresh seed** from a reputable seed supplier and not saving seed from the previous crop as it may be infected. Tomato cultivars which are **grafted** on

Figure 18.22 Tomato mosaic virus symptoms on tomato leaves (source: Shutterstock, Plant Pathology)

Figure 18.23 Leaf roll virus on potatoes; the leaves are thickened and roll upwards

to the rootstock 'Estamino' F1 can benefit from its high resistance to ToMV. Using a **growing bag** or **nutrient-film** method of cultivation can be effective in helping to avoid infection from contaminated roots of the previous crop. For soil-grown crops, **crop rotation** can help to break the infection cycle. Potentially infected tools such as knives should be **sterilized** with a horticultural disinfectant such as peroxyacetic acid between use on different plants and after use. Hands of growers should be **cleaned** with a sanitizer before, between and after working on susceptible plants. **Physical control** involves the immediate **removal** and **burning** of infected plants as they will not recover and can act as source of further infection.

Potato leaf roll virus (PLRV)

PLRV is a significant global problem in potato production. The occurrence of this important viral disease in a crop that is vegetatively propagated presents problems for control.

Damage. The leaves of the plant show an upward **rolling** (Figure 18.23). These leaves often are **light green** in colour and may turn red on the upper side and purple on the lower side. The leaves may also cause a rattling sound when shaken against each other. Leaf roll virus does not show any visible symptoms in the **tubers**, but can cause a serious **reduction** in tuber yield, especially if the grower reuses tubers from the previous year's crop and when aphid numbers are high.

Life cycle and spread. The main **vector** of leaf roll virus is the peach-potato aphid. The aphid, while feeding on the sugars present in phloem tissues of an infected potato plant, will take up the virus into its body and the insect remains infective for the **whole** of its life. The aphid may move from plant to plant spreading the virus and can be blown considerable distances by wind currents. A high incidence of peach-potato aphid usually results in increased levels of leaf roll virus.

Control. **Cultural control** includes selection of cultivars which show good **resistance** to PLRV such as 'Celine', 'Pentland Crown' and 'Rooster'. **Avoid** growing potato crops from the previous year's tubers as they may be infected with PLRV. Purchase 'seed' potatoes which are certified under the Seed Potato Classification Scheme to ensure the tubers are healthy. **Scout** the crop regularly to check for aphids and take action to control aphids when identified. **Physical control** – remove from site and burn any plants showing leaf roll symptoms.

Blackcurrant reversion

This virus can seriously reduce blackcurrant plant health and crop yields.

Damage. Flower buds on infected blackcurrant bushes are almost **hairless** and appear brighter in **colour** than healthy buds. Infected leaves often have fewer main veins than healthy ones (Figure 18.24). After infection, the bush may grow vigorously but fruit production decreases significantly and the bush has to be removed. The virus is spread by the **blackcurrant gall mite** and reversion-infected plants are particularly susceptible to attack by this pest. *Control*. **Cultural control** – most cultivars are **susceptible** but 'Ben Gairn' does show **resistance** to the virus. Purchase planting material which is certified by the Fruit Propagation Certification Scheme. **Physical control** includes digging up and burning or removal from site of bushes infected with the virus. Any stems which contain **swollen buds** associated with big bud mites should be cut out and burnt or removed from site when carrying out routine **winter pruning**. It should be remembered that the presence of big bud mite does not necessarily indicate the plant is infected with the virus.

Figure 18.24 Reversion disease on blackcurrant; note that the infected lower leaf has fewer main veins and leaf lobes than the healthy upper leaf

Tulip breaking virus (TBV)

When infected with TBV, the tepals of infected tulips produce **irregular coloured streaks** and may appear distorted. Leaves may become light green and the plants become stunted.

Aphid vectors are significant in the spread of this virus, and it can also be spread by **infected tools** such as knives. Removal and burning of infected plants in the garden prevents a source of virus for aphid transmission.

Plum pox

This viral disease, also known as '**Sharka**', has increased in significance in Britain and Ireland since 1970, after its introduction from mainland Europe. Plums, damsons, greengages, apricots and peaches can be affected. Cherries were previously thought not to be susceptible; however, a strain of plum pox virus which affects cherries has been confirmed on cherries in mainland Europe. **Leaf symptoms** are interveinal yellow blotches, spots and rings. These symptoms are most obvious in late spring and early summer. The most reliable symptoms, however, are found on **fruit**, where uneven ripening and sunken dark blotches are seen with necrotic tissue developing in the flesh through to the stone. The fruit has a low sugar content and is inedible. It is spread on infected rootstocks or scion material, which can lead to very fast movement of the disease over long distances in plant transport. The virus is also spread by **aphids** from infected trees or susceptible wild hosts such as blackthorn (*Prunus spinosa*). Spread by this method is usually more localized. *Control.* Plum pox is a **quarantine (**notifiable) disease, and suspected infections must be reported to the relevant plant health authorities. In commercial orchards removal of wild hosts such as blackthorn and programmes to control aphid vectors will help to reduce the risk of the disease occurring. **Cultural control** – planting stock certified by the Fruit Propagation Certification Scheme should be purchased. **Physical control** involves removing and burning infected trees.

Arabis mosaic virus (ArMV)

ArMV infects a wide range of plants including **edible and ornamental species**. On **strawberries**, yellow spots or mottling is produced on the leaves and the plants become severely stunted. On ornamental plants such as *Daphne odora*, yellow rings and lines are seen on infected leaves and the plants may slowly die back, particularly when this virus is associated with cucumber mosaic inside the plant. Some **weeds** such as chickweed can harbour this disease. The virus is spread by a common soil-inhabiting **nematode**, *Xiphinema diversicaudatum*, which may retain the virus in its body for several months. *Control.* **Cultural control** includes purchasing fruit planting, material which is certified by the Fruit Propagation Certification Scheme. **Rotation** of fruit crops should take place; for example, do not plant successive strawberry crops on the same area. Control of susceptible weeds should routinely take place. **Physical control** involves the immediate removal and burning of all infected plants.

Plant disorders

There are many symptoms of poor health which can appear on plants that are not caused by pests or diseases. A **plant disorder** is a damaging condition resulting from an **abiotic** (non-living) factor such as environmental conditions. The causes of disorders include excessively low or high temperatures, shade, drought, waterlogging, high humidity, unsuitable soil pH, nutrient deficiencies, excess fertilizer and fasciation.

Frost damage

Plant species differ greatly in their **tolerance** (hardiness) to frost (see p. 28). It often causes the

Figure 18.25 Frost-damaged *Magnolia* flowers (source: Shutterstock, futo david)

above ground parts of sensitive plants to become brown or blackened (Figure 18.25) and then collapse into a mass of dead tissue (after ice has formed inside the plant and fractured all the cells). For example, cultivars of *Begonia semperflorens* Cultorum Group left to grow outside in autumn will be severely damaged when a frost occurs. At the beginning of the growing season, early **potatoes** may be damaged by severe frosts in late spring. Some hardy plants such as apple, *Malus* spp., tolerate low temperatures very well, but their flowers are tender and frost can damage or destroy the flowers and young fruitlets. The risk of frost damage can be reduced by the selection of suitable species and cultivars with the flowering period being suitable for the location's frost susceptibility. Tender species in containers can be moved in to frost-free protected structures such as cold frames, polytunnels and greenhouses until the risk of frost has passed. Outdoor plants can be protected over winter by horticultural fleece or straw, which helps to keep the temperature higher around the plant.

Low temperature

Lower temperatures reduce rates of photosynthesis and respiration which in turn results in slower **growth and development** including seed germination which may be delayed. Pollinating insects tend to be less active in low temperatures, and pollination of fruit crops can be adversely affected. Plant species vary in their temperature requirements; **hardy** species such as *Cyclamen coum* grows well in cold, winter conditions and snapdragon (*Antirrhinum majus*) grows well when night temperatures are no higher than 12°C, while most **tender or half-hardy** bedding plants prefer a higher night temperature. Avoid planting out tender and half-hardy species until the danger of low temperatures has passed and ensure they are adequately **acclimatized** ('hardened off') prior to planting outside. On a small scale, **overnight protection** can be provided by covering individual plants with hessian, horticultural fleece or straw which can be effective in reducing the effects of low temperatures. In vegetable gardens, cloches can be used to protect early sowings of seeds; higher temperature under the cloche will help to facilitate germination of the seeds and healthy early growth of the seedlings.

High temperature

The effects of high temperature will differ depending on species. As the plant approaches its upper level of tolerance, 36°C for temperate species, **photosynthesis** slows down due to enzymes breaking down and becoming ineffective. Plants, such as sweetcorn (*Zea mays*) which originates from the tropics, have different photosynthesis systems to help cope with this problem (see p. 115). The direct effects of high temperature are varied, but it causes leaf margins to dry off before the whole leaf dies (Figure 18.26 a,b). Flowers and fruits can also be scorched. This problem can be managed in greenhouses and other protected structures by deactivating heating systems and providing shading and adequate ventilation to reduce temperatures. Seed germination in some species can be affected by high temperatures (thermodormancy). Lettuce seed germination is inhibited in temperatures over approximately 25°C.

Shade

Houseplant species are sometimes placed in areas of a **building** unsuitable for their ideal growth. For example, poinsettia (*Euphorbia pulcherrima*) needs high, indirect light levels to thrive and will grow poorly in a dark room. Plants **outdoors** may suffer from the same problems. *Pelargonium zonale* cultivars used as summer bedding plants should be given full sunlight and will develop a pale yellowing of foliage if grown in a shady location. However, *Impatiens* spp. can withstand shade and maintain their rich, dark green foliage and flower production. Species which are not adapted to grow in shade can become **etiolated** and produce weak, spindly growth. Variegated cultivars commonly suffer from 'reversion' in shade with their leaves reverting to wholly green (see p. 311). These problems can be avoided by thoughtfully selecting plant species which can tolerate the prevailing low light levels of shady locations.

Drought

A plant needs sufficient water to maintain **turgidity** in plant tissues (see p. 115), carry nutrients around the plant, be present as a requirement for photosynthesis and transpire from the leaf in order

Figure 18.26 (a) A healthy *Salvia splendens*; (b) high temperature effect on *Salvia splendens*: the leaves are wilting and going dark brown

to keep a desirable leaf temperature and draw water up the plant in xylem tissue by 'transpiration pull' (see p. 127). In some plant species such as lawn grasses, leaves change from shiny to dull as a first signal of water stress, and also may change from bright green to a grey green. New leaves wilt, but in some **woody** species such as holly and conifers only the very youngest leaves wilt. Premature flowering (**bolting**) may occur in some **annual** species such as lettuce or **biennial** species such as cabbage. Flowers may fade quickly and fall prematurely. Older leaves often turn brown, dry and fall off. Seed germination is also inhibited by lack of water as it is an essential requirement for germination. Digging a few centimetres into the soil may indicate the need for watering with shallow-rooted perennials and annual border plants. Established woody species with deep roots rarely need watering, although transplanted older shrubs may show summer water stress for a number of years. The problem can be managed by providing a **well-structured soil** or growing medium which is able to retain moisture adequately. Water inputs should be managed according to **plant species**, but aim to ensure that available water (see p. 174) is present in the soil or growing medium.

Waterlogging

Waterlogging can be caused by a high natural water table, poor soil structure, excessive rainfall or overwatering. When the soil is **saturated** and all the pore spaces in the soil or growing medium are full of water, oxygen is excluded, creating **anaerobic conditions**. In these conditions, inefficient anaerobic respiration (see p. 119) takes place in root tissues leading to symptoms such as slow and stunted growth, the whole plant wilting in herbaceous species and the lower leaves turning yellow and dropping. In prolonged periods of waterlogging, oedema may occur and the roots of susceptible species may die; some pathogens which cause diseases such as *Phytophthora* root rot, damping off in seedlings and clubroot of brassicas may become more prevalent. Soil structure can be destroyed and there will also be detrimental effects on **soil organisms** such as earthworms, insects and other arthropods, beneficial bacteria and fungi; this will have a long-term impact on soil health affecting organic matter breakdown, soil structure and nutrient recycling including denitrification in the soil (see nitrogen cycle p. 187). If the soil has structural problems such as compaction or a perched water table, then action such as appropriate cultivation practices will be necessary to remedy the underlying problems. If there is a naturally high water table, the installation of a drainage system may be necessary. Overwatering can be rectified by using irrigation practices which are appropriate for the plant species, soil type and environmental conditions. Plant selection should always be appropriate for the prevailing site and soil conditions (see Table 15.1), and in these conditions species which are tolerant

Figure 18.27 Oedema symptoms on *Pelargonium peltatum*; note the raised, corky spots on the lower leaf surface

of a soil with a high soil moisture content for extended periods of time should be selected.

Oedema

Oedema is seen as raised **corky** spots on the lower leaf surfaces in species such as *Pelargonium peltatum* (Figure 18.27). *Peperomia* spp. and orchids in greenhouses and *Camellia* spp. and *Eucalyptus* spp. outdoors are susceptible to this disorder.

Oedema is most commonly seen in **greenhouses**, where overwatering and poor ventilation leads to high humidity levels and can result in this disorder. It occurs when the roots' ability to take in and supply water to the leaves exceeds the leaves' ability to lose water by transpiration. Excessively **high** water levels are retained in the leaf cells which results in them rupturing, causing damage that results in disfiguring corky spots developing on the lower leaf surface. Conditions which favour oedema occur most commonly during extended periods of cool, cloudy weather when a combination of ample water being **available** to the roots and the cool, humid air brings on the condition most severely; the symptoms are commonly seen in unheated greenhouses. The problem can be greatly reduced in greenhouses by growing susceptible plants in a **free-draining** growing medium, avoiding overwatering and providing adequate **ventilation** to prevent excessively high humidity occurring. Affected leaves should not be removed, as this will further reduce the plants' ability to expel water and results in further problems developing in the remaining leaves.

Figure 18.28 Flower balling with papery, dried petals on the outside of the rose flower

Flower balling

This is a condition which is commonly seen on **roses** but can also affect paeonies and camellias. It can occur when the outer petals of **developing flowers** become wet from rain or high humidity but then dry out rapidly when exposed to sunlight and higher temperatures. The outer petals become dry and fuse together, preventing the inner petals from emerging to produce an open bloom. The affected flower buds remain globular in shape (Figure 18.28) and are often subsequently infected with **grey mould** (see p. 296), which can lead to stem dieback. Rose cultivars which have many slender petals ('**double flowers**') are particularly prone to this problem. The withered blooms eventually drop off the plant.

Affected flowers should be **removed** by cutting back to just above a bud on the stem. Growing roses in locations which have good **airflow** can reduce moisture on flowers, and pruning to develop open plants can also help reduce the problem. Care should be taken not to deposit water on flowers by **avoiding** overhead irrigation; water should be directed to the soil. Selection of 'single flowered' cultivars will help avoid the incidence of flower balling.

Soil pH and lime-induced chlorosis

Cultivated plants have developed from species in many different habitats around the world. Most plant species and soil organisms do well in a pH between 6 and 7, but some species do have a **tolerance** of more acid or alkaline conditions. Outside their limits of tolerance, plants may be subjected to problems such as **lime-induced chlorosis**. This is a common condition that can occur when calcifuge plants such *Pieris* spp., *Rhododendron* spp. and *Camellia* spp. are grown in soils with a pH above 7. The high level of alkalinity affects the plant's ability to absorb iron, and a deficiency of iron results in young leaves showing an unhealthy yellow appearance, especially between the veins (interveinal chlorosis) (Figure 18.29), and stunting of the plant. Soil pH levels below 6 can detrimentally affect the activity of **soil organisms** such as earthworms, bacteria and fungi which influence decomposition of organic matter, soil structure and nutrient recycling. In some cases, the activity of microorganisms which are plant **pathogens** and cause disease such as brassica clubroot are affected. The principal and most sustainable approach to this problem should be to select plant species according to their ability to tolerate the existing soil pH and other prevailing conditions in any given location. Alternatively, where appropriate, consideration may be given to growing plants in containers or raised planting areas where a growing medium with an appropriate pH can be provided.

Nutrient deficiencies

Each essential **major** and **minor** nutrient is required in the correct amounts to enable the plant to carry out healthy growth and development. When nutrients are present in too low amounts, **deficiencies** develop. Care should be taken to ensure soil fertility is maintained at appropriate levels by appropriately managing soil structure, water and air levels, organic matter content, pH and fertilizer applications, where necessary.

Symptoms of common nutrient deficiencies include **blossom end rot** (Figure 18.30) which occurs in some members of the **Solanaceae** family. It is most common in tomato but can also affect aubergine and pepper. It produces a symptom of a black, concave lesion on the end of fruit that looks at first sight like a fungal disease. It is caused by a lack of **calcium** in the developing fruit. Most soils and container growing media have sufficient calcium for healthy plant growth; **deficiency** in the plant is most likely to occur when the soil or growing medium is allowed to dry out while the fruits are developing. The resulting **irregular** water uptake and transport within the plant limits the amount of calcium available to the fruit and the disorder develops. It is seen most often in **container-grown** plants rather than in plants growing in the open garden or greenhouse borders and is most common when plants are grown in **growing bags**, where they have a limited, shallow root run that dries out easily. Although there is no cure for blossom end rot once the symptoms begin to appear, the recommendation is that **susceptible** crops should never be allowed to have **dry roots** and the soil water content should be maintained as close to **field capacity** (see p. 174) as possible.

Bitter pit in **apples** (Figure 18.31) occurs when the fruits develop many small, dark brown, **sunken** pits. The tissues below are stained to a depth of about 2 mm. Cultivars such as 'Bramley's Seedling' and 'Egremont Russet' are **susceptible**. Young, heavy-bearing trees show the worst effects. The disorder is caused by insufficient **calcium** levels in the fruit and, like blossom end rot, is influenced by **irregular** water supply in the tree, most commonly in dry summers. Actions to help prevent this problem include ensuring a regular **water** supply to the tree during dry spells; **mulching** around trees to help moisture retention; **summer pruning** young, vigorous trees to help available calcium to be transported to the **developing fruits**; and applying sprays of a water-soluble **calcium oxide**

Figure 18.29 Lime-induced chlorosis on raspberry; note the pale areas between the main veins

Figure 18.30 Blossom end rot on tomato fruits (source: Shutterstock, PKDAENG)

fertilizer directly on to **foliage** and developing **fruits** every **two weeks** from petal fall to **one week** before apple harvesting at application rates specified by the manufacturer.

Figure 18.31 Bitter pit in apple fruit with numerous small brown spots appearing on the fruit surface in late summer

Excess fertilizer

When soluble fertilizers are present in **excess** levels, roots are damaged and therefore unable to absorb **nutrients and water** efficiently and consequently cannot provide nutrients for the other parts of the plant. This condition is sometimes described as **high soil conductivity**. Careful consideration of the appropriate **type**, **amount** and **frequency** of fertilizer will help prevent this situation.

Fasciation

Fasciation is a disorder in which the **stems** and **flowers** of susceptible plants grow a**bnormally** because of an unknown disturbance. Fasciated stems are often **wider and flatter** than normal stems (Figure 18.32a) and **appear** as though a number of stems are **fused** together. Flowers can be flattened and elongated; sometimes, fasciation results in abnormal growth of a whole inflorescence (Figure 18.32b) as in members of Asteraceae family. It is commonly seen on **woody plants** such as *Forsythia* spp. and *Fraxinus* spp. and **herbaceous plants** such as *Veronicastrum virginicum* and *Digitalis* spp. This disorder is not well understood but some species show a natural tendency to develop it and it may be caused by random genetic disturbances, undiagnosed viral or bacterial infections or physical damage caused by pests, environmental conditions or mechanical disturbance. Horticulturists have used the genetic tendency towards fasciation shown in plants such as the fasciated Japanese fantail willow *Salix udensis* 'Sekka' and have perpetuated it as a desirable selection by means of vegetative propagation. Cockscomb (*Celosia argentea* var. *cristata*), however, is a garden annual from tropical Africa origin that carries the genetic tendency to fasciation from generation to generation through its **seed**. Fasciated parts of plants, if deemed to be undesirable, should be removed by cutting back to healthy tissue.

Reversion

Many plant species have been selected and cultivated by horticulturists for their interesting leaf

Figure 18.32 (a) Fasciation is affecting the lower specimen which has a flattened stem. The upper specimen is normal; (b) a fasciated inflorescence

Table 18.2 Some symptoms of diseases and physiological disorders

Symptom	Cause	Other cause
Leaf spot	Fungus, e.g. apple scab	Bacterial canker (*Prunus*)
Raised leaf spots	Rusts	Oedema (corky spots)
White covering on leaf	Powdery mildew – upper	Spraying hard water
	Downy mildew – lower	
Chlorosis	Nitrogen deficiency (in older leaves)	Root disease/waterlogged soil
Brown edge to the leaves	Potassium deficiency	
Leaves curl and go brown	Lack of water	
Dry, crumbling leaves	Plants overheated	Excess fertilizer
Interveinal chlorosis	Magnesium deficiency (symptoms in older leaves) or iron deficiency (symptoms in younger leaves)	
Dark-coloured leaves	Phosphorus deficiency	
Lower leaves yellow	Wilt fungus/nitrogen deficiency	Overwatering
Yellow/green leaf mottle	Virus mosaic	Mutation/chimaera, variegated cultivar
Fruit spots	Fungus	Bitter pit (apple)
Sunken fruit lesions	Blossom end rot (tomatoes)	
Bud or leaf drop	Sudden change of temperature	Irregular watering
Stems etiolated	Lack of light	
Whole plant wilts	Severe underwatering	Wilt disease, vine weevil larva damage
Fluffy mould	*Botrytis* (grey mould)	
Brown stem lesions	Tomato mosaic virus	
Sunken lesions on woody stems	Fungal/bacterial canker, fireblight	
Oozing from woody stems	Bacterial canker/fireblight	
Brown roots	Root rot	Overfertilizing

features, especially **leaf variegation** (see p. 117). Common examples of these include the variegated cultivars of common species of *Euonymus*, *Hedera*, *Cornus*, *Daphne*, *Ilex* and *Elaeagnus*. This condition commonly occurs when a variegated cultivar is grown in a location with insufficient light and the foliage reverts to a green, non-variegated form. Plants with variegated leaves should be planted in locations with adequate light intensity and duration to prevent this disorder occurring. If the disorder does occur, the reverted parts of the plant should be cut out immediately.

Symptoms of common disease and plant disorders

Table 18.2 presents a summary of common symptoms to help to determine some common plant disease problems and physiological disorders.

Further reading

Adams, C., Brook, J. and Early, M. (2015) *Principles of Horticulture Level 3*. Routledge.

Beales, P. et al. (2019) *Plant Diseases and Biosecurity*. Oxford University Press.

British Crop Production Council. (2025) *UK Pesticide Guide*. BCPC.

Brown, L.V. (2008) *Applied Principles of Horticulture*. 3rd ed. Butterworth Heinemann.

Buczacki, S. and Harris, K. (2014) *Pests, Diseases and Disorders of Garden Plants*. HarperCollins.

Costello, L.R. (2003) *Abiotic Disorders of Landscape Plants*. University of California, Agriculture and Natural Resources.

Greenwood, P. and Halstead, A. (2018) *Pests and Diseases*. Dorling Kindersley.

Hessayon, D.G. (2007) *Pest and Weed Expert*. Transworld Publishers.

Ingram, D. and Robertson, N. (2013) *Plant Diseases*. Collins.

Oliver, R. (2024) *Agrios' Plant Pathology*. Academic Press.

RHS. (2024) *Fungicides for Home Gardeners*. Online: RHS Advisory Service www.rhs.org.uk/advice/pdfs/fungicides-for-home-gardeners.pdf

www.daera-ni.gov.uk/topics/plant-and-tree-health

https://forestrycommission.blog.gov.uk/category/tree-health/

https://planthealthportal.defra.gov.uk/

www.sasa.gov.uk/plant-health

www.teagasc.ie/crops/horticulture/plant-health/

The online material is accessible via the QR code and includes further information on many of the topics in the book, such as

- Sudden oak death
- Fusarium patch of turf

Glossary

abscission shedding of leaves, flowers and fruits

active transport movement of a substance into a cell across the cell membrane against a concentration gradient. It requires energy

adhesion the attraction of water molecules to charged surfaces due to hydrogen bonding

adult growth sexually reproductive (flowering) growth

adventitious buds buds which do not arise from a stem

adventitious roots roots which do not derive from the radicle of the embryo. Often found on other plant parts such as stems

aerobic respiration process by which sugars are broken down in the presence of oxygen to yield energy, the end products being carbon dioxide and water

aggregation hormone a chemical (included in sticky traps) that induces particular flying pests to fly towards the trap

air-filled porosity (AFP) percentage of air in a growing medium immediately after it has stopped draining having been saturated with water

allelopathy where one organism influences the growth of another through the release of chemicals

alternate host plants that some pests or pathogens alternate between as different host plants in different seasons

alternative host plants that some pest or pathogens sometimes use as a host in addition to their preferred host

respiration. anaerobic respiration in the absence of oxygen, with the release of energy, carbon dioxide and ethanol

androecium the male part of the flower consisting of the stamens

angiosperms seed bearing flowering plants

annual a plant that completes its life cycle within a growing season

apical meristem area of cell division at the root and shoot tips responsible for increase in length

apoplast all the cell walls and the spaces in between the cells linked together

apomixis production of viable seed without fertilisation

asexual reproduction the formation of new individuals without fusion of gametes

authority citation (naming authority) the name of the person who first published the plant name

autotroph see producer

available water content (AWC) the water held in the soil between field capacity and the permanent wilting point

base dressing fertilizer that is applied to the soil and worked into the seedbed, or incorporated in composts, before sowing/planting

binomial a two-part name made up of a generic epithet and a specific epithet

biodiversity encompasses the genetic variation within a species, the total number of species present and the variety of habitats they live in

biological control the use of natural predators, parasites and pathogens to control pests and diseases

biomass the weight or volume of living plant and animal material in an ecosystem

biome large geographical area or global community of distinctive plants and animals, containing many ecosystems whose communities have adapted to a particular climate

bitter pit small spots on apple fruits resulting from insufficient calcium
blossom end rot black lesions produced on tomato fruit, caused by insufficient calcium
boundary layer the layer of still air next to the leaf surface
bryophytes non-vascular non-seed- bearing plants
buffering capacity the ability of the soil to retain nutrients including lime against loss by leaching
bulb an underground modified shoot which mainly consists of food storage leaves (scale leaves)

calcicoles plants that are adapted to grow on calcareous soils
calcifuges plants that are adapted to grow on acid soils below pH 5.5
calyx collective name for all the sepals
canker a swollen stem caused by a disease (or pest)
capillarity the movement of water against gravity in narrow tubes such as xylem vessels
capillary action/capillary rise capillary rise of water is its movement against gravity within thin tubes as seen with water moving up into paper towels
carbon diioxide fixation the conversion of carbon dioxide from the atmosphere in carbohydrates such as glucose
carnivore an organism that feeds on animal tissue
carpel the female reproductive unit made up of a stigma, style and ovary. Individual carpels are often fused together
centre of origin a geographical area where plant diversity is high
chemical control the use of a pesticide to prevent or control a weed, pest or disease
chimaera see graft hybrid
clay soil particle less than 0.002 mm in diameter
clay soil soil with more than 35% clay particles
climacteric rise burst of respiration in some fruits on ripening
climax community the final community in the successional sequence
cohesion the attraction of water molecules to each other due to hydrogen bonding
community a group of populations in a given area or habitat
companion planting an association of plant species that derives benefit of some kind from each other
compatible pollen pollen which germinates and grows enabling delivery of male gametes
complex tissue tissue made up of more than one type of cell
compost a growing medium used for growing plants in containers OR a product of composting
composting the decomposition of organic matter in a pile before it is applied to soils
compound fertilizers those that supply two or more of the nutrients nitrogen, phosphorus and potassium
consumers organisms which feed on other living organisms
corm a swollen and compressed underground stem which acts as a food storage organ
corolla collective name for all the petals
COSHH Control of Substances Hazardous to Health Regulations (2002)
critical daylength the period of daylength which initiates flowering or vegetative growth
cross-pollination pollen transfer between flowers on different plants
CSS Countryside Stewardship Scheme (1991–2014)
cultivar a variation within a species which has usually arisen and has been maintained in cultivation.
cultivated plant a plant that does not exist in the wild anywhere
cultural control a process in which management of the growing environment results in conditions which prevents or controls weeds, pests and diseases
cymose an inflorescence where the stem terminates in a flower

damping down applying water to floor and bench surfaces in a glasshouse to raise humidity and reduce temperature

day neutral plants flowering is unaffected by daylength
deciduous a plant that sheds all its leaves at once, often at the end of the growing season
decomposer an organism that breaks down dead organic matter (see saprophyte)
dehiscent fruit a fruit that splits open to release the seeds
detritivore an organism that feeds on decomposed organic matter
development, plant the changes in structure, form and behaviour throughout its life cycle
differentiation the changes that take place in a cell, tissue or organ enabling it to perform a specific function
diffusion the movement of a substance from a high concentration to a lower concentration
dioecious plants with female flowers on one plant and male flowers on another plant
dormancy the condition when viable seed fails to germinate even when all germination requirements are met
double fertilization in angiosperms where one male gamete fuses with the ovum and one fuses with the endosperm nucleus
drainage the removal of gravitational (excess) water from the growing medium

economic threshold the level of pest or disease damage allowable before economic loss in a crop is observed
ecosystem a community of living organisms together with their non-living environment functioning together as a unit
endospermic seed where the endosperm is the main food store for the embryo
ephemerals a plant that has several life cycles in a growing season
epigeal germination germination in which the cotyledon/s emerge above the ground, initially enclosed inside the testa
epinasty the downward bending of petals, petioles and branches due to uneven growth
erosion the movement of rock fragments and soil
ESS Environmental Stewardship Scheme (2005)
essential minerals inorganic substances necessary to maintain unrestricted growth and development (often referred to as 'nutrients')
etiolated growth characteristic growth of plants in the absence of light
etiolation abnormally long and spindly plant growth which occurs in low light levels
evergreen a plant retaining leaves in all seasons

false fruit a fruit containing structures derived from flower parts other than the ovary (a pseudocarp)
fasciation a disorder which produces abnormally flattened stems and inflorescences
fatty acids a contact insecticide and herbicide which contains potassium salts of fatty acids and is derived from vegetable oils
fertigation describes irrigation when it is used to deliver nutrients as well as water
fertilization the fusion of a male sex cell (the male gamete) from a pollen grain with a female sex cell (the female gamete) in the ovule to produce an embryo
fertilizer a concentrated source of plant nutrients that is added to growing media
fibrous root many roots growing from the base of the stem with no dominant root
field capacity (FC) the amount of water the soil can hold against the force of gravity
flower balling the outer petals of a flower become dry and fuse together, preventing the inner petals from emerging
foliar feed nutrients supplied in a spray to the foliage
forma a genetic variant of a species which shows a minor degree of difference from the species (f.)

formulation a pesticide product containing an active ingredient and other necessary ingredients

friable the consistency of the soil when it is easily cultivated i.e. readily forms crumbs

fruit the structure formed from the ovary wall usually after fertilization

fungicide, systemic a fungicide that travels through vascular tissues

gamete sex cell

garden health plan a method of managing plant health which considers all potential impacts on plant health, not just biotic factors such as pests, diseases and weeds

genus a group of species within a family which have characteristics in common

geotropism growth movement in response to gravity

germination of seeds the emergence of the young root or radicle through the testa, usually at the micropyle

glue band a permanently sticky plastic strip wound round a tree trunk to trap flightless pests

graft hybrid a non-sexual hybrid resulting from a graft union and containing genetic material from both parents

gravitational water the water that can be removed from the soil by the force of gravity (also known as 'excess water')

gravitropism see geotropism

group name a name given to a group of cultivars with similar characteristics

growth habit the general appearance of a weed (upright or prostrate)

growth, plant the increase in the size of cells, organs and the whole plant due to cell division and cell expansion

growth the increase in size of cells, organs or the whole plant

guttation the exudation of water from leaves and plant surfaces due to root pressure

gymnosperms seed bearing, non-flowering plants

gynoecium the female part of the flower consisting of the carpels

habitat an area occupied by a community of organisms

half-hardy a plant able to survive temperatures between 1°C to –5°C

hardiness a plant's ability to survive low temperatures

hardy a plant able to survive below –5°C

hazard something with the potential to cause injury or illness (see risk)

heated glasshouse (hardiness) a plant requiring temperatures above 5°C to survive

hemiparasite a plant which is parasitic on other plants and also photosynthesises

herbaceous perennial a perennial that is non-woody and generally loses its stems and foliage at the end of the growing season

herbicide, contact a herbicide which is used on annual and seedling weeds which scorches the weed's foliage and leads to its death

herbicide, residual a herbicide which is applied to the soil where it remains active over a period of weeks to months

herbicide, selective a herbicide which kills certain groups of weeds, leaving other species unharmed

herbicide, total a herbicide which will kill all eudicot and monocotyledon plants they are applied to

herbicide, translocated a herbicide which enters the leaves, stems or roots and then moves around the plant via the vascular system

herbivore an organism that feeds on plant tissue

hermaphrodite flower a flower with both male and female organs

heterostyly where flower structure varies within a species e.g. *Primula*

heterotroph see consumers

honeydew an insect secretion containing sugar

hormones, plant a group of plant growth regulators produced within the plant: auxin, abscisic acid, gibberellin, cytokinin and ethylene

hydathodes specialized groups of cells which exude water due to guttation
hydroponics the cultivation of plants in nutrient solution without soils
hypogeal germination germination in which the cotyledon/s remain below the ground

ICN (ICBN) the International Code of Nomenclature (formerly the International Code of Botanical Nomenclature)
ICNCP the International Code of Nomenclature for Cultivated Plants
incompatible pollen pollen which fails to germinate or grow
indehiscent fruit a fruit that does not split open to release the seeds
inflorescence the arrangement of flowers on a stem
insecticide, contact an insecticide which kills on contact rather than ingestion by the pest
insecticide, systemic an insecticide which is absorbed into the plant and kills when the pest feeds on the plant
integrated pest management (IPM) the use of a range of compatible cultural, physical, biological and chemical methods of pest and disease control
intergeneric hybrid a sexual hybrid between two species from different genera
interspecific competition competition for resources between individuals of different species
interspecific hybrid a sexual hybrid between two different species in the same genus
intraspecific competition competition for resources between individuals of the same species

juvenile growth non-reproductive (vegetative) growth

larva immature stage of some insects which have complete lifecycles
lateral meristem area of cell division within the stem and root responsible for increase in width
lateral root roots branching from a taproot
Law of limiting factors states that the factor in least supply will limit the rate of a process, for example, in photosynthesis
light compensation point the light level at which the carbon dioxide fixed by photosynthesis is all lost by respiration
light saturation point the light level above which there is no further increase in the rate of photosynthesis
lime requirement the quantity of calcium carbonate required to raise the soil pH to 6.5
liquid feed fertilizers dissolved and watered on to soils or composts
long day plants flower if the daylength is more than their critical daylength

manure a source of bulky organic matter comprising animal faeces (dung) and bedding
misting spraying a fine mist of water over plants in a greenhouse
molluscicide a pesticide effective against slugs and snails
monoculture where only one species of plant is present
monoecious plants with separate male and female flowers on the same plant
mosaic a mottling symptom seen on leaves, often caused by a virus
mulches materials applied to the surface of the soil to suppress weeds, modify soil temperatures, reduce water loss, protect the soil surface and/or reduce erosion
mutualism a relationship between two organisms which is beneficial to both
mycelium a mass of fungal tissue

native plant present at the end of the last Ice Age when Britain and Ireland separated from the rest of Europe
naturalized plant introduced by humans and now reproduces in the wild
niche the position or role of each species within a habitat
non-endospermic seed where the cotyledons are the main food stores for the embryo
nymph immature stage of some insects, which have incomplete lifecycles, and mites

oedema a dry, corky spotting on leaves and petals of some greenhouse plants caused by excess root water pressure

organ a collection of plant tissues carrying out a particular function

osmosis the movement of water from a high water (low solute) concentration to a low water (high solute) concentration across a selectively permeable membrane

osmoregulation the active regulation of solutes within a plant's cell

Parasitism an extreme form of predation where the predator often lives within the host

parent material the rock from which a soil is made

parthenocarpy production of fruit without fertilisation

parthenogenesis a process in which female animals give birth without fertilization by a male

perennating organ a plant organ that stores food (usually as starch), enabling the plant to survive unfavourable conditions e.g. bulbs, corms, tubers and rhizomes

perennial a plant living through several growing seasons

permanent wilting point (PWP) the water content of the soil when a wilted plant does not recover overnight

pesticide a chemical product used to control weeds, pests and diseases

phenology the timing of stages in plant and animal life cycles

pheromone trap a sticky trap that contains an aromatic substance attractive to specific flying pest species

phloem the tissue which transports sugars made in photosynthesis from the leaves to other plant organs

photosynthesis the process in the chloroplasts by which green plants trap light energy from the sun, convert it into chemical energy stored it in glucose and use it to produce food in the form of carbohydrates such as sugars and starch

phototropism growth movement in response to light

phyllosphere a community of bacteria and fungi that live on the leaf surface

physical control use of hand or mechanical means to remove and destroy weeds, pests or diseases or the use of materials to block or trap them

phytochrome a pigment which mediates photoperiodic and other growth responses in plants

pioneer community the first community in the successional sequence

plant breeders' rights legal rights granted to plant breeders giving them exclusive control over new cultivars

plant disorder a plant condition caused by a non-living factor (such as low temperature)

plant growth regulator (PGR) a chemical produced within the plant or applied to it which influences plant development

plant hardiness a plant's ability to survive low temperatures

pollination the transfer of pollen from stamen to stigma of a flower or flowers

polyploid an organism containing three or more sets of chromosomes in its cells (triploid = three sets, tetraploid = four sets etc.)

population a group of individuals of one species which interbreed

predation a relationship between two organisms which is beneficial to one partner (the predator) and harmful to the other (the prey)

predator an animal species that eats another animal species

prickle a sharp pointed outgrowth of the epidermis modified for defence

primary growth growth which arises from the primary meristems (apical meristem) giving rise to primary tissues

primary root see taproot

producer, primary an organism which manufactures its own food from simple molecules e.g. green plants through the process of photosynthesis

propagule any part of a plant that can be used to create a new plant

protandry where the stamens mature before the stigma

protogyny where the carpels mature before the stamens

pteridophytes vascular non-seed-bearing plants

pupa resting stage of some insects such as moths

pure breeding lines plants that are genetically identical and produce the same traits in their offspring. They are produced by self fertilisation and selection over many generations

quiescent seed a viable seed which does not germinate because the environmental requirements (water, oxygen, a suitable temperature) are not present

racemose an inflorescence where the apex of the stem continues to grow

resistance, chemical an ability in a pest or pathogen to tolerate a specific pesticide (group)

resistance, plant an ability in a plant that leads to it resisting a pathogen or a pest

respiration anaerobic the process by which glucose (food) is broken down to yield energy, the end products being carbon dioxide and water

response periods when the plant/crop benefits from the addition of water

reversion a return to a green-leaf condition in a plant selected for its variegated foliage

rhizome a horizontal, generally underground stem which may act as a food storage organ

rhizomorph 'bootlaces' produced by some toadstool fungi such as honey fungus that are able to infect plants underground

risk how likely and how seriously someone could be harmed (see hazard)

Root pressure the osmotic pressure that builds up in the root systems of plants which forces water upwards in the xylem tissue

rotation planting particular crops in different plots each year

runner a stem that runs along the ground and produces new plantlets at the nodes (e.g. *Ranunculus repens*) or at the tip (e.g. *Fragaria* spp.)

sand a soil particle between 0.06 mm and 2.0 mm in diameter

saprophytes organisms that live on dead plant material

saturation point of soils when water has filled all the soil pores (i.e. no air in the pores); waterlogging

scarification physically damaging the seed coat to break dormancy OR raking turf to remove moss and thatch

sclerotia small dark-coloured clumps of fungal tissue that enable some fungi to survive for long periods

secondary growth (thickening) growth which arises from the secondary meristems (vascular cambium and phellogen) giving rise to secondary tissues

secondary root see lateral root

seed dressing an insecticide or fungicide coating applied to seeds

seed germination the emergence of the young root or radicle through the testa, usually at the micropyle

seed the structure that develops from the ovule after fertilization

seismotropism see thigmotropism

self-pollination pollen transfer between flowers on the same plant or within the same flower

semi-evergreen a plant that retains some of its leaves through the year but may shed most leaves under severe weather conditions

sexual reproduction the formation of new individuals through fusion of male and female gametes (sex cells)

short day plants flower if the daylength is less than their critical daylength

shrub a multistemmed woody perennial plant having side branches emerging from near ground level

silt a soil particle between 0.002 mm and 0.06 mm in diameter

simple tissue tissue made up of one type of cell

soil horizon a specific layer in the soil revealed by digging a soil pit

soil moisture deficit (SMD) the amount of water required to return the soil to field capacity (FC)

soil sterilization, partial the use of heat to kill weed seeds, pests and diseases in soil, while leaving beneficial bacteria unharmed

soil structure the arrangement of particles in the soil

soil texture the relative proportion of the sand, silt and clay particles in the soil
species a group of individuals within a genus which have characteristics in common and are able to breed among themselves
spermatophytes vacular seed bearing plants
spine a sharp pointed leaf modified for defence and to reduce water loss
spore-case a small fungal structure that contains spores
sticky traps a sticky yellow or blue card hung in glasshouses to catch flying insect pests
stolon a long arching branch which roots at its tip to produce a new plant e.g. in *Rubus fruticosus*
stones soil particles larger than 2 mm in diameter
straights fertilizers that supply only one of the major nutrients
stratification moist storage of seeds in cold or warm temperatures to break dormancy
subsoil the layer (horizon) below the cultivated layer (topsoil) and lighter in colour because of its low organic matter level
subspecies a genetic variant of a species which shows a large degree of difference from the species (subsp.)
succession a sequence of changes in the composition of plants and animals in an area over time
succulent fruit a soft, fleshy fruit
sustainable an action or process which provides the best for the environment and people, socially and economically, both new and indefinitely into the future
symplast all the cell protoplasts linked together through plasmodesmata
symptoms a visible feature that is characteristic of a disease-causing organism or agent

taproot (primary root) a single large root derived from the radicle
taproot a single large root, derived from the radicle
tender a plant able to survive temperatures 1°C–5°C. Tolerant of low temperatures but will not survive frost
tepal a fused petal and sepal found in monocotyledonous flowers
thigmotropism growth movement in response to touch
thorn a sharp pointed branch or shoot modified for defence
tilth the crumb structure of the seedbed
tissue a collection of cells carrying out a specific function
top dressing a fertilizer applied to the surface of soils
topsoil an upper layer (horizon) of soil (above the subsoil and below the litter layer) normally moved during cultivation. It is typically 10–40 cm deep and darkened by the decomposed organic matter it contains.
trade designation a 'selling name' which replaces the cultivar name at the point of sale
transpiration the evaporation of water vapour from the leaves and other plant surfaces
transpiration pull the main mechanism by which water moves in the xylem due to transpiration
tree a large woody perennial unbranched for some distance above ground, on a single stem
tropism a directional growth response to an environmental stimulus
tuber a swollen underground food storage organ which may be formed from a root or stem
tylose balloon-like outgrowths of the cytoplasm of adjacent parenchyma cells protruding into xylem vessels through pits in their cell wall walls

ultrasonic deterrent an electric device that creates sound waves intolerable to moles

varietas a genetic variant of a species which shows a moderate degree of difference from the species (var.)

variety, horticultural a general, non-botanical, term for plants which vary from the species
vascular plants plants with conducting tissue i.e. xylem and phloem
vector an organism (such as an aphid) that transfers pathogens from plant to plant
vegetative propagation involves asexual reproduction and results in a clone
viable seed seed that has the potential for germination when the required external conditions are supplied
vivipary giving birth to live young rather than laying eggs

water-holding capacity (WHC) the amount of water held by the soil at field capacity
weathering the breakdown of rocks (or soil particles)
weed, annual a weed with one life cycle in a growing season
weed, ephemeral a weed with several life cycles in a growing season
weed, perennial a weed that lives for more than 2 years
wettable powder a pesticide formulation that needs constant agitation in the spray tank
woody perennial a perennial that maintains a live woody framework of stems at the end of the growing season

xerophyte a plant that has adapted to survive in an environment where there is only a small amount of liquid water available
xeromorphic adaptations structural adaptations which enable xerophytes to survive
xylem the tissue which transports water and dissolved mineral nutrients from the roots, up the stem to the leaves and other plant organs

zygote cell resulting from the fusion of male and female gametes

Index

Note: page numbers in *italics* indicate a figure and page numbers in **bold** indicate a table on the corresponding page.